Fabrication and Welding Engineering

Fabrication and Welding Engineering

Roger Timings

ELSEVIER

AMSTERDAM • BOSTON • HEIDELBERG • LONDON • NEW YORK • OXFORD
PARIS • SAN DIEGO • SAN FRANCISCO • SINGAPORE • SYDNEY • TOKYO
Newnes is an imprint of Elsevier

Newnes

Newnes is an imprint of Elsevier
Linacre House, Jordan Hill, Oxford OX2 8DP, UK
30 Corporate Drive, Suite 400, Burlington, MA 01803, USA

British Library Cataloguing in Publication Data
A catalogue record for this book is available from the British Library

Library of Congress Cataloguing in Publication Data
A catalogue record for this book is available from the Library of Congress

ISBN: 978-0-7506-6691-6

For information on all Newnes publications visit our
web site at http://books.elsevier.com

Typeset by Charon Tec Ltd., A Macmillan Company. (www.macmillansolutions.com)

Printed and bound in Hungary

08 09 10 11 12   10 9 8 7 6 5 4 3 2 1

Contents

Preface

This book is designed to provide an accessible course in the basic sheet-metal fabrication and welding engineering principles and applications required in a wide range of vocational courses. No prior knowledge of sheet-metal fabrication and welding is assumed.

I trust that *Fabrication and Welding Engineering* will be found to be a worthy complement to my previous books on the fundamentals of general and mechanical engineering. As well as offering up-to-date best practice and technical information, this new title has been fully matched to the latest courses, in particular Level 2 NVQs in Performing Engineering Operations from EAL and the City & Guilds. Guidance on the depth of treatment has been taken from the SEMTA National Occupational Standards on which all NVQs are based. The book is especially useful for trainees who are involved in a SEMTA Apprenticeship in Engineering.

All of the chapters end with a selection of exercises. These will help with assessing the trainees' performance criteria for the underpinning knowledge and understanding that is an essential part of their training. The book ends with an Appendix of useful websites on the more advanced aspects of welding and profile cutting involving: plasma arc welding; electron beam welding; laser welding and cutting and automated (robotic) welding and cutting (CNC).

I also wish to pay tribute to the late Mr F.J.M. Smith who collaborated with me on matters of sheet-metal working and welding where this topic area occurred in my previous books. I have drawn heavily on his expertise in this current book.

Finally, the author and publishers are grateful to SEMTA, Pearson Educational Ltd for allowing me to quote extensively from F.J.M. Smith's book *Fundamentals of Fabrication and Welding Engineering* , Henkel Loctite, Tucker Fasteners Ltd and all the other companies and organizations listed under *acknowledgements* for allowing the reproduction and adaptation of their illustrations and material in this text.

Roger Timings

Acknowledgements of figures and tables

The author is indebted to Mr Bob Goss, Senior Technology Specialist at Henkel Loctite Adhesives Limited for his assistance in compiling the information on adhesive bonding in Chapter 12. The author also wishes to acknowledge Emhart Technologies (Tucker Fasteners Ltd.) for permission to use illustrations in the extracts from *Mechanical Engineer's Pocket Book*, 3rd Edn, Newnes, 2005, pp. 295 to 300, 320 to 324 and 341 to 349.

Further acknowledgement is for the Figures and Tables is given on the following pages.

Figure No.	Source	Copyright holder
1.1	R. Timings, *Engineering Fundamentals*, Newnes, 2002	Training Publications Ltd
1.2	R. Timings, *Engineering Fundamentals*, Newnes, 2002	Federation of Building Trades Employers
1.3	R. Timings, *Engineering Fundamentals*, Newnes, 2002	Training Publications Ltd
1.4	R. Timings, *Engineering Fundamentals*, Newnes, 2002	Training Publications Ltd
1.5	R. Timings, *Engineering Fundamentals*, Newnes, 2002	Pearson Education Ltd
1.6	R. Timings, *Engineering Fundamentals*, Newnes, 2002	British Standards Institution
1.7	R. Timings, *Engineering Fundamentals*, Newnes, 2002	British Standards Institution
1.8	R. Timings, *Engineering Fundamentals*, Newnes, 2002	British Standards Institution
1.9	R. Timings, *Engineering Fundamentals*, Newnes, 2002	British Standards Institution
1.10	R. Timings, *Engineering Fundamentals*, Newnes, 2002	Training Publications Ltd
1.11	R. Timings, *Engineering Fundamentals*, Newnes, 2002	Pearson Education Ltd
1.12	R. Timings, *Engineering Fundamentals*, Newnes, 2002	Pearson Education Ltd
1.13	R. Timings, *Engineering Fundamentals*, Newnes, 2002	Pearson Education Ltd
1.14	R. Timings, *Engineering Fundamentals*, Newnes, 2002	Pearson Education Ltd
1.15	R. Timings, *Engineering Fundamentals*, Newnes, 2002	Pearson Education Ltd
1.16	R. Timings, *Engineering Fundamentals*, Newnes, 2002	Pearson Education Ltd
1.17	R. Timings, *Engineering Fundamentals*, Newnes, 2002	Pearson Education Ltd
1.18	R. Timings, *Engineering Fundamentals*, Newnes, 2002	Training Publications Ltd
1.19	R. Timings, *Engineering Fundamentals*, Newnes, 2002	Training Publications Ltd
1.20	R. Timings, *Engineering Fundamentals*, Newnes, 2002	Training Publications Ltd
1.21	R. Timings, *Engineering Fundamentals*, Newnes, 2002	Training Publications Ltd
1.22	R. Timings, *Engineering Fundamentals*, Newnes, 2002	Pearson Education Ltd
1.23	R. Timings, *Engineering Fundamentals*, Newnes, 2002	Training Publications Ltd
1.24	R. Timings, *Engineering Fundamentals*, Newnes, 2002	Training Publications Ltd
1.25	R. Timings, *General Engineering*, Longman, 1995	Federation of Building Trades Employers
1.26	R. Timings, *General Engineering*, Longman, 1995	Federation of Building Trades Employers
1.27	R. Timings, *General Engineering*, Longman, 1995	Federation of Building Trades Employers
1.28	R. Timings, *General Engineering*, Longman, 1995	Federation of Building Trades Employers
1.29	R. Timings, *General Engineering*, Longman, 1995	Federation of Building Trades Employers
1.30	R. Timings, *General Engineering*, Longman, 1995	Training Publications Ltd
1.31	R. Timings, *Engineering Fundamentals*, Newnes, 2002	Training Publications Ltd/Pearson Education Ltd
1.32	R. Timings, *General Engineering*, Longman, 1995	British Standards Institution
1.33	R. Timings, *General Engineering*, Longman, 1995	Pearson Education Ltd
1.34	R. Timings, *Engineering Fundamentals*, Newnes, 2002	Pearson Education Ltd
1.35	R. Timings, *Engineering Fundamentals*, Newnes, 2002	Pearson Education Ltd
1.36	R. Timings, *Engineering Fundamentals*, Newnes, 2002	Pearson Education Ltd
1.37	R. Timings, *General Engineering*, Longman, 1995	Pearson Education Ltd
1.38	R. Timings, *Engineering Fundamentals*, Newnes, 2002	Pearson Education Ltd
1.39	F.J.M. Smith, *Fundamentals of Fabrication and Welding Engineering*, Longman, 1992	Training Publications Ltd
1.40	F.J.M. Smith, *Fundamentals of Fabrication and Welding Engineering*, Longman, 1992	British Oxygen Company
1.41	F.J.M. Smith, *Fundamentals of Fabrication and Welding Engineering*, Longman, 1992	British Oxygen Company
1.42	F.J.M. Smith, *Fundamentals of Fabrication and Welding Engineering*, Longman, 1992	Training Publications Ltd
1.43	F.J.M. Smith, *Fundamentals of Fabrication and Welding Engineering*, Longman, 1992	Training Publications Ltd

(Continued)

Figure No.	Source	Copyright holder
1.44	F.J.M. Smith, *Fundamentals of Fabrication and Welding Engineering*, Longman, 1992	Pearson Education Ltd
1.45	F.J.M. Smith, *Fundamentals of Fabrication and Welding Engineering*, Longman, 1992	Pearson Education Ltd
1.46	F.J.M. Smith, *Fundamentals of Fabrication and Welding Engineering*, Longman, 1992	Pearson Education Ltd
1.47	F.J.M. Smith, *Fundamentals of Fabrication and Welding Engineering*, Longman, 1992	Pearson Education Ltd
1.48	F.J.M. Smith, *Fundamentals of Fabrication and Welding Engineering*, Longman, 1992	Pearson Education Ltd
1.49	F.J.M. Smith, *Fundamentals of Fabrication and Welding Engineering*, Longman, 1992	Training Publications Ltd
1.50	F.J.M. Smith, *Fundamentals of Fabrication and Welding Engineering*, Longman, 1992	Training Publications Ltd
1.51	F.J.M. Smith, *Fundamentals of Fabrication and Welding Engineering*, Longman, 1992	Federation of Building Trades Employers
1.52	F.J.M. Smith, *Fundamentals of Fabrication and Welding Engineering*, Longman, 1992	Federation of Building Trades Employers
1.53	F.J.M. Smith, *Fundamentals of Fabrication and Welding Engineering*, Longman, 1992	Federation of Building Trades Employers
1.54	F.J.M. Smith, *Fundamentals of Fabrication and Welding Engineering*, Longman, 1992	Federation of Building Trades Employers
1.55	R. Timings, *Engineering Fundamentals*, Newnes, 2002	Pearson Education Ltd
1.56	R. Timings, *Engineering Fundamentals*, Newnes, 2002	Pearson Education Ltd
2.1	R. Timings, *Engineering Fundamentals*, Newnes, 2002	Pearson Education Ltd
2.2	R. Timings, *Engineering Fundamentals*, Newnes, 2002	Pearson Education Ltd
2.3	R. Timings, *Engineering Fundamentals*, Newnes, 2002	Pearson Education Ltd
3.1	R. Timings, *Engineering Fundamentals*, Newnes, 2002	Pearson Education Ltd
3.2	R. Timings, *Engineering Fundamentals*, Newnes, 2002	Pearson Education Ltd
3.3	R. Timings, *Engineering Fundamentals*, Newnes, 2002	Pearson Education Ltd
3.4	R. Timings, *Engineering Fundamentals*, Newnes, 2002	Pearson Education Ltd
3.5	R. Timings, *Engineering Fundamentals*, Newnes, 2002	Pearson Education Ltd
3.6	R. Timings, *Engineering Fundamentals*, Newnes, 2002	Pearson Education Ltd
3.7	R. Timings, *Engineering Fundamentals*, Newnes, 2002	Pearson Education Ltd
3.8	R. Timings, *Engineering Fundamentals*, Newnes, 2002	Pearson Education Ltd
3.9	R. Timings, *Engineering Fundamentals*, Newnes, 2002	Pearson Education Ltd
3.10	R. Timings, *Engineering Fundamentals*, Newnes, 2002	Pearson Education Ltd
3.11	R. Timings, *Engineering Fundamentals*, Newnes, 2002	Pearson Education Ltd
3.12	R. Timings, *Engineering Fundamentals*, Newnes, 2002	Pearson Education Ltd
3.13	R. Timings, *Engineering Fundamentals*, Newnes, 2002	Pearson Education Ltd
3.14	R. Timings, *Engineering Fundamentals*, Newnes, 2002	Pearson Education Ltd
3.15	R. Timings, *Engineering Fundamentals*, Newnes, 2002	Pearson Education Ltd
3.16	R. Timings, *Engineering Fundamentals*, Newnes, 2002	Pearson Education Ltd
3.17	R. Timings, *Engineering Fundamentals*, Newnes, 2002	Pearson Education Ltd
3.18	R. Timings, *Engineering Fundamentals*, Newnes, 2002	Pearson Education Ltd
3.19	R. Timings, *Engineering Fundamentals*, Newnes, 2002	Pearson Education Ltd
3.20	R. Timings, *Engineering Fundamentals*, Newnes, 2002	Pearson Education Ltd

(*Continued*)

Figure No.	Source	Copyright holder
4.1	R. Timings, *Engineering Fundamentals*, Newnes, 2002	Pearson Education Ltd
4.2	R. Timings, *Engineering Fundamentals*, Newnes, 2002	Pearson Education Ltd
4.3	R. Timings, *Engineering Fundamentals*, Newnes, 2002	Pearson Education Ltd
4.4	R. Timings, *Engineering Fundamentals*, Newnes, 2002	Pearson Education Ltd
4.5	R. Timings, *Engineering Fundamentals*, Newnes, 2002	Pearson Education Ltd
4.6	R. Timings, *Engineering Fundamentals*, Newnes, 2002	Pearson Education Ltd
4.7	R. Timings, *Engineering Fundamentals*, Newnes, 2002	Pearson Education Ltd
4.8	R. Timings, *Engineering Fundamentals*, Newnes, 2002	Pearson Education Ltd
4.9	R. Timings, *Engineering Fundamentals*, Newnes, 2002	Pearson Education Ltd
4.10	R. Timings, *Engineering Fundamentals*, Newnes, 2002	Pearson Education Ltd
4.11	R. Timings, *Engineering Fundamentals*, Newnes, 2002	Pearson Education Ltd
4.12	R. Timings, *Engineering Fundamentals*, Newnes, 2002	Pearson Education Ltd
4.13	R. Timings, *Engineering Fundamentals*, Newnes, 2002	Pearson Education Ltd
4.14	R. Timings, *Engineering Fundamentals*, Newnes, 2002	Pearson Education Ltd
4.15	R. Timings, *Engineering Fundamentals*, Newnes, 2002	British Standards Institution
4.16	R. Timings, *Engineering Fundamentals*, Newnes, 2002	British Standards Institution
4.17	R. Timings, *Engineering Fundamentals*, Newnes, 2002	British Standards Institution
4.18	R. Timings, *Engineering Fundamentals*, Newnes, 2002	British Standards Institution
4.19	R. Timings, *Engineering Fundamentals*, Newnes, 2002	British Standards Institution
4.20	R. Timings, *Engineering Fundamentals*, Newnes, 2002	British Standards Institution
4.21	F.J.M. Smith, *Fundamentals of Fabrication and Welding Engineering*, Longman, 1992	Pearson Education Ltd
4.22	F.J.M. Smith, *Fundamentals of Fabrication and Welding Engineering*, Longman, 1992	Pearson Education Ltd
4.23	F.J.M. Smith, *Fundamentals of Fabrication and Welding Engineering*, Longman, 1992	Pearson Education Ltd
4.24	F.J.M. Smith, *Fundamentals of Fabrication and Welding Engineering*, Longman, 1992	Pearson Education Ltd
4.25	F.J.M. Smith, *Fundamentals of Fabrication and Welding Engineering*, Longman, 1992	Pearson Education Ltd
4.26	F.J.M. Smith, *Fundamentals of Fabrication and Welding Engineering*, Longman, 1992	Pearson Education Ltd
4.27	F.J.M. Smith, *Fundamentals of Fabrication and Welding Engineering*, Longman, 1992	Pearson Education Ltd
4.28	F.J.M. Smith, *Fundamentals of Fabrication and Welding Engineering*, Longman, 1992	Pearson Education Ltd
4.29	F.J.M. Smith, *Fundamentals of Fabrication and Welding Engineering*, Longman, 1992	Pearson Education Ltd
4.30	F.J.M. Smith, *Fundamentals of Fabrication and Welding Engineering*, Longman, 1992	Pearson Education Ltd
4.31	F.J.M. Smith, *Fundamentals of Fabrication and Welding Engineering*, Longman, 1992	Pearson Education Ltd
4.32	F.J.M. Smith, *Fundamentals of Fabrication and Welding Engineering*, Longman, 1992	Pearson Education Ltd
4.33	F.J.M. Smith, *Fundamentals of Fabrication and Welding Engineering*, Longman, 1992	Pearson Education Ltd
4.34	F.J.M. Smith, *Fundamentals of Fabrication and Welding Engineering*, Longman, 1992	Pearson Education Ltd
4.35	F.J.M. Smith, *Fundamentals of Fabrication and Welding Engineering*, Longman, 1992	Pearson Education Ltd

(Continued)

Figure No.	Source	Copyright holder
4.36	F.J.M. Smith, *Fundamentals of Fabrication and Welding Engineering*, Longman, 1992	Pearson Education Ltd
4.37	E. Jackson, Jackson's Geometrical and Engineering Drawing, Longman, 1969	Pearson Education Ltd
4.38	E. Jackson, Jackson's Geometrical and Engineering Drawing, Longman, 1969	Pearson Education Ltd
4.39	R. Timings, *Engineering Fundamentals*, Newnes, 2002	Pearson Education Ltd
4.40	R. Timings, *Engineering Fundamentals*, Newnes, 2002	Pearson Education Ltd
4.41	R. Timings, *Engineering Fundamentals*, Newnes, 2002	Pearson Education Ltd
4.42	R. Timings, *Engineering Fundamentals*, Newnes, 2002	Pearson Education Ltd
4.43	R. Timings, *Engineering Fundamentals*, Newnes, 2002	Pearson Education Ltd
5.1	F.J.M. Smith, *Fundamentals of Fabrication and Welding Engineering*, Longman, 1992	Pearson Education Ltd
5.2	R. Timings, *Engineering Fundamentals*, Newnes, 2002	Pearson Education Ltd
5.3	R. Timings, *Engineering Fundamentals*, Newnes, 2002	Moore and Wright
5.4	R. Timings, *Engineering Fundamentals*, Newnes, 2002	Moore and Wright
5.5	R. Timings, *Engineering Fundamentals*, Newnes, 2002	Pearson Education Ltd
5.6	R. Timings, *Engineering Fundamentals*, Newnes, 2002	Pearson Education Ltd
5.7	R. Timings, *Engineering Fundamentals*, Newnes, 2002	Pearson Education Ltd
5.8	R. Timings, *Engineering Fundamentals*, Newnes, 2002	Pearson Education Ltd
5.9	R. Timings, *Engineering Fundamentals*, Newnes, 2002	Pearson Education Ltd
5.10	R. Timings, *Engineering Fundamentals*, Newnes, 2002	Pearson Education Ltd
5.11	R. Timings, *Engineering Fundamentals*, Newnes, 2002	Pearson Education Ltd
5.12	R. Timings, *Engineering Fundamentals*, Newnes, 2002/ F.J.M. Smith, *Fundamentals of Fabrication and Welding Engineering*, Longman, 1992	Pearson Education Ltd Pearson Education Ltd
5.13	R. Timings, *Engineering Fundamentals*, Newnes, 2002/ F.J.M. Smith, *Fundamentals of Fabrication and Welding Engineering*, Longman, 1992	Pearson Education Ltd
5.14	R. Timings, *Engineering Fundamentals*, Newnes, 2002	Moore and Wright
5.15	R. Timings, *Engineering Fundamentals*, Newnes, 2002	Moore and Wright
5.16	R. Timings, *Engineering Fundamentals*, Newnes, 2002	Moore and Wright
5.17	R. Timings, *Engineering Fundamentals*, Newnes, 2002	Pearson Education Ltd
5.18	R. Timings, *Engineering Fundamentals*, Newnes, 2002	Pearson Education Ltd
5.19	R. Timings, *Engineering Fundamentals*, Newnes, 2002	Moore and Wright
5.20	R. Timings, *Engineering Fundamentals*, Newnes, 2002	Pearson Education Ltd
5.21	F.J.M. Smith, *Fundamentals of Fabrication and Welding Engineering*, Longman, 1992	Pearson Education Ltd
5.22	F.J.M. Smith, *Fundamentals of Fabrication and Welding Engineering*, Longman, 1992	Training Publications Ltd
5.23	F.J.M. Smith, *Fundamentals of Fabrication and Welding Engineering*, Longman, 1992	Pearson Education Ltd
5.24	F.J.M. Smith, *Fundamentals of Fabrication and Welding Engineering*, Longman, 1992	Pearson Education Ltd
5.25	F.J.M. Smith, *Fundamentals of Fabrication and Welding Engineering*, Longman, 1992	PELFOFWE/TPL
5.26	F.J.M. Smith, *Fundamentals of Fabrication and Welding Engineering*, Longman, 1992	Pearson Education Ltd

(Continued)

Figure No.	Source	Copyright holder
5.27	F.J.M. Smith, *Fundamentals of Fabrication and Welding Engineering*, Longman, 1992	Pearson Education Ltd
5.28	F.J.M. Smith, *Fundamentals of Fabrication and Welding Engineering*, Longman, 1992	Pearson Education Ltd
5.29	F.J.M. Smith, *Fundamentals of Fabrication and Welding Engineering*, Longman, 1992	Pearson Education Ltd
5.30	F.J.M. Smith, *Fundamentals of Fabrication and Welding Engineering*, Longman, 1992	Pearson Education Ltd
5.31	F.J.M. Smith, *Fundamentals of Fabrication and Welding Engineering*, Longman, 1992	Pearson Education Ltd
5.32	F.J.M. Smith, *Fundamentals of Fabrication and Welding Engineering*, Longman, 1992	Pearson Education Ltd
5.33	F.J.M. Smith, *Fundamentals of Fabrication and Welding Engineering*, Longman, 1992	Pearson Education Ltd
5.34	F.J.M. Smith, *Fundamentals of Fabrication and Welding Engineering*, Longman, 1992	Pearson Education Ltd
5.35	F.J.M. Smith, *Fundamentals of Fabrication and Welding Engineering*, Longman, 1992	Pearson Education Ltd
5.36	F.J.M. Smith, *Fundamentals of Fabrication and Welding Engineering*, Longman, 1992	Pearson Education Ltd
5.37	F.J.M. Smith, *Fundamentals of Fabrication and Welding Engineering*, Longman, 1992	Pearson Education Ltd
5.38	F.J.M. Smith, *Fundamentals of Fabrication and Welding Engineering*, Longman, 1992	Pearson Education Ltd
5.39	F.J.M. Smith, *Fundamentals of Fabrication and Welding Engineering*, Longman, 1992	Pearson Education Ltd
5.40	R. Timings, *Engineering Fundamentals*, Newnes, 2002	Pearson Education Ltd
6.1	R. Timings, *Engineering Fundamentals*, Newnes, 2002	Pearson Education Ltd
6.2	R. Timings, *Engineering Fundamentals*, Newnes, 2002	Pearson Education Ltd
6.3	R. Timings, *Engineering Fundamentals*, Newnes, 2002	Pearson Education Ltd
6.4	R. Timings, *Engineering Fundamentals*, Newnes, 2002	Pearson Education Ltd
6.5	R. Timings, *Engineering Fundamentals*, Newnes, 2002	Pearson Education Ltd
6.6	R. Timings, *Engineering Fundamentals*, Newnes, 2002	Pearson Education Ltd
6.7	R. Timings, *Engineering Fundamentals*, Newnes, 2002	Pearson Education Ltd
6.8	F.J.M. Smith, *Fundamentals of Fabrication and Welding Engineering*, Longman, 1992	Pearson Education Ltd
6.9	R. Timings, *Engineering Fundamentals*, Newnes, 2002	Pearson Education Ltd
6.10	R. Timings, *Engineering Fundamentals*, Newnes, 2002	Pearson Education Ltd
6.11	R. Timings, *Engineering Fundamentals*, Newnes, 2002	Pearson Education Ltd
6.12	F.J.M. Smith, *Fundamentals of Fabrication and Welding Engineering*, Longman, 1992	Pearson Education Ltd
6.13	R. Timings, *Engineering Fundamentals*, Newnes, 2002	Pearson Education Ltd
6.14	R. Timings, *Engineering Fundamentals*, Newnes, 2002	Pearson Education Ltd
6.15	R. Timings, *Engineering Fundamentals*, Newnes, 2002	Pearson Education Ltd
6.16	R. Timings, *Engineering Fundamentals*, Newnes, 2002	Pearson Education Ltd
6.17	R. Timings, *Engineering Fundamentals*, Newnes, 2002	Pearson Education Ltd
6.18	R. Timings, *General Engineering*, Longman, 1995	Pearson Education Ltd
6.19	R. Timings, *Engineering Fundamentals*, Newnes, 2002	Pearson Education Ltd
6.20	R. Timings, *Engineering Fundamentals*, Newnes, 2002	Pearson Education Ltd

(Continued)

Figure No.	Source	Copyright holder
6.21	R. Timings, *Workshop Processes and Materials, Level 1*, 2nd edn, Longman, 1984	Pearson Education Ltd
6.22	R. Timings, *Workshop Processes and Materials, Level 1*, 2nd edn, Longman, 1984	Pearson Education Ltd
6.23	R. Timings, *Engineering Fundamentals*, Newnes, 2002	British Standards Institution
6.24	R. Timings, *Engineering Fundamentals*, Newnes, 2002	Pearson Education Ltd
6.25	R. Timings, *Engineering Fundamentals*, Newnes, 2002	Pearson Education Ltd
6.26	R. Timings, *Engineering Fundamentals*, Newnes, 2002	Pearson Education Ltd
6.27	F.J.M. Smith, *Fundamentals of Fabrication and Welding Engineering*, Longman, 1992	Pearson Education Ltd
6.28	F.J.M. Smith, *Fundamentals of Fabrication and Welding Engineering*, Longman, 1992	Pearson Education Ltd
6.29	R. Timings, *Engineering Fundamentals*, Newnes, 2002	Pearson Education Ltd
6.30	R. Timings, *Engineering Fundamentals*, Newnes, 2002	Pearson Education Ltd
6.31	R. Timings, *Engineering Fundamentals*, Newnes, 2002	Pearson Education Ltd
6.32	F.J.M. Smith, *Fundamentals of Fabrication and Welding Engineering*, Longman, 1992	Pearson Education Ltd
6.33	F.J.M. Smith, *Fundamentals of Fabrication and Welding Engineering*, Longman, 1992	Pearson Education Ltd
6.34	F.J.M. Smith, *Fundamentals of Fabrication and Welding Engineering*, Longman, 1992	Pearson Education Ltd
6.35	F.J.M. Smith, *Fundamentals of Fabrication and Welding Engineering*, Longman, 1992	Pearson Education Ltd
6.36	F.J.M. Smith, *Fundamentals of Fabrication and Welding Engineering*, Longman, 1992	Pearson Education Ltd
6.37	F.J.M. Smith, *Fundamentals of Fabrication and Welding Engineering*, Longman, 1992	Pearson Education Ltd
6.38	F.J.M. Smith, *Fundamentals of Fabrication and Welding Engineering*, Longman, 1992	Pearson Education Ltd
6.39	F.J.M. Smith, *Fundamentals of Fabrication and Welding Engineering*, Longman, 1992	Pearson Education Ltd
6.40	F.J.M. Smith, *Fundamentals of Fabrication and Welding Engineering*, Longman, 1992	Pearson Education Ltd
6.41	F.J.M. Smith, *Fundamentals of Fabrication and Welding Engineering*, Longman, 1992	Pearson Education Ltd
6.42	F.J.M. Smith, *Fundamentals of Fabrication and Welding Engineering*, Longman, 1992	Pearson Education Ltd
6.43	F.J.M. Smith, *Fundamentals of Fabrication and Welding Engineering*, Longman, 1992	Pearson Education Ltd
6.44	F.J.M. Smith, *Fundamentals of Fabrication and Welding Engineering*, Longman, 1992	Pearson Education Ltd
6.45	R. Timings, *Workshop Processes and Materials, Level 1*, 2nd edn, Longman, 1984	Pearson Education Ltd
6.46	F.J.M. Smith, *Fundamentals of Fabrication and Welding Engineering*, Longman, 1992	Pearson Education Ltd
6.47	F.J.M. Smith, *Fundamentals of Fabrication and Welding Engineering*, Longman, 1992	Pearson Education Ltd
6.48	F.J.M. Smith, *Fundamentals of Fabrication and Welding Engineering*, Longman, 1992	Pearson Education Ltd
6.49	R. Timings, *Engineering Fundamentals*, Newnes, 2002	Pearson Education Ltd

(Continued)

Figure No.	Source	Copyright holder
6.50	R. Timings, *Engineering Fundamentals*, Newnes, 2002	Pearson Education Ltd
6.51	R. Timings, *Engineering Fundamentals*, Newnes, 2002	Pearson Education Ltd
6.52	F.J.M. Smith, *Fundamentals of Fabrication and Welding Engineering*, Longman, 1992	Pearson Education Ltd
6.53	R. Timings, *Workshop Processes and Materials, Level 1*, 2nd edn, Longman, 1984	Pearson Education Ltd
6.54	F.J.M. Smith, *Fundamentals of Fabrication and Welding Engineering*, Longman, 1992	Pearson Education Ltd
6.55	F.J.M. Smith, *Fundamentals of Fabrication and Welding Engineering*, Longman, 1992	Pearson Education Ltd
6.56	F.J.M. Smith, *Fundamentals of Fabrication and Welding Engineering*, Longman, 1992	Pearson Education Ltd
6.57	F.J.M. Smith, *Fundamentals of Fabrication and Welding Engineering*, Longman, 1992	British Oxygen Company
6.58	F.J.M. Smith, *Fundamentals of Fabrication and Welding Engineering*, Longman, 1992	British Oxygen Company/Training Publications Ltd
6.59	F.J.M. Smith, *Fundamentals of Fabrication and Welding Engineering*, Longman, 1992	Training Publications Ltd
6.60	F.J.M. Smith, *Fundamentals of Fabrication and Welding Engineering*, Longman, 1992	Training Publications Ltd
6.61	F.J.M. Smith, *Fundamentals of Fabrication and Welding Engineering*, Longman, 1992	British Oxygen Company
7.1	F.J.M. Smith, *Fundamentals of Fabrication and Welding Engineering*, Longman, 1992	Pearson Education Ltd
7.2	F.J.M. Smith, *Fundamentals of Fabrication and Welding Engineering*, Longman, 1992	Pearson Education Ltd
7.3	F.J.M. Smith, *Fundamentals of Fabrication and Welding Engineering*, Longman, 1992	Pearson Education Ltd
7.4	F.J.M. Smith, *Fundamentals of Fabrication and Welding Engineering*, Longman, 1992	Pearson Education Ltd
7.5	F.J.M. Smith, *Fundamentals of Fabrication and Welding Engineering*, Longman, 1992	Pearson Education Ltd
7.6	F.J.M. Smith, *Fundamentals of Fabrication and Welding Engineering*, Longman, 1992	Pearson Education Ltd
7.7	F.J.M. Smith, *Fundamentals of Fabrication and Welding Engineering*, Longman, 1992	Pearson Education Ltd
7.8	F.J.M. Smith, *Fundamentals of Fabrication and Welding Engineering*, Longman, 1992	Pearson Education Ltd
7.9	F.J.M. Smith, *Fundamentals of Fabrication and Welding Engineering*, Longman, 1992	Pearson Education Ltd
7.10	F.J.M. Smith, *Fundamentals of Fabrication and Welding Engineering*, Longman, 1992	Pearson Education Ltd
7.11	F.J.M. Smith, *Fundamentals of Fabrication and Welding Engineering*, Longman, 1992	Pearson Education Ltd
7.12	F.J.M. Smith, *Fundamentals of Fabrication and Welding Engineering*, Longman, 1992	Pearson Education Ltd
7.13	F.J.M. Smith, *Fundamentals of Fabrication and Welding Engineering*, Longman, 1992	Pearson Education Ltd

(Continued)

Figure No.	Source	Copyright holder
7.14	F.J.M. Smith, *Fundamentals of Fabrication and Welding Engineering*, Longman, 1992	Pearson Education Ltd
7.15	F.J.M. Smith, *Fundamentals of Fabrication and Welding Engineering*, Longman, 1992	Pearson Education Ltd
7.16	F.J.M. Smith, *Fundamentals of Fabrication and Welding Engineering*, Longman, 1992	Pearson Education Ltd
7.17	F.J.M. Smith, *Fundamentals of Fabrication and Welding Engineering*, Longman, 1992	Pearson Education Ltd
7.18	F.J.M. Smith, *Fundamentals of Fabrication and Welding Engineering*, Longman, 1992	Training Publications Ltd
7.19	F.J.M. Smith, *Fundamentals of Fabrication and Welding Engineering*, Longman, 1992	Pearson Education Ltd
7.20	F.J.M. Smith, *Fundamentals of Fabrication and Welding Engineering*, Longman, 1992	Pearson Education Ltd
7.21	F.J.M. Smith, *Fundamentals of Fabrication and Welding Engineering*, Longman, 1992	Pearson Education Ltd
7.22	F.J.M. Smith, *Fundamentals of Fabrication and Welding Engineering*, Longman, 1992	Pearson Education Ltd
7.23	F.J.M. Smith, *Fundamentals of Fabrication and Welding Engineering*, Longman, 1992	Pearson Education Ltd
7.24	F.J.M. Smith, *Fundamentals of Fabrication and Welding Engineering*, Longman, 1992	Pearson Education Ltd
7.25	F.J.M. Smith, *Fundamentals of Fabrication and Welding Engineering*, Longman, 1992	Pearson Education Ltd
7.26	F.J.M. Smith, *Fundamentals of Fabrication and Welding Engineering*, Longman, 1992	Pearson Education Ltd
7.27	F.J.M. Smith, *Fundamentals of Fabrication and Welding Engineering*, Longman, 1992	Pearson Education Ltd
7.28	F.J.M. Smith, *Fundamentals of Fabrication and Welding Engineering*, Longman, 1992	Pearson Education Ltd
7.29	F.J.M. Smith, *Fundamentals of Fabrication and Welding Engineering*, Longman, 1992	Pearson Education Ltd
7.30	F.J.M. Smith, *Fundamentals of Fabrication and Welding Engineering*, Longman, 1992	Pearson Education Ltd
7.31	F.J.M. Smith, *Fundamentals of Fabrication and Welding Engineering*, Longman, 1992	Pearson Education Ltd
7.32	F.J.M. Smith, *Fundamentals of Fabrication and Welding Engineering*, Longman, 1992	Pearson Education Ltd
7.33	F.J.M. Smith, *Fundamentals of Fabrication and Welding Engineering*, Longman, 1992	Pearson Education Ltd
7.34	F.J.M. Smith, *Fundamentals of Fabrication and Welding Engineering*, Longman, 1992	Pearson Education Ltd
7.35	F.J.M. Smith, *Fundamentals of Fabrication and Welding Engineering*, Longman, 1992	Pearson Education Ltd
7.36	F.J.M. Smith, *Fundamentals of Fabrication and Welding Engineering*, Longman, 1992	Pearson Education Ltd
7.37	F.J.M. Smith, *Fundamentals of Fabrication and Welding Engineering*, Longman, 1992	Pearson Education Ltd
7.38	F.J.M. Smith, *Fundamentals of Fabrication and Welding Engineering*, Longman, 1992	Pearson Education Ltd

(Continued)

Figure No.	Source	Copyright holder
7.39	F.J.M. Smith, *Fundamentals of Fabrication and Welding Engineering*, Longman, 1992	Pearson Education Ltd
7.40	F.J.M. Smith, *Fundamentals of Fabrication and Welding Engineering*, Longman, 1992	Pearson Education Ltd
7.41	F.J.M. Smith, *Fundamentals of Fabrication and Welding Engineering*, Longman, 1992	Pearson Education Ltd
7.42	F.J.M. Smith, *Fundamentals of Fabrication and Welding Engineering*, Longman, 1992	Pearson Education Ltd
7.43	F.J.M. Smith, *Fundamentals of Fabrication and Welding Engineering*, Longman, 1992	Pearson Education Ltd
7.44	F.J.M. Smith, *Fundamentals of Fabrication and Welding Engineering*, Longman, 1992	Pearson Education Ltd
7.45	F.J.M. Smith, *Fundamentals of Fabrication and Welding Engineering*, Longman, 1992	Pearson Education Ltd
7.46	F.J.M. Smith, *Fundamentals of Fabrication and Welding Engineering*, Longman, 1992	Pearson Education Ltd
7.47	F.J.M. Smith, *Fundamentals of Fabrication and Welding Engineering*, Longman, 1992	Pearson Education Ltd
7.48	F.J.M. Smith, *Fundamentals of Fabrication and Welding Engineering*, Longman, 1992	Pearson Education Ltd
7.49	F.J.M. Smith, *Fundamentals of Fabrication and Welding Engineering*, Longman, 1992	Pearson Education Ltd
7.50	F.J.M. Smith, *Fundamentals of Fabrication and Welding Engineering*, Longman, 1992	Pearson Education Ltd
7.51	F.J.M. Smith, *Fundamentals of Fabrication and Welding Engineering*, Longman, 1992	Pearson Education Ltd
7.52	F.J.M. Smith, *Fundamentals of Fabrication and Welding Engineering*, Longman, 1992	Pearson Education Ltd
7.53	F.J.M. Smith, *Fundamentals of Fabrication and Welding Engineering*, Longman, 1992	Pearson Education Ltd
7.54	F.J.M. Smith, *Fundamentals of Fabrication and Welding Engineering*, Longman, 1992	Pearson Education Ltd
7.55	F.J.M. Smith, *Fundamentals of Fabrication and Welding Engineering*, Longman, 1992	Pearson Education Ltd
7.56	F.J.M. Smith, *Fundamentals of Fabrication and Welding Engineering*, Longman, 1992	Pearson Education Ltd
7.57	F.J.M. Smith, *Fundamentals of Fabrication and Welding Engineering*, Longman, 1992	Pearson Education Ltd
7.58	F.J.M. Smith, *Fundamentals of Fabrication and Welding Engineering*, Longman, 1992	Pearson Education Ltd
7.59	F.J.M. Smith, *Fundamentals of Fabrication and Welding Engineering*, Longman, 1992	Pearson Education Ltd
7.60	F.J.M. Smith, *Fundamentals of Fabrication and Welding Engineering*, Longman, 1992	Pearson Education Ltd
8.1	F.J.M. Smith, *Fundamentals of Fabrication and Welding Engineering*, Longman, 1992	Pearson Education Ltd
8.2	F.J.M. Smith, *Fundamentals of Fabrication and Welding Engineering*, Longman, 1992	Pearson Education Ltd

(Continued)

Figure No.	Source	Copyright holder
8.3	F.J.M. Smith, *Fundamentals of Fabrication and Welding Engineering*, Longman, 1992	Pearson Education Ltd
8.4	F.J.M. Smith, *Fundamentals of Fabrication and Welding Engineering*, Longman, 1992	Pearson Education Ltd
8.5	F.J.M. Smith, *Fundamentals of Fabrication and Welding Engineering*, Longman, 1992	Pearson Education Ltd
8.6	F.J.M. Smith, *Fundamentals of Fabrication and Welding Engineering*, Longman, 1992	Pearson Education Ltd
8.7	F.J.M. Smith, *Fundamentals of Fabrication and Welding Engineering*, Longman, 1992	Pearson Education Ltd
8.8	F.J.M. Smith, *Fundamentals of Fabrication and Welding Engineering*, Longman, 1992	Pearson Education Ltd
8.9	F.J.M. Smith, *Fundamentals of Fabrication and Welding Engineering*, Longman, 1992	Pearson Education Ltd
8.10	F.J.M. Smith, *Fundamentals of Fabrication and Welding Engineering*, Longman, 1992	Pearson Education Ltd
8.11	F.J.M. Smith, *Fundamentals of Fabrication and Welding Engineering*, Longman, 1992	Pearson Education Ltd
8.12	F.J.M. Smith, *Fundamentals of Fabrication and Welding Engineering*, Longman, 1992	Pearson Education Ltd
8.13	F.J.M. Smith, *Fundamentals of Fabrication and Welding Engineering*, Longman, 1992	Pearson Education Ltd
8.14	F.J.M. Smith, *Fundamentals of Fabrication and Welding Engineering*, Longman, 1992	Pearson Education Ltd
8.15	F.J.M. Smith, *Fundamentals of Fabrication and Welding Engineering*, Longman, 1992	Pearson Education Ltd
8.16	F.J.M. Smith, *Fundamentals of Fabrication and Welding Engineering*, Longman, 1992	Pearson Education Ltd
8.17	F.J.M. Smith, *Fundamentals of Fabrication and Welding Engineering*, Longman, 1992	Pearson Education Ltd
8.18	F.J.M. Smith, *Fundamentals of Fabrication and Welding Engineering*, Longman, 1992	Pearson Education Ltd
8.19	F.J.M. Smith, *Fundamentals of Fabrication and Welding Engineering*, Longman, 1992	Pearson Education Ltd
8.20	F.J.M. Smith, *Fundamentals of Fabrication and Welding Engineering*, Longman, 1992	Pearson Education Ltd
8.21	F.J.M. Smith, *Fundamentals of Fabrication and Welding Engineering*, Longman, 1992	Pearson Education Ltd
8.22	F.J.M. Smith, *Fundamentals of Fabrication and Welding Engineering*, Longman, 1992	Pearson Education Ltd
8.23	F.J.M. Smith, *Fundamentals of Fabrication and Welding Engineering*, Longman, 1992	Pearson Education Ltd
8.24	F.J.M. Smith, *Fundamentals of Fabrication and Welding Engineering*, Longman, 1992	Pearson Education Ltd
8.25	F.J.M. Smith, *Fundamentals of Fabrication and Welding Engineering*, Longman, 1992	British Standards Institution
8.26	F.J.M. Smith, *Fundamentals of Fabrication and Welding Engineering*, Longman, 1992	Pearson Education Ltd
8.27	F.J.M. Smith, *Fundamentals of Fabrication and Welding Engineering*, Longman, 1992	Pearson Education Ltd

(Continued)

Figure No.	Source	Copyright holder
8.28	F.J.M. Smith, *Fundamentals of Fabrication and Welding Engineering*, Longman, 1992	Pearson Education Ltd
8.29	F.J.M. Smith, *Fundamentals of Fabrication and Welding Engineering*, Longman, 1992	Pearson Education Ltd
8.30	F.J.M. Smith, *Fundamentals of Fabrication and Welding Engineering*, Longman, 1992	Pearson Education Ltd
8.31	F.J.M. Smith, *Fundamentals of Fabrication and Welding Engineering*, Longman, 1992	Pearson Education Ltd
8.32	F.J.M. Smith, *Fundamentals of Fabrication and Welding Engineering*, Longman, 1992	Pearson Education Ltd
9.1	F.J.M. Smith, *Fundamentals of Fabrication and Welding Engineering*, Longman, 1992	Pearson Education Ltd
9.2	F.J.M. Smith, *Fundamentals of Fabrication and Welding Engineering*, Longman, 1992	Pearson Education Ltd
9.3	F.J.M. Smith, *Fundamentals of Fabrication and Welding Engineering*, Longman, 1992	Pearson Education Ltd
9.4	F.J.M. Smith, *Fundamentals of Fabrication and Welding Engineering*, Longman, 1992	Pearson Education Ltd
9.5	F.J.M. Smith, *Fundamentals of Fabrication and Welding Engineering*, Longman, 1992	Pearson Education Ltd
9.6	F.J.M. Smith, *Fundamentals of Fabrication and Welding Engineering*, Longman, 1992	Pearson Education Ltd
9.7	F.J.M. Smith, *Fundamentals of Fabrication and Welding Engineering*, Longman, 1992	Pearson Education Ltd
9.8	F.J.M. Smith, *Fundamentals of Fabrication and Welding Engineering*, Longman, 1992	Pearson Education Ltd
9.9	F.J.M. Smith, *Fundamentals of Fabrication and Welding Engineering*, Longman, 1992	Pearson Education Ltd
9.10	F.J.M. Smith, *Fundamentals of Fabrication and Welding Engineering*, Longman, 1992	Pearson Education Ltd
9.11	F.J.M. Smith, *Fundamentals of Fabrication and Welding Engineering*, Longman, 1992	Pearson Education Ltd
9.12	F.J.M. Smith, *Fundamentals of Fabrication and Welding Engineering*, Longman, 1992	Pearson Education Ltd
9.13	F.J.M. Smith, *Fundamentals of Fabrication and Welding Engineering*, Longman, 1992	Pearson Education Ltd
9.14	R. Timings, *Engineering Fundamentals*, Newnes, 2002	Pearson Education Ltd
9.15	R. Timings, *Engineering Fundamentals*, Newnes, 2002	Pearson Education Ltd
9.16	R. Timings, *Engineering Fundamentals*, Newnes, 2002	Pearson Education Ltd
9.17	R. Timings, *Engineering Fundamentals*, Newnes, 2002	Pearson Education Ltd
9.18	R. Timings, *Engineering Fundamentals*, Newnes, 2002	British Standards Institution
9.19	F.J.M. Smith, *Fundamentals of Fabrication and Welding Engineering*, Longman, 1992	Pearson Education Ltd
9.20	F.J.M. Smith, *Fundamentals of Fabrication and Welding Engineering*, Longman, 1992	Pearson Education Ltd
9.21	F.J.M. Smith, *Fundamentals of Fabrication and Welding Engineering*, Longman, 1992	Pearson Education Ltd
9.22	F.J.M. Smith, *Fundamentals of Fabrication and Welding Engineering*, Longman, 1992	Pearson Education Ltd

(Continued)

Figure No.	Source	Copyright holder
9.23	F.J.M. Smith, *Fundamentals of Fabrication and Welding Engineering*, Longman, 1992	Pearson Education Ltd
9.24	F.J.M. Smith, *Fundamentals of Fabrication and Welding Engineering*, Longman, 1992	Pearson Education Ltd
9.25	F.J.M. Smith, *Fundamentals of Fabrication and Welding Engineering*, Longman, 1992	Pearson Education Ltd
9.26	F.J.M. Smith, *Fundamentals of Fabrication and Welding Engineering*, Longman, 1992	Pearson Education Ltd
9.27	F.J.M. Smith, *Fundamentals of Fabrication and Welding Engineering*, Longman, 1992	Pearson Education Ltd
9.28	F.J.M. Smith, *Fundamentals of Fabrication and Welding Engineering*, Longman, 1992	Pearson Education Ltd
9.29	F.J.M. Smith, *Fundamentals of Fabrication and Welding Engineering*, Longman, 1992	Pearson Education Ltd
9.30	F.J.M. Smith, *Fundamentals of Fabrication and Welding Engineering*, Longman, 1992	Pearson Education Ltd
9.31	F.J.M. Smith, *Fundamentals of Fabrication and Welding Engineering*, Longman, 1992	Pearson Education Ltd
9.32	F.J.M. Smith, *Fundamentals of Fabrication and Welding Engineering*, Longman, 1992	Pearson Education Ltd
10.1	F.J.M. Smith, *Fundamentals of Fabrication and Welding Engineering*, Longman, 1992	Pearson Education Ltd
10.2	F.J.M. Smith, *Fundamentals of Fabrication and Welding Engineering*, Longman, 1992	Pearson Education Ltd
10.3	F.J.M. Smith, *Fundamentals of Fabrication and Welding Engineering*, Longman, 1992	TRI
10.4	F.J.M. Smith, *Fundamentals of Fabrication and Welding Engineering*, Longman, 1992	Pearson Education Ltd
10.5	F.J.M. Smith, *Fundamentals of Fabrication and Welding Engineering*, Longman, 1992	Pearson Education Ltd
10.6	F.J.M. Smith, *Fundamentals of Fabrication and Welding Engineering*, Longman, 1992	Pearson Education Ltd
10.7	F.J.M. Smith, *Fundamentals of Fabrication and Welding Engineering*, Longman, 1992	Pearson Education Ltd
10.8	F.J.M. Smith, *Fundamentals of Fabrication and Welding Engineering*, Longman, 1992	Pearson Education Ltd
10.9	F.J.M. Smith, *Fundamentals of Fabrication and Welding Engineering*, Longman, 1992	Pearson Education Ltd
10.10	F.J.M. Smith, *Fundamentals of Fabrication and Welding Engineering*, Longman, 1992	Pearson Education Ltd
10.11	F.J.M. Smith, *Fundamentals of Fabrication and Welding Engineering*, Longman, 1992	Pearson Education Ltd
10.12	F.J.M. Smith, *Fundamentals of Fabrication and Welding Engineering*, Longman, 1992	Pearson Education Ltd
10.13	F.J.M. Smith, *Fundamentals of Fabrication and Welding Engineering*, Longman, 1992	Pearson Education Ltd
10.14	F.J.M. Smith, *Fundamentals of Fabrication and Welding Engineering*, Longman, 1992	Pearson Education Ltd

(Continued)

Figure No.	Source	Copyright holder
10.15	F.J.M. Smith, *Fundamentals of Fabrication and Welding Engineering*, Longman, 1992	Pearson Education Ltd
10.16	F.J.M. Smith, *Fundamentals of Fabrication and Welding Engineering*, Longman, 1992	Pearson Education Ltd
10.17	F.J.M. Smith, *Fundamentals of Fabrication and Welding Engineering*, Longman, 1992	Pearson Education Ltd
10.18	F.J.M. Smith, *Fundamentals of Fabrication and Welding Engineering*, Longman, 1992	Pearson Education Ltd
10.19	F.J.M. Smith, *Fundamentals of Fabrication and Welding Engineering*, Longman, 1992	Pearson Education Ltd
10.20	F.J.M. Smith, *Fundamentals of Fabrication and Welding Engineering*, Longman, 1992	Pearson Education Ltd
10.21	F.J.M. Smith, *Fundamentals of Fabrication and Welding Engineering*, Longman, 1992	Pearson Education Ltd
10.22	F.J.M. Smith, *Fundamentals of Fabrication and Welding Engineering*, Longman, 1992	British Oxygen Company
10.23	F.J.M. Smith, *Fundamentals of Fabrication and Welding Engineering*, Longman, 1992	British Oxygen Company
10.24	F.J.M. Smith, *Fundamentals of Fabrication and Welding Engineering*, Longman, 1992	British Oxygen Company
10.25	F.J.M. Smith, *Fundamentals of Fabrication and Welding Engineering*, Longman, 1992	Pearson Education Ltd
11.1	F.J.M. Smith, *Fundamentals of Fabrication and Welding Engineering*, Longman, 1992	Pearson Education Ltd
11.2	F.J.M. Smith, *Fundamentals of Fabrication and Welding Engineering*, Longman, 1992	Pearson Education Ltd
11.3	F.J.M. Smith, *Fundamentals of Fabrication and Welding Engineering*, Longman, 1992	Pearson Education Ltd
11.4	F.J.M. Smith, *Fundamentals of Fabrication and Welding Engineering*, Longman, 1992	British Oxygen Company
11.5	F.J.M. Smith, *Fundamentals of Fabrication and Welding Engineering*, Longman, 1992	British Oxygen Company
11.6	F.J.M. Smith, *Fundamentals of Fabrication and Welding Engineering*, Longman, 1992	British Oxygen Company
11.7	F.J.M. Smith, *Fundamentals of Fabrication and Welding Engineering*, Longman, 1992	British Oxygen Company
11.8	F.J.M. Smith, *Fundamentals of Fabrication and Welding Engineering*, Longman, 1992	Pearson Education Ltd
11.9	F.J.M. Smith, *Fundamentals of Fabrication and Welding Engineering*, Longman, 1992	British Oxygen Company
11.10	F.J.M. Smith, *Fundamentals of Fabrication and Welding Engineering*, Longman, 1992	British Oxygen Company
11.11	F.J.M. Smith, *Fundamentals of Fabrication and Welding Engineering*, Longman, 1992	Training Publications Ltd
11.12	F.J.M. Smith, *Fundamentals of Fabrication and Welding Engineering*, Longman, 1992	Pearson Education Ltd
11.13	F.J.M. Smith, *Fundamentals of Fabrication and Welding Engineering*, Longman, 1992	Pearson Education Ltd

(Continued)

Figure No.	Source	Copyright holder
11.14	F.J.M. Smith, *Fundamentals of Fabrication and Welding Engineering*, Longman, 1992	Pearson Education Ltd
11.15	F.J.M. Smith, *Fundamentals of Fabrication and Welding Engineering*, Longman, 1992	British Oxygen Company
11.16	F.J.M. Smith, *Fundamentals of Fabrication and Welding Engineering*, Longman, 1992	British Oxygen Company
11.17	F.J.M. Smith, *Fundamentals of Fabrication and Welding Engineering*, Longman, 1992	Pearson Education Ltd
11.18	F.J.M. Smith, *Fundamentals of Fabrication and Welding Engineering*, Longman, 1992	Pearson Education Ltd
11.19	F.J.M. Smith, *Fundamentals of Fabrication and Welding Engineering*, Longman, 1992	Pearson Education Ltd
11.20	F.J.M. Smith, *Fundamentals of Fabrication and Welding Engineering*, Longman, 1992	Pearson Education Ltd
11.22	F.J.M. Smith, *Fundamentals of Fabrication and Welding Engineering*, Longman, 1992	Pearson Education Ltd
11.23	F.J.M. Smith, *Fundamentals of Fabrication and Welding Engineering*, Longman, 1992	Training Publications Ltd
11.24b	F.J.M. Smith, *Fundamentals of Fabrication and Welding Engineering*, Longman, 1992	Pearson Education Ltd
11.25	F.J.M. Smith, *Fundamentals of Fabrication and Welding Engineering*, Longman, 1992	Pearson Education Ltd
11.26	F.J.M. Smith, *Fundamentals of Fabrication and Welding Engineering*, Longman, 1992	Pearson Education Ltd
11.27	F.J.M. Smith, *Fundamentals of Fabrication and Welding Engineering*, Longman, 1992	Pearson Education Ltd
11.28	F.J.M. Smith, *Fundamentals of Fabrication and Welding Engineering*, Longman, 1992	Pearson Education Ltd
11.29	F.J.M. Smith, *Fundamentals of Fabrication and Welding Engineering*, Longman, 1992	Pearson Education Ltd
11.31	F.J.M. Smith, *Fundamentals of Fabrication and Welding Engineering*, Longman, 1992	Pearson Education Ltd
11.32	R. Timings, *Workshop Processes and Materials, Level 1*, 2nd edn, Longman, 1984	Pearson Education Ltd
11.33	F.J.M. Smith, *Fundamentals of Fabrication and Welding Engineering*, Longman, 1992	Pearson Education Ltd
11.34	F.J.M. Smith, *Fundamentals of Fabrication and Welding Engineering*, Longman, 1992	Pearson Education Ltd
11.35	F.J.M. Smith, *Fundamentals of Fabrication and Welding Engineering*, Longman, 1992	Pearson Education Ltd
11.36	F.J.M. Smith, *Fundamentals of Fabrication and Welding Engineering*, Longman, 1992	Pearson Education Ltd
11.37	F.J.M. Smith, *Fundamentals of Fabrication and Welding Engineering*, Longman, 1992	Pearson Education Ltd
11.38	F.J.M. Smith, *Fundamentals of Fabrication and Welding Engineering*, Longman, 1992	Pearson Education Ltd
11.39	F.J.M. Smith, *Fundamentals of Fabrication and Welding Engineering*, Longman, 1992	Pearson Education Ltd

(Continued)

Figure No.	Source	Copyright holder
11.40	F.J.M. Smith, *Fundamentals of Fabrication and Welding Engineering*, Longman, 1992	Pearson Education Ltd
11.41	F.J.M. Smith, *Fundamentals of Fabrication and Welding Engineering*, Longman, 1992	Pearson Education Ltd
11.42	F.J.M. Smith, *Fundamentals of Fabrication and Welding Engineering*, Longman, 1992	Pearson Education Ltd
11.43	F.J.M. Smith, *Fundamentals of Fabrication and Welding Engineering*, Longman, 1992	Pearson Education Ltd
11.44	F.J.M. Smith, *Fundamentals of Fabrication and Welding Engineering*, Longman, 1992	Pearson Education Ltd
12.1	R. Timings, *Materials Technology, Level 3*, Longman, 1985	Pearson Education Ltd
12.2	R. Timings, *Engineering Materials, Volume 1*, 2nd edn, Pearson Education Ltd, 1998	Pearson Education Ltd
12.3	R. Timings, *Engineering Materials, Volume 1*, 2nd edn, Pearson Education Ltd, 1998	Pearson Education Ltd
12.4	R. Timings, *Engineering Materials, Volume 1*, 2nd edn, Pearson Education Ltd, 1998	Pearson Education Ltd
12.5	R. Timings, *Engineering Materials, Volume 1*, 2nd edn, Pearson Education Ltd, 1998	Pearson Education Ltd
12.6	R. Timings, *Engineering Materials, Volume 1*, 2nd edn, Pearson Education Ltd, 1998	Pearson Education Ltd
12.7	R. Timings, *Engineering Materials, Volume 1*, 2nd edn, Pearson Education Ltd, 1998	Pearson Education Ltd
12.8	R. Timings, *Engineering Materials, Volume 1*, 2nd edn, Pearson Education Ltd, 1998	Pearson Education Ltd
12.9	R. Timings, *Engineering Materials, Volume 1*, 2nd edn, Pearson Education Ltd, 1998	Pearson Education Ltd
12.10	R. Timings, *Engineering Materials, Volume 1*, 2nd edn, Pearson Education Ltd, 1998	Pearson Education Ltd
12.11	R. Timings, *Engineering Materials, Volume 1*, 2nd edn, Pearson Education Ltd, 1998	Pearson Education Ltd
12.12	R. Timings, *Engineering Materials, Volume 1*, 2nd edn, Pearson Education Ltd, 1998	Pearson Education Ltd
12.13	R. Timings, *Workshop Processes and Materials, Level 1*, 2nd edn, Longman, 1984	Pearson Education Ltd
12.14	R. Timings, *Engineering Materials, Volume 1*, 2nd edn, Pearson Education Ltd, 1998	Pearson Education Ltd
12.15	R. Timings, *Workshop Processes and Materials, Level 1*, 2nd edn, Longman, 1984	Pearson Education Ltd
12.16	R. Timings, *Workshop Processes and Materials, Level 1*, 2nd edn, Longman, 1984	Pearson Education Ltd
12.17	R. Timings, *Workshop Processes and Materials, Level 1*, 2nd edn, Longman, 1984	Pearson Education Ltd
12.18	R. Timings, *Workshop Processes and Materials, Level 1*, 2nd edn, Longman, 1984	Pearson Education Ltd
12.19	R. Timings, *Workshop Processes and Materials, Level 1*, 2nd edn, Longman, 1984	Pearson Education Ltd
12.20	R. Timings, *Workshop Processes and Materials, Level 1*, 2nd edn, Longman, 1984	Pearson Education Ltd

(*Continued*)

Figure No.	Source	Copyright holder
12.21	R. Timings, *Workshop Processes and Materials, Level 1*, 2nd edn, Longman, 1984	Pearson Education Ltd
12.22	R. Timings, *Workshop Processes and Materials, Level 1*, 2nd edn, Longman, 1984	Pearson Education Ltd
12.23	R. Timings, *Workshop Processes and Materials, Level 1*, 2nd edn, Longman, 1984	Pearson Education Ltd
12.24	R. Timings, *Engineering Materials, Volume 1*, 2nd edn, Pearson Education Ltd, 1998	Pearson Education Ltd
12.25	R. Timings, *Engineering Materials, Volume 1*, 2nd edn, Pearson Education Ltd, 1998	Pearson Education Ltd
12.26	R. Timings, *Engineering Materials, Volume 1*, 2nd edn, Pearson Education Ltd, 1998	Pearson Education Ltd
12.27	R. Timings, *Engineering Materials, Volume 1*, 2nd edn, Pearson Education Ltd, 1998	Pearson Education Ltd
12.28	R. Timings, *Engineering Materials, Volume 1*, 2nd edn, Pearson Education Ltd, 1998	Pearson Education Ltd

Table No.	Source	Copyright holder
1.1	R. Timings, *General Engineering*, Longman, 1995	Pearson Education Ltd
1.2	F.J.M. Smith, *Fundamentals of Fabrication and Welding Engineering*, Longman, 1992	Pearson Education Ltd
1.3	F.J.M. Smith, *Fundamentals of Fabrication and Welding Engineering*, Longman, 1992	Pearson Education Ltd
1.4	F.J.M. Smith, *Fundamentals of Fabrication and Welding Engineering*, Longman, 1992	Pearson Education Ltd
1.5	F.J.M. Smith, *Fundamentals of Fabrication and Welding Engineering*, Longman, 1992	Pearson Education Ltd
1.6	R. Timings, *Engineering Fundamentals*, Newnes, 2002	Pearson Education Ltd
1.7	R. Timings, *Engineering Fundamentals*, Newnes, 2002	Pearson Education Ltd
1.8	R. Timings, *Engineering Fundamentals*, Newnes, 2002	Pearson Education Ltd
3.1	R. Timings, *Engineering Fundamentals*, Newnes, 2002	Pearson Education Ltd
3.2	R. Timings, *Engineering Fundamentals*, Newnes, 2002	Pearson Education Ltd
3.3	R. Timings, *Engineering Fundamentals*, Newnes, 2002	Pearson Education Ltd
3.4	R. Timings, *Engineering Fundamentals*, Newnes, 2002	Pearson Education Ltd
3.5	R. Timings, *Engineering Fundamentals*, Newnes, 2002	Pearson Education Ltd
3.6	R. Timings, *Engineering Fundamentals*, Newnes, 2002	Pearson Education Ltd
3.7	R. Timings, *Engineering Fundamentals*, Newnes, 2002	Pearson Education Ltd
3.8	R. Timings, *Engineering Fundamentals*, Newnes, 2002	Pearson Education Ltd
3.9	R. Timings, *Engineering Fundamentals*, Newnes, 2002	Pearson Education Ltd
3.10	R. Timings, *Engineering Fundamentals*, Newnes, 2002	Pearson Education Ltd
3.11	R. Timings, *Engineering Fundamentals*, Newnes, 2002	Pearson Education Ltd
3.12	R. Timings, *Engineering Fundamentals*, Newnes, 2002	Pearson Education Ltd
3.13	R. Timings, *Engineering Fundamentals*, Newnes, 2002	Pearson Education Ltd
3.14	R. Timings, *Engineering Fundamentals*, Newnes, 2002	Pearson Education Ltd
3.15	R. Timings, *Engineering Fundamentals*, Newnes, 2002	

(Continued)

Table No.	Source	Copyright holder
4.1	R. Timings, *Engineering Fundamentals*, Newnes, 2002	Pearson Education Ltd
4.2	R. Timings, *Engineering Fundamentals*, Newnes, 2002	Pearson Education Ltd
4.3	R. Timings, *Engineering Fundamentals*, Newnes, 2002	Pearson Education Ltd
4.4	R. Timings, *Engineering Fundamentals*, Newnes, 2002	Pearson Education Ltd
4.5	R. Timings, *Engineering Fundamentals*, Newnes, 2002	Pearson Education Ltd
5.1	F.J.M. Smith, *Fundamentals of Fabrication and Welding Engineering*, Longman, 1992	Pearson Education Ltd
5.2	F.J.M. Smith, *Fundamentals of Fabrication and Welding Engineering*, Longman, 1992	Pearson Education Ltd
5.3	F.J.M. Smith, *Fundamentals of Fabrication and Welding Engineering*, Longman, 1992	Pearson Education Ltd
6.1	F.J.M. Smith, *Fundamentals of Fabrication and Welding Engineering*, Longman, 1992	Pearson Education Ltd
6.2	R. Timings, *Engineering Fundamentals*, Newnes, 2002	Pearson Education Ltd
6.3	R. Timings, *Engineering Fundamentals*, Newnes, 2002	Pearson Education Ltd
6.4	F.J.M. Smith, *Fundamentals of Fabrication and Welding Engineering*, Longman, 1992	Pearson Education Ltd
6.5	R. Timings, *Engineering Fundamentals*, Newnes, 2002	Pearson Education Ltd
6.6	R. Timings, *Engineering Fundamentals*, Newnes, 2002	Pearson Education Ltd
6.7	R. Timings, *Engineering Fundamentals*, Newnes, 2002	Pearson Education Ltd
6.8	R. Timings, *Engineering Fundamentals*, Newnes, 2002	Pearson Education Ltd
6.9	R. Timings, *Engineering Fundamentals*, Newnes, 2002	Pearson Education Ltd
6.10	R. Timings, *Workshop Processes and Materials, Level 1*, 2nd edn, Longman, 1984	Pearson Education Ltd
7.1	F.J.M. Smith, *Fundamentals of Fabrication and Welding Engineering*, Longman, 1992	Pearson Education Ltd
7.2	F.J.M. Smith, *Fundamentals of Fabrication and Welding Engineering*, Longman, 1992	Pearson Education Ltd
7.3	F.J.M. Smith, *Fundamentals of Fabrication and Welding Engineering*, Longman, 1992	Pearson Education Ltd
7.4	F.J.M. Smith, *Fundamentals of Fabrication and Welding Engineering*, Longman, 1992	Pearson Education Ltd
7.5	F.J.M. Smith, *Fundamentals of Fabrication and Welding Engineering*, Longman, 1992	Pearson Education Ltd
8.1	F.J.M. Smith, *Fundamentals of Fabrication and Welding Engineering*, Longman, 1992	Pearson Education Ltd
8.2	F.J.M. Smith, *Fundamentals of Fabrication and Welding Engineering*, Longman, 1992	Pearson Education Ltd
8.3	F.J.M. Smith, *Fundamentals of Fabrication and Welding Engineering*, Longman, 1992	Pearson Education Ltd
9.1	R. Timings, *Engineering Fundamentals*, Newnes, 2002	Pearson Education Ltd
9.2	F.J.M. Smith, *Fundamentals of Fabrication and Welding Engineering*, Longman, 1992	Pearson Education Ltd
10.1	F.J.M. Smith, *Fundamentals of Fabrication and Welding Engineering*, Longman, 1992	Pearson Education Ltd

(Continued)

Table No.	Source	Copyright holder
10.2	F.J.M. Smith, *Fundamentals of Fabrication and Welding Engineering*, Longman, 1992	Pearson Education Ltd
10.3	F.J.M. Smith, *Fundamentals of Fabrication and Welding Engineering*, Longman, 1992	Pearson Education Ltd
10.4	F.J.M. Smith, *Fundamentals of Fabrication and Welding Engineering*, Longman, 1992	Pearson Education Ltd
10.5	F.J.M. Smith, *Fundamentals of Fabrication and Welding Engineering*, Longman, 1992	Pearson Education Ltd
10.6	F.J.M. Smith, *Fundamentals of Fabrication and Welding Engineering*, Longman, 1992	Pearson Education Ltd
10.7	F.J.M. Smith, *Fundamentals of Fabrication and Welding Engineering*, Longman, 1992	Pearson Education Ltd
10.8	F.J.M. Smith, *Fundamentals of Fabrication and Welding Engineering*, Longman, 1992	Pearson Education Ltd
10.9	F.J.M. Smith, *Fundamentals of Fabrication and Welding Engineering*, Longman, 1992	Pearson Education Ltd
11.1	F.J.M. Smith, *Fundamentals of Fabrication and Welding Engineering*, Longman, 1992	Pearson Education Ltd
11.2	F.J.M. Smith, *Fundamentals of Fabrication and Welding Engineering*, Longman, 1992	British Oxygen Company
11.3	F.J.M. Smith, *Fundamentals of Fabrication and Welding Engineering*, Longman, 1992	British Oxygen Company
11.4	F.J.M. Smith, *Fundamentals of Fabrication and Welding Engineering*, Longman, 1992	British Oxygen Company
11.5	F.J.M. Smith, *Fundamentals of Fabrication and Welding Engineering*, Longman, 1992	Pearson Education Ltd
12.1	R. Timings, *Engineering Materials, Volume 1*, 2nd edn, Pearson Education Ltd, 1998	Pearson Education Ltd
12.2	R. Timings, *Engineering Materials, Volume 1*, 2nd edn, Pearson Education Ltd, 1998	Pearson Education Ltd
12.3	R. Timings, *Engineering Materials, Volume 1*, 2nd edn, Pearson Education Ltd, 1998	Pearson Education Ltd
12.4	R. Timings, *Engineering Materials, Volume 1*, 2nd edn, Pearson Education Ltd, 1998	Pearson Education Ltd
12.5	R. Timings, *Engineering Materials, Volume 1*, 2nd edn, Pearson Education Ltd, 1998	

1

Health and safety

When you have read this chapter, you should understand:

- The statutory requirements for general health and safety at work
- Accident and first aid procedures
- Fire precautions and procedures
- Protective clothing and equipment
- Correct manual lifting and carrying techniques
- How to use lifting equipment
- Safe working practices – cutting and forming tools
- Safe working practices – portable power tools
- Safe working practices – oxy-fuel-gas brazing and welding
- Safe working practices – electric arc welding
- Safe working practices – thermal cutting processes
- Safe working practices – working on site

1.1 Health, safety and the law

1.1.1 Health and Safety at Work etc. Act

It is essential to observe safe working practices not only to safeguard yourself, but also to safeguard the people with whom you work. The Health and Safety at Work etc. Act provides a comprehensive and integrated system of law for dealing with the health, safety and welfare of work-people and the general public as affected by industrial, commercial and associated activities.

The Act places the responsibility for safe working equally upon:

1. The employer.
2. The employee (that means you).
3. The manufacturers and suppliers of materials, goods, equipment and machinery.

1.1.2 Health and Safety Commission

The Act provides for a full-time, independent chairman and between six and nine part-time commissioners. The commissioners are made up of three trade union members appointed by the TUC, three management members appointed by the CBI, two Local Authority members, and one independent member. The commission has taken over the responsibility previously held by various Government Departments for the control of most occupational health and safety matters. The commission is also responsible for the organization and functioning of the Health and Safety Executive.

1.1.3 Health and Safety Executive

The inspectors of the Health and Safety Executive (HSE) have very wide powers. Should an inspector find a contravention of one of the provisions of earlier Acts or Regulations still in force, or a contravention of the Health and Safety at Work etc. Act, the inspector has three possible lines of action available.

Prohibition notice

If there is a risk of serious personal injury, the inspector can issue a *Prohibition Notice*. This immediately stops the activity that is giving rise to the risk until the remedial action specified in the notice has been taken to the Inspector's satisfaction. The prohibition notice can be served upon the person undertaking the dangerous activity, or it can be served upon the person in control of the activity at the time the notice is served.

Improvement notice

If there is a legal contravention of any of the relevant statutory provisions, the inspector can issue an *Improvement Notice*. This notice requires the infringement to be remedied within a specified time. It can be served on any person on whom the responsibilities are placed. The latter person can be an employer, employee or a supplier of equipment or materials.

Prosecution

In addition to serving a Prohibition Notice or an Improvement Notice, the inspector can prosecute any person (including an employee – you) for contravening a relevant statutory provision. Finally, the inspector can seize, render harmless or destroy any substance or article which the inspector considers to be the cause of imminent danger or personal injury.

Thus every employee must be a fit and trained person capable of carrying out his or her assigned task properly and safely. Trainees must work under the supervision of a suitably trained, experienced worker or instructor. By law, every employee must:

1. Obey all the safety rules and regulations of his or her place of employment.
2. Understand and use, as instructed, the safety practices incorporated in particular activities or tasks.
3. Not proceed with his or her task if any safety requirement is not thoroughly understood, guidance must be sought.
4. Keep his or her working area tidy and maintain his or her tools in good condition.
5. Draw the attention of his or her immediate supervisor or the safety officer to any potential hazard.
6. Report all accidents or incidents (even if injury does not result from the incident) to the responsible person.
7. Understand emergency procedures in the event of an accident or an alarm.
8. Understand how to give the alarm in the event of an accident or an incident such as fire.
9. Co-operate promptly with the senior person in charge in the event of an accident or an incident such as fire.

Therefore, safety health and welfare are very personal matters for any worker, such as you, who is just entering the engineering industry. This chapter sets out to identify the main hazards and suggests how they may be avoided. Factory life, and particularly fabrication and welding engineering, is potentially dangerous and you must take a positive approach towards safety, health and welfare.

1.1.4 Further legislation and regulations concerning safety

In addition to the Health and Safety at Work etc. Act, the following are examples of legislation and regulations that also control the conditions under which you work and the way in which you work. Such legislation is updated from time to time and new legislation concerning health and safety is introduced. Ensure that you are familiar with the latest issue of such legislation.

1. Factories Act.
2. Safety Representatives and Safety Committees Regulations.
3. Notification of Accidents and General Occurrences Regulations.
4. The Management of Health and Safety at Work Regulations.
5. The Protection of Eyes Regulations.
6. Electricity at Work Regulations.
7. Low Voltage Electrical Equipment (Safety) Regulations. This includes voltage ranges of 50 volts to 1000 volts (AC) and 75 volts to 1500 volts (DC).

8. Abrasive Wheels Regulations.

9. Noise at Work Regulations.

You are not expected to have a detailed knowledge of all this legislation, but you are expected to know of its existence, the main topic areas that it covers, and how it affects your working conditions, your responsibilities, and the way in which you work.

There are many other Laws and Regulations that you will come across depending upon the branch of the fabrication and welding engineering industry in which you work and the equipment you use.

1.2 Employers' responsibilities

All employers must, by law, maintain a safe place to work. To fulfil all the legal obligations imposed upon them, employers must ensure that:

1. The workplace is provided with a safe means of access and exit so that in the case of an emergency (such as fire) no one will be trapped. This is particularly important when the workplace is not at ground level. Pedestrian access and exits should be segregated away from trucks that are delivering materials or collecting finished work. The premises must be kept in good repair – worn floor coverings and stair treads are a major source of serious falls.

2. All plant and equipment is safe so that it not only complies with the *Machinery Directive* but also complies with British Standards Institution (BSI) and CE requirements. It must be correctly installed and properly maintained. The plant and any associated cutters and tools must also be properly guarded.

3. Working practices and systems are safe and that, where necessary, protective clothing is provided.

4. A safe, healthy and comfortable working environment is provided. That the temperature and humidity is maintained at the correct levels for the work being undertaken.

5. There is an adequate supply of fresh air and that fumes and dust are either eliminated altogether, or are reduced to an acceptable and safe level.

6. There is adequate and suitable natural and artificial lighting in the workplace, particularly over stairways.

7. There is adequate and convenient provision for washing and sanitation.

8. There are adequate first aid facilities under the supervision of a qualified person. This can range from a first aid box under the supervision of a person trained in basic first aid procedures for a small firm, to a full-scale ambulance room staffed by professionally qualified medical personnel in a large firm.

9. Provision is made for the safe handling, storing, and transportation of raw materials, work-in-progress and finished goods awaiting delivery.

10. Provision is made for the safe handling, storing, transportation and use of dangerous substances such as compressed gases (e.g. oxygen and acetylene), and toxic and flammable solvents.

11. There is a correct and legal system for the reporting of accidents and the logging of such accidents in the *accident register*.

12. There is a company policy for adequate instruction, training and supervision of employees. This must not only be concerned with safety procedures but also

with good working practices. Such instruction and training to be updated at regular intervals.

13. There is a safety policy in force. This safety policy must be subject to regular review. One of the more important innovations of the Health and Safety at Work etc. Act is contained in section 2(4) which provides for the appointment of safety representatives from amongst the employees, who will represent them in consultation with the employers, and have other prescribed functions.

14. Where an employer receives a written request from at least two safety representatives to form a *safety committee* the employer shall, after consulting with the applicants and representatives of other recognized unions (if applicable) whose members work in the workplace concerned, establish a safety committee within the period of three months after the request. The employer must post a notice of the composition of the committee and the workplaces covered. The notice must be positioned where it may be easily read by the employees concerned.

15. Membership of the safety committee should be settled by consultation. The number of management representatives should not exceed the number of safety representatives. Where a company doctor, industrial hygienist or safety officer/adviser is employed they should be ex-officio members of the committee.

16. Management representation is aimed at guaranteeing the necessary knowledge and expertise to provide accurate information on company policy, production needs and technical matters in relation to premises, processes, plant, machinery and equipment.

1.3 Employees' responsibilities

All employees (including you) are as equally responsible for safety as are their employers. Under the Health and Safety at Work etc. Act, employees are expected to take reasonable care for their own health and safety together with the health and safety of other people with whom they work, and members of the public who are affected by the work being performed.

Further, the misuse of, or interference with, equipment provided by an employer for health and safety purposes is a *criminal offence*. It is up to all workers to develop a sense of *safety awareness* by following the example set by their instructors. Regrettably not all older workers observe the safety regulations as closely as they should. Take care who you choose for your 'role model'.

The basic requirements for safe working are to:

1. Learn the safe way of doing each task. This is also the correct way.
2. Use the safe way of carrying out the task in practice.
3. Ask for instruction if you do not understand a task or have not received previous instruction.
4. Be constantly on your guard against careless actions by yourself or by others.
5. Practice good housekeeping at all times, e.g. keep your working area clean and tidy and your tools and equipment in good condition.
6. Co-operate promptly in the event of an accident or a fire.
7. Report all accidents to your instructor or supervisor.
8. Draw your instructor's or your supervisor's attention to any potential hazard you have noticed.

1.4 Electrical hazards

The most common causes of electrical shock are shown in Fig. 1.1. The installation and maintenance of electrical equipment must be carried out only by a fully trained and registered electrician. The installation and equipment must conform to international standards and regulations as laid down in Safety Legislation and the Codes of Practice and Regulations, published by the Institution of Electrical Engineers (IEE).

An electric shock from a 240 volt single-phase supply (lighting and office equipment) or a 415 volt three-phase supply (most factory machines) can easily kill you. Even if the shock is not sufficiently severe to cause death, it can still cause serious injury. The sudden convulsion caused by the shock can throw you from a ladder or against moving machinery. To reduce the risk of shock, all electrical equipment should be *earthed* or *double insulated*. Further, portable power tools should be fed from a low-voltage transformer at 110 volts. The power tool must be suitable for operating at such a voltage. The transformer itself should be protected by a circuit breaker containing a residual current detector.

The fuses and circuit breakers designed to protect the supply circuitry to the transformer react too slowly to protect the user from electric shock. The electrical supply to a portable power tool should, therefore, be additionally protected by a residual current detector (RCD). Such a device compares the magnitudes of the current flowing in the live and neutral conductors supplying the tool. Any leakage to earth through the body of the user or by any other route will upset the balance between these two currents. This results in the supply being immediately disconnected. The sensitivity of residual current detectors is such that a difference of only a few milliamperes is sufficient to cut off the supply and the time delay is only a few microseconds. Such a small current applied for such a short time is not normally dangerous.

In the event of rendering first aid to the victim of electrical shock, great care must be taken when pulling the victim clear of the fault which caused the shock. The victim can act as a conductor and thus, in turn, electrocute the rescuer. If the supply cannot be quickly and completely disconnected, always pull the victim clear by his or her clothing which, if dry, will act as an insulator. If in doubt, hold the victim with a plastic bag or cloth known to be dry. Never touch the victim's bare flesh

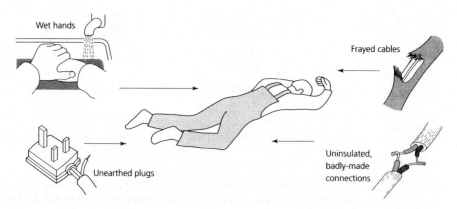

Figure 1.1 Causes of electric shock

until the victim is clear of the electrical fault. Artificial respiration must be started immediately the victim has been pulled clear of the fault or the live conductor.

1.5 Fire fighting

Fire fighting is a highly skilled operation and most medium and large firms have properly trained teams who can contain the fire locally until the professional brigade arrives. The best way you can help is to learn the correct fire drill, both how to give the alarm and how to leave the building. It only requires one person to panic and run in the wrong direction to cause a disaster.

In an emergency never lose your head and panic

Smoke is the main cause of panic. It spreads quickly through a building, reducing visibility and increasing the risk of falls down stairways. It causes choking and even death by asphyxiation. Smoke is less dense near the floor: as a last resort – crawl. To reduce the spread of smoke and fire, keep fire doors closed at all times but never locked. The plastic materials used in the finishes and furnishings of modern buildings give off highly toxic fumes. Therefore it is best to leave the building as quickly as possible and leave the fire-fighting to the professionals who have breathing apparatus.

If you do have to fight a fire there are some basic rules to remember. A fire is the rapid oxidation (burning) of flammable materials at relatively high temperatures. Figure 1.2 shows that removing the air (oxygen), or the flammable materials (fuel), or lowering the temperature will result in the fire ceasing to burn. It will go out. It can also be seen from Fig. 1.2 that different fires require to be dealt with in different ways.

Saving human life is more important than saving property

1.5.1 Fire extinguishers

The normally available fire extinguishers and the types of fire they can be used for are as follows.

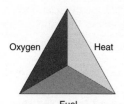

Oxygen   Heat

Fuel

The 3 essentials to start a fire

Note: Once the fire has started it produces sufficient heat to maintain its own combustion reactions and sufficient surplus heat to spread the fire

Remove heat

When solids are on fire remove heat by applying water

Remove oxygen

Liquids, such as petrol etc. on fire can be extinguished by removing oxygen with a foam or dry powder extinguisher

Remove fuel

Electrical or gas fires can usually be extinguised by turning off the supply of energy

Figure 1.2 How to remove each of the three items necessary to start a fire (Note: once the fire has started it provides sufficient heat to maintain its own combustion reaction and sufficient surplus heat spread the fire)

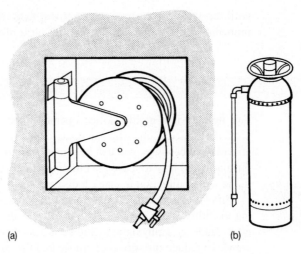

(a)   (b)

Figure 1.3 Hose point (a) and pressurized water extinguisher (b)

Water

Used in large quantities, water reduces the temperature and puts out the fire. The steam generated also helps to smother the flames as it displaces the air and therefore the oxygen essential to the burning process. However, for various technical reasons, water should only be used on burning solids such as wood, paper and some plastics. A typical hose point and a typical pressurized water extinguisher is shown in Fig. 1.3.

Foam extinguishers

These are used for fighting oil and chemical fires. The foam smothers the flames and prevents the oxygen in the air from reaching the burning materials at the seat of the fire. Water alone cannot be used because oil floats on the water and this spreads the area of the fire. A typical foam extinguisher is shown in Fig. 1.4(a).

Note: Since both water and foam are electrically conductive, do not use them on fires associated with electrical equipment or the person wielding the hose or the extinguisher could be electrocuted.

Carbon dioxide (CO_2) extinguishers

These are used on burning gases and vapours. They can also be used for oil and chemical fires in confined places. The carbon dioxide gas replaces the air and smothers the fire. It can only be used in confined places, where it cannot be displaced by draughts.

Note: If the fire cannot breathe, neither can you breathe so care must be taken to evacuate all living creatures from the vicinity before operating the extinguisher. Back away from the bubble of CO_2 gas as you operate the extinguisher, do not advance towards it. Fig. 1.4(b) shows a typical CO_2 extinguisher.

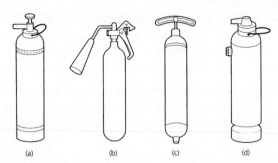

(a)   (b)   (c)   (d)

Figure 1.4 Fire extinguishers: (a) foam; (b) CO_2; (c) vaporizing liquid; (d) dry powder

Vaporizing liquid extinguishers

These include CTC, CBM and BCF extinguishers. The heat from the fire causes rapid vaporization of the liquid sprayed from the extinguisher and this vapour displaces the air and smothers the fire. Since a small amount of liquid produces a very large amount of vapour, this is a very efficient way of producing the blanketing vapour. Any vapour that will smother the fire will also smother all living creatures which must be evacuated before using such extinguishers. As with CO_2 extinguishers always back away from the bubble of vapour, never advance into it. Vaporizing liquid extinguishers are suitable for oil, gas, vapour and chemical fires. Like CO_2 extinguishers, vaporizing liquid extinguishers are safe to use on fires associated with electrical equipment. A typical example of a vaporizing liquid extinguisher is shown in Fig. 1.4(c).

Dry powder extinguishers

These are suitable for small fires involving flammable liquids and small quantities of solids such as paper. They are also useful for fires in electrical equipment, offices and kitchens since the powder is not only non-toxic, it can be easily removed by vacuum cleaning and there is no residual mess. The active ingredient is powdered sodium bicarbonate (baking powder) which gives off carbon dioxide when heated. A typical example of a dry powder extinguisher is shown in Fig. 1.4(d).

1.5.2 General rules governing the use of portable extinguishers

- Since fire spreads quickly, a speedy attack is essential if the fire is to be contained.
- Sound the alarm immediately the fire is discovered.
- Send for assistance before attempting to fight the fire.
- Remember:
  (a) Extinguishers are only provided to fight small fires.
  (b) Take up a position between the fire and the exit, so that your escape cannot be cut off.
  (c) **Do not** continue to fight the fire if:
    (i) It is dangerous to do so
    (ii) There is any possibility of your escape route being cut off by fire, smoke, or collapse of the building

(iii) The fire spreads despite your efforts

(iv) Toxic fumes are being generated by the burning of plastic furnishings and finishes

(v) There are gas cylinders or explosive substances in the vicinity of the fire.

If you have to withdraw from the fire, wherever possible close windows and doors behind you but not if such actions endanger your escape. Finally, ensure that all extinguishers are recharged immediately after use.

1.6 Fire precautions and prevention

1.6.1 Fire precautions

It is the responsibility of employers and their senior management (duty of care) to ensure the safety of their employees in the event of fire. However, the following precautions should be taken to prevent fire breaking out in the first place:

- Ensure ease of exit from the premises at all times – emergency exits must not be locked or obstructed.
- Easy access for fire appliances from the local, professional Brigade.
- Regular inspection of the plant, premises and processes by the Local Authority Fire Brigade's fire prevention officer. No new plant or processes involving flammable substances should be used without prior notification and inspection by the fire prevention officer.
- The above note also applies to the Company's insurance inspector.
- Regular and frequent fire drills must be carried out and a log kept of such drills including the time taken to evacuate the premises. A roll call of all persons present should be taken immediately the evacuation is complete. A meeting of the safety committee should be called as soon as possible after a fire drill to discuss any problems, improve procedures and to learn lessons from the exercise.

1.6.2 Fire prevention

Prevention is always better than cure, and fire prevention is always better than fire fighting. Tidiness is of paramount importance in reducing the possibility of outbreaks of fire. Fires have small beginnings and it is usually amongst accumulated rubbish that many fires originate. So you should make a practice of constantly removing rubbish, shavings, off-cuts, cans, bottles, waste paper, oily rags, and other unwanted materials to a safe place at regular intervals. Discarded foam plastic packing is not only highly flammable, but gives off highly dangerous toxic fumes when burnt.

Highly flammable materials, such as the gases used for welding, brazing and flame-cutting processes, should be stored in specially designed and equipped compounds away from the main working areas. In the case of paints, adhesives and solvents only minimum quantities of such materials for immediate use should be allowed into the workshop at any one time, and then only into *non-smoking zones*. The advice of the Local Authority Fire Brigade's fire prevention officer should also be sought.

It is good practice to provide metal containers with air-tight hinged lids with proper markings as to the type of rubbish they should contain since some types of rubbish will ignite spontaneously when mixed. The lids of the bins should be kept closed so that, if a fire starts, it will quickly use up the air in the bin and go out of its own accord without doing any damage.

1.7 Accidents

Accidents do not happen, they are caused. There is not a single accident that could not have been prevented by care and forethought on somebody's part. Accidents can and must be prevented. They cost millions of lost man-hours of production every year, but this is of little importance compared with the immeasurable cost in human suffering. On average, nearly one hundred workers are the victims of industrial accidents in every eight-hour shift. Many of these will be blinded, maimed for life, or confined to a hospital bed for months. At least two of them will die. Figure 1.5 shows the main causes of accidents.

Figure 1.5 Average national causes of industrial accidents (by per cent of all accidents)

1.7.1 Accident procedure

You must learn and obey the accident procedures for your company.

1. Report all accidents, no matter how small and trivial they seem, to your supervisor, instructor or tutor. Record your report and details of the incident on an accident form.
2. Receive first-aid treatment from a *qualified* person, or your company's medical centre, depending upon the size of your company and its policy.

> Failure to log all accidents is an offence under the Health and Safety at Work etc. Act and can lead to prosecution in the courts

It is important that you follow the procedures laid down by your company since the accident register has to be produced on request by any HSE inspector visiting your company.

Also if at some future date you had to seek compensation as a result of the accident, your report is important evidence.

1.7.2 Warning signs and labels

Let's now consider some examples of the signs and labels you may come across in your place of work, commencing with *warning signs*. You must be aware of the warning signs and their meanings. You must also obey such signs. To disregard them is an offence under the Health and Safety at Work etc. Act.

Warning signs are triangular in shape and all the sides are the same length. The background colour is *yellow* and there is a *black* border. In addition to warning signs there are also *warning labels*. Figure 1.6 shows some typical warning signs and warning labels. It also gives their meanings.

Prohibition signs

You can recognize these signs as they have a red circular band and a red crossbar on a white background. Figure 1.7 shows five typical prohibition signs. These signs indicate activities that are *prohibited* at all times. They *must be obeyed*, you have no

Caution, risk of fire	Caution, risk of explosion	Caution, toxic hazard	Caution, corrosive substance	Caution, risk of ionizing radiation

Caution, overhead load	Caution, industrial trucks	Caution, risk of electric shock	General warning, caution, risk of danger	Caution, laser beam

Figure 1.6 Warning signs

option in the matter. To disregard them in an *offence in law*, as you would be putting yourself and others at considerable risk.

Mandatory signs

You can recognize these signs as they have a blue background colour. The symbol must be white. Figure 1.8 shows five typical mandatory signs. These signs indicate things that *you must do* and precautions that *you must take*. These signs *must be obeyed*, you have no option in the matter. To disregard them in an *offence in law* as, again, you would be putting yourself at considerable risk.

Safe condition signs

In addition to the signs discussed so far that tell you what to look out for, what you should do and what you should not do, there are also signs that tell what is safe. These have a white symbol on a green background. The example shown in Fig. 1.9(a)

| No smoking | Smoking and naked flames prohibited | Pedestrians prohibited | Do not extinguished with water | Not drinking water (that is, do not drink) |

Figure 1.7 Prohibition signs

| Eye protection must be worn | Head protection must be worn | Hearing protection must be worn | Foot protection must be worn | Hand protection must be worn |

Figure 1.8 Mandatory signs

First aid

(a)

- Background colour shall be green.
- The symbol or text shall be white. The shape of the sign shall be oblong or square as necessary to accommodate the symbol or text.
- Green shall cover at least 50% of the area of the safety sign.

Indication of direction

(b)

Figure 1.9 Safe condition signs

indicates a first aid post or an ambulance room. The example shown in Fig. 1.9(b) indicates a safe direction in which to travel, e.g. towards an emergency exit.

1.8 First aid

Accidents can happen anywhere at any time. The injuries caused by such accidents can range from minor cuts and bruises to broken bones and life threatening injuries. It is a very good idea to know what to do in an emergency.

- You must be aware of the accident procedure.
- You must know where to find your nearest first aid post.
- You must know the quickest and easiest route to the first aid post.
- You must know who is the *qualified* first-aid person on duty and where he/she can be found).

First aid should only be adminis-tered by a quali-fied person

Unfortunately in this day and age, more and more people are being encouraged to seek compensation through the courts of law. Complications resulting from amateur-ish but well-intentioned and well-meaning attempts at first aid on your part could result in you being sued for damages. If a qualified person is not immediately availa-ble, summon the paramedics by telephoning for the emergency services. Remember that workshop materials and equipment are, medically speaking, essentially dirty and can easily carry dirt into a wound causing it to become infected. Hence the need for any wound, however minor, to be assessed and treated by a qualified person.

1.8.1 In the event of an emergency

If you are first on the scene of a serious incident, but you are not qualified to render first aid:

- Remain calm.
- Get help quickly by sending for the appropriate skilled and qualified personnel.
- Act and speak in a calm and confident manner to give the casualty confidence.
- Do not attempt to move the casualty.
- Do not administer fluids.
- Hand over to the experts as quickly as possible.

Minor wounds

Prompt first aid can help nature heal small wounds and deal with germs. If you have to treat yourself then wash the wound clean and apply a plaster. However, you must seek medical advice if:

- There is a foreign body embedded in the wound.
- There is a special risk of infection (such as a dog bite or the wound has been caused by a dirty object).
- A wound shows signs of becoming infected.

Sometimes there can be *foreign bodies* in minor wounds. Small pieces of glass or grit lying on a wound can be picked off with tweezers or rinsed off with cold water before treatment. However, you MUST NOT try to remove objects that are embedded in the wound; you may cause further tissue damage and bleeding.

- Control any bleeding by applying firm pressure on either side of the object, and raising the wounded part.
- Drape a piece of gauze lightly over the wound to minimize the risk of germs entering it, then build up padding around the object until you can bandage with out pressing down upon it.
- Take or send the casualty to hospital.

Bruises

These are caused by internal bleeding that seeps through the tissues to produce the discoloration under the skin. Bruising may develop very slowly and appear hours, even days, after injury. Bruising that develops rapidly and seems to be the main problem will benefit from first aid. Caution, bruises may indicate deeper injury. Seek professional advice.

Minor burns and scalds

These are treated to stop the burning, to relieve pain and swelling and to minimize the risk of infection. If you are in any doubt as to the severity of the injury seek the advice of a doctor.

Do Not

- Break blisters or interfere with the injured area; you are likely to introduce an infection.
- Use adhesive dressings or strapping.
- Apply lotions, ointments, creams or fats to the injury.

Note: Chemical burns to the skin and particularly the eyes require immediate and specialist treatment. Expert attention must be obtained *immediately*.

Foreign bodies in the eye

Foreign bodies in the eye can lead to blurred vision with pain or discomfort. They can also lead to redness and watering of the eye. A speck of dust or grit, or a loose eyelash floating on the white of the eye can generally be removed easily. However, a foreign body that adheres to, or penetrates the eyeball, or rests on the coloured part of the eye, should NOT be removed, even by a person qualified in First Aid. DO NOT touch anything sticking to, or embedded in, the eyeball or the coloured part of the eye. Cover the affected eye with an eye pad, bandage both eyes, then take or send the casualty to hospital.

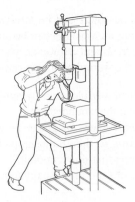

Figure 1.10 The hazard of long hair

1.9 Personal protection

1.9.1 Appearance

Clothing

For general workshop purposes a boiler suit is the most practical and safest form of clothing. Whatever form of overall you wear it should be changed and cleaned regularly. Dirty overalls not only look scruffy and are a poor advert for the company which employs you; they can also cause your skin to become infected. See also section 1.10.2.

Long hair

Long hair is liable to be caught in moving machinery such as drilling machines. This can result in the hair and scalp being torn away which is extremely dangerous and painful. Permanent disfigurement could result and brain damage can even occur

Long hair is a health hazard, as it is almost impossible to keep clean and free from infection in a workshop environment. It is also a serious hazard in a workshop. If it becomes entangled in a machine, as shown in Fig. 1.10, the operator can be scalped. If you wish to retain a long hairstyle in the interests of fashion, then your hair must be contained in a close fitting cap. This also helps to keep your hair and scalp clean and healthy.

1.9.2 Head and eye protection

When working on site, or in a heavy engineering erection shop involving the use of overhead cranes, all persons should wear a safety helmet complying with BS 2826. Even small objects such as nuts and bolts can cause serious head injuries when dropped from a height. Safety helmets (hard hats) are made from high impact resistant plastics or from fibre-glass reinforced polyester mouldings. Figure 1.11(a) shows such a helmet. Such helmets can be colour-coded for personnel identification and are light and comfortable to wear. Despite their lightweight construction, they have a high resistance to impact and penetration. To eliminate the possibility of electric shock, safety helmets have no metal parts. The harness inside a safety helmet should be adjusted so as to provide ventilation and a fixed safety clearance between the

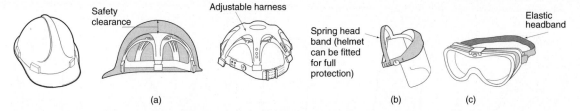

Figure 1.11 Head protection: (a) a typical fibre-glass safety helmet made to BS 2826; (b) plastic face safety visor for complete protection against chemical and salt-bath splashes; (c) transparent plastic goggles suitable for machining operations

outer shell of the helmet and the wearer's skull. This clearance must be maintained at 32 millimetres. The entire harness is removable for regular cleaning and sterilizing. It is fully adjustable for size, fit and angle to suit the individual wearer's head.

Whilst it is possible to walk about on an artificial leg, nobody has ever seen out of a glass eye. Therefore, eye protection is possibly the most important precaution you can take in a workshop. Eye protection is provided by wearing suitable visors as shown in Fig. 1.11(b) or goggles as shown in Fig. 1.11(c). Where eye safety is concerned, prevention is better than cure. There may be no cure! Eye injuries fall into three main categories:

- Pain and inflammation due to abrasive grit and dust getting between the lid and the eye.
- Damage due to exposure to ultraviolet radiation (arc-welding) and high intensity visible light (oxy-acetylene welding). Particular care is required when using laser equipment for plate cutting.
- Loss of sight due to the eyeball being pierced or the optic nerve cut by flying splinters of metal (swarf), or by the blast of a compressed air jet.

1.9.3 Hand protection

The edges of thin sheet metal can be razor sharp and can cause deep and serious cuts. Gloves and 'palms' of a variety of styles and types of materials are available to protect your hands whatever the nature of the work. Some examples are shown in Fig. 1.12. In general terms, plastic gloves are impervious to liquids and should be worn when handling oils, greases and chemicals. However, they are unsuitable and even dangerous for handling hot materials. Leather gloves should be used when handling sharp, rough and hot materials. NEVER handle hot metal with plastic gloves. These could melt onto and into your flesh causing serious burns that would be difficult to treat.

1.9.4 Hand cleansing

DO NOT use solvents to clean your hands. As well as removing oils, greases, paints and adhesives, solvents also remove the natural protective oils from your skin. This leaves the skin open to infection and can lead to cracking and sores. It can also result in sensitization of the skin and the onset of industrial dermatitis or worse.

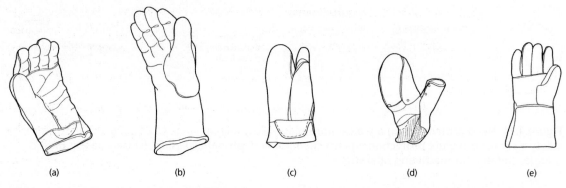

Figure 1.12 Gloves suitable for industrial purposes: (a) leather glove with reinforced palm – ideal for handling sheet steel and sections; (b) gauntlet – available in rubber, neoprene or PVC for handling chemical, corrosive or oily materials; (c) heat resistant leather glove – can be used for handling objects heated up to 300°C; (d) chrome leather hand pad or 'palm' – very useful for handling sheet steel, sheet glass, etc.; (e) industrial gauntlets – usually made of leather because of its heat resistance: gauntlets nor only protect the hands but also the wrists and forearms from slashes from molten salts and quenching media

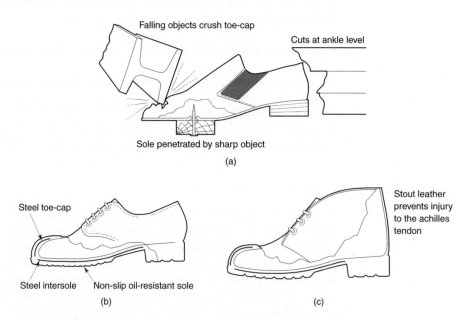

Figure 1.13 Safety footware: (a) lightweight shoes offer no protection: (b) industrial safety shoes; (c) industrial safety boot

Where gloves are inappropriate, but your hands still need to be protected from oil and dirt rather that from cuts and abrasions, then you should use a barrier cream. This is a mildly antiseptic cream that you can rub into your hands before work. It fills the pores of your skin and prevents the entry of oils and dirt that could cause infection. The cream is water-soluble and can be easily removed by washing your hand with ordinary soap and water at the end of the shift. Removal of the barrier cream also carries away the dirt and sources of infection.

1.9.5 Foot protection

The injuries that you can suffer when wearing lightweight, casual shoes in a workshop environment are shown in Fig. 1.13. This figure also shows some examples of safety footwear as specified in BS 1870. Such safety footwear is available in a variety of styles and prices. It looks as smart as normal footwear and is equally as comfortable.

1.10 Hazards in the workplace

1.10.1 Health Hazards

Noise

Excessive noise can be a dangerous pollutant of the working environment. The effects of noise can result in:

- Fatigue leading to careless accidents.
- Mistaken communications between workers leading to accidents.
- Ear damage leading to deafness.
- Permanent nervous disorders.

Noise is energy and it represents waste since it does not do useful work. Ideally, it should be suppressed at source to avoid waste of energy and to improve the working environment. If this is not possible then you should be insulated from the noise by sound absorbent screens and/or ear-protectors (ear-muffs).

Narcotic (anaesthetic) effects

Exposure to small concentrations of narcotic substances causes headaches, giddiness and drowsiness. Under such conditions you are obviously prone to accidents since your judgement and reactions are adversely affected.

A worker who has become disorientated by the inhalation of narcotics is a hazard to him or herself and a hazard to other workers

Examples of narcotic substances are to be found amongst the many types of solvent used in industry. Solvents are used in paints, adhesives, polishes and degreasing agents. Careful storage and use is essential and should be carefully supervised by qualified persons. Fume extraction and adequate ventilation of the workplace must be provided when working with these substances. Suitable respirators should be available for use in emergencies.

Irritant effects

Many substances cause irritation to the skin both externally and internally. External irritants can cause industrial dermatitis by coming into contact with your skin. The main irritants met with in a workshop are oils (particularly cutting oils and coolants), adhesive, degreasing solvents, and electro-plating chemicals. Internal irritants are the more dangerous as they may have long term and deep-seated effects on the major organs of the body. They may cause inflammation, ulceration, internal bleeding, poisoning and the growth of cancerous tumours. Internal irritants are usually air pollutants in the form of dusts (asbestos fibres), fumes and vapours. As well as being inhaled, they may also be carried into your body on food handled without washing. Even the cutting oils used on machine tools can be dangerous if you allow your overalls to become impregnated with the spray. Change your overalls regularly.

Systemic effects

Toxic substances, also known as *systemics*, affect the fundamental organs and bodily functions. They damage your brain, heart, lungs, kidneys, liver, your central nervous system and your bone marrow. Their effects cannot be reversed and thus lead to chronic ill-health and, ultimately, early death. These toxic substances may enter the body in various ways.

- Dust and vapour can be breathed in through your nose. Observe the safety codes when working with such substances and wear the respirator provided, no matter how inconvenient or uncomfortable.
- Liquids and powders contaminating your hands can be transferred to the digestive system by handling food or cigarettes with dirty hands. Always wash before eating or smoking. Never smoke in a prohibited area. Not only may there be a fire risk, but some vapours change chemically and become highly toxic (poisonous) when inhaled through a cigarette.
- Liquids, powders, dusts and vapours may all enter the body through the skin:
  a) directly through the pores
  b) by destroying the outer tough layers of the skin and attacking the sensitive layers underneath
  c) by entering through undressed wounds.

Again it must be stressed that regular washing with soap and water, the use of a barrier cream, the use of suitable protective gloves, and the immediate dressing of cuts (no matter how small) is essential to proper hand care.

1.10.2 Personal hygiene

Personal hygiene is most important. It ensures good health and freedom from industrial diseases and can go a long way towards preventing skin diseases, both irritant and infectious. Your employer's safety policy should make recommendations on dress and hygiene and they should provide suitable protective measures. As previously mentioned, dirty and oil-soaked overalls are a major source of skin infection. Correct dress not only makes you look smart and feel smart, it helps you to avoid accidents and industrial diseases. This is why overalls should be regularly changed and cleaned. Finally, you must always wash your hands thoroughly before handling and eating any food, and when going to the toilet. If your hands are dirty and oily it is essential to wash them *before* as well as *after*.

1.10.3 Behaviour in workshops

In an industrial environment reckless, foolish and boisterous behaviour such as pushing, shouting, throwing objects and practical joking by a person or a group of persons cannot be tolerated. Such actions can distract a worker's attention and break his or her concentration which could lead to:

- Scrapped work, serious accidents and even fatalities.
- Incorrect operation of equipment or inadvertent contact with moving machinery or cutters.
- Someone being pushed against moving machinery or factory transport.
- Someone being pushed against ladders and trestles upon which people are working at heights.
- Someone being pushed against and dislodging heavy, stacked components.
- Contact with electricity, compressed air or dangerous chemicals.

1.10.4 Hazards associated with hand tools

Newcomers to industry often overlook the fact that, as well as machine tools, badly maintained and incorrectly used hand tools can also represent a serious safety hazard

The time and effort taken to fetch the correct tool from the stores or to service a worn tool is considerably less than the time taken to recover from injury. Figure 1.14 shows some badly maintained and incorrectly used hand tools. Chipping screens, as shown in Fig. 1.15, should be used when removing metal or arc welding slag with a cold chisel to prevent injury from the pieces of metal flying from the cutting edge of the chisel. For this reason, goggles should also be worn and you should never chip towards another worker.

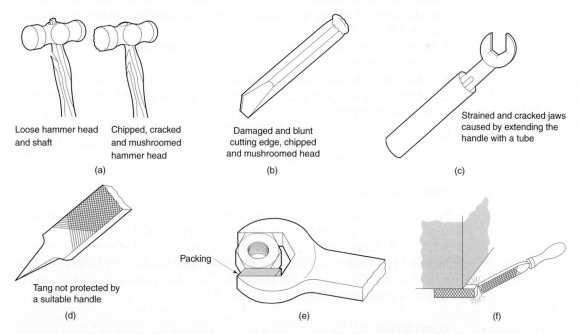

Loose hammer head and shaft   Chipped, cracked and mushroomed hammer head

Damaged and blunt cutting edge, chipped and mushroomed head

Strained and cracked jaws caused by extending the handle with a tube

(a)   (b)   (c)

Tang not protected by a suitable handle

Packing

(d)   (e)   (f)

Figure 1.14 Hand tools in a dangerous condition and misused: (a) hammer faults; (b) chisel faults; (c) spanner faults; (d) file faults (e) do not use oversize spanner and packing – use the correct size spanner for both the nut or bolt head; (f) do not use a file as a lever

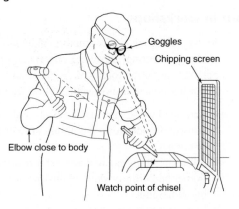

Figure 1.15 Safety when using a cold chisel

1.10.5 Hazards associated with machine tools

Metal cutting and forming machines are potentially dangerous.

- Before operating any machinery make sure that you have been fully instructed in how to use it, the dangers associated with it, and that you have been given permission to use it.
- Do not operate a machine unless all the guards and safety devices are in position and are operating correctly. Guards should only be fitted and adjusted by a qualified person.
- Make sure you understand any special rules and regulations applicable to the particular machine you are about to use, even if you have been trained on machines in general.
- Never clean or adjust a machine whilst it is in motion. Stop the machine and isolate it from the supply.
- Report any dangerous aspect of the machine you are using, or are about to use, immediately and do not use it until it has been made safe by a suitably qualified and authorized person.
- A machine may have to be stopped in an emergency. Learn how to make an emergency stop without having to pause and think about it and without having to search for the emergency stop switch.

Transmission guards

By law, no machine can be sold or hired out unless all gears, belts, shafts and couplings making up the power transmission system are guarded so that they cannot be touched when they are in motion. Figure 1.16 shows a typical transmission guard fitted to a motorized guillotine shear. Sometimes guards have to be removed in order to replace, adjust or service the components they are covering. This must only be done by a qualified maintenance mechanic.

Cutter guards

The machine manufacturer does not normally provide cutter guards because of the wide range of work a machine may have to do.

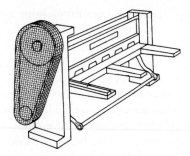

Figure 1.16 Typical transmission guard for the belt drive of a motorised guillotine shear

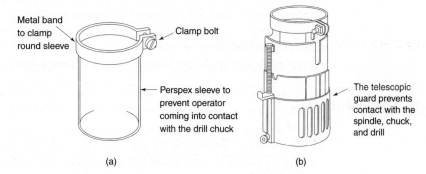

Figure 1.17 Drill chuck guards: (a) simple; (b) telescopic

- It is the responsibility of the owner or the hirer of the machine to supply cutter guards that are appropriate for the machine and for the process being performed on it.
- It is the responsibility of the setter and/or the operator to make sure that the guards are fitted and working correctly before operating the machine, and to use the guards as instructed. It is an offence in law for the operator to remove or tamper with the guards provided.
- If ever you are doubtful about the adequacy of a guard or the safety of a process, consult your instructor or your safety officer without delay.

The simple drilling machine guard shown in Fig. 1.17(a) only covers the chuck and is only suitable for jobbing work when small diameter drills are being used. The drill chuck shown in Fig. 1.17(b) is used for larger drills and for drills which are mounted directly into the drilling machine spindle. It covers the whole length of the drill and telescopes up as the drill penetrates into the work-piece.

1.11 Manual lifting

1.11.1 Individual lifting

In the engineering industry it is often necessary to lift fairly heavy loads. As a general rule, loads lifted manually should not exceed 20 kg.

Mechanical lifting equipment should be used for loads in excess of 20 kg

However, even lifting loads less than 20 kg can cause strain and lifting loads incorrectly is one of the major causes of back trouble. If the load is obviously too heavy or bulky for one person to handle, you should ask for assistance. Even a light load can be dangerous if it obscures your vision as shown in Fig. 1.18. All moveable objects that form hazardous obstructions should be moved to a safe place before movement of the load commences.

The correct way to lift a load manually is shown in Fig. 1.19. You should start the lift in a balanced squatting position with your legs at hip width apart and one foot slightly ahead of the other. The load to be lifted should be held close to your body. Make sure that you have a safe and secure grip on the load. Before taking the weight of the load, your back should be straightened and as near to the vertical as possible. Keep your head up and your chin drawn in; this helps to keep your spine straight and rigid as shown in Fig. 1.19(a). To raise the load, first straighten your legs; this ensures that the load is being raised by your powerful thigh muscles and bones, as shown in Fig. 1.19(b), and not by your back. To complete the lift, raise the upper part of your body to a vertical position as shown in Fig. 1.19(c). To carry the load, keep your body upright and hold the load close to your body as shown in Fig. 1.20.

1.11.2 Team lifting

When a lifting party is formed to move a particularly large or heavy load the team leader is solely responsible for the safe completion of the task. The team leader should not take part in the actual lifting but should ensure that:

■ Everyone understands what the job involves and the method chosen for its completion.

Clear movable objects

Figure 1.18 Obstructions to safe movement must be removed

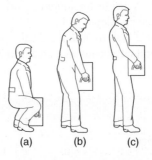

(a)   (b)   (c)

Figure 1.19 Correct manual lifting: (a) keep back straight and near vertical; (b) keep your spine straight; (c) straighten your legs to raise the load

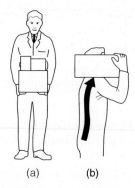

(a)   (b)

Figure 1.20 Correct carrying: (a) keep body upright and load close to body; (b) let your bone structure support the load

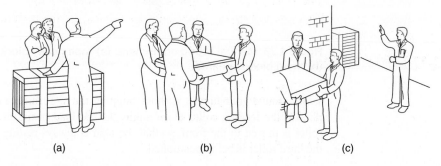

(a)   (b)   (c)

Figure 1.21 Team lifting

- The area is clear of obstructions and that the floor is safe and will provide a good foothold.
- The members of the lifting team are of similar height and physique and that they are wearing any necessary protective clothing. Each person should be positioned so that the weight is evenly distributed.
- The team leader should take up a position which gives the best all round view of the area and will permit the development of any hazardous situation to be seen so that appropriate action can be taken in time to prevent an accident.
- Any equipment moved in order to carry out the operation is replaced in its original position when the task has been completed.

The sequence of events is shown in Fig. 1.21.

1.11.3 Moving loads that are too heavy to lift

Loads that are too heavy to be lifted or carried can still be moved manually by using a crowbar and rollers as shown in Fig. 1.22. The rollers should be made from thick walled tubes so that there is no danger of trapping your fingers if the load should move whilst positioning the rollers. Never close your fingers around a roller when moving it. Always put your fingers inside the tubular roller so if it moves, it rotates

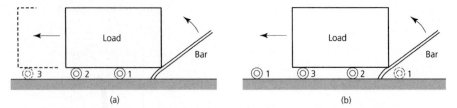

Figure 1.22 Use of rollers: (a) load is rolled forward on rollers 1 and 2 until it is on rollers 2 and 3; (b) roller 1 is moved to the front ready for the next move

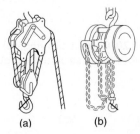

Figure 1.23 Manual lifting equipment: (a) rope pulley blocks (snatch blocks); (b) chain blocks (geared)

safely around your fingers without trapping them. Turning a corner is achieved by placing the leading roller at an angle. As the load clears the rearmost roller, this roller is moved to the front, so that the load is always resting on two rollers, whilst the third roller is being positioned.

1.12 Mechanical lifting equipment

Mechanical lifting equipment can be classified according to the motive power used to operate it.

1.12.1 Manual (muscle power)

Examples of this type of equipment are shown in Fig. 1.23. Rope pulley blocks (snatch blocks) are light and easily mounted. However, the tail rope has to be tied off to prevent the load falling when the effort force is removed. Some rope blocks have an automatic brake which is released by giving the tail rope a sharp tug before lowering the load. They are suitable for loads up to 250 kg.Chain pulley blocks are portable and are used for heavier loads from 250 kg to 1 tonne. They have the advantage that they do not run back (overhaul) when the effort raising the load is removed.

1.12.2 Powered

An example of an electrically powered chain block is shown in Fig. 1.24. Powered lifting equipment can raise greater loads more quickly than manually operated chain blocks.

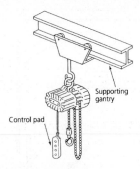

Figure 1.24 Powered lifting equipment

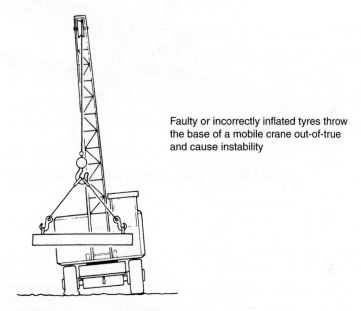

Faulty or incorrectly inflated tyres throw the base of a mobile crane out-of-true and cause instability

Figure 1.25 Importance of stability

1.12.3 Mobile cranes

Fabrication and welding engineers often have to work on site where any heavy lifting is performed with the aid of mobile cranes. These are powered by diesel, petrol or LPG fuelled engines. Because of the exhaust fumes they are unsuitable for use indoors. Adequate ventilation is natural and normal out of doors on construction sites. The engine is coupled to the mechanism of the crane either mechanically or hydraulically.

Special precautions must be taken when using mobile cranes

■ Never stand under suspended loads – this applies to all forms of hoist.
■ Check the condition of the ground. Soft ground and faulty or incorrectly inflated tyres can have a destabilizing effect as shown in Fig. 1.25.
■ Beware of overhead, high-voltage cables as shown in Fig. 1.26.
■ Do not swing the load to increase the reach of the jib. Loads should only be lifted vertically. It is a hazard to swing the load out manually to gain additional radius

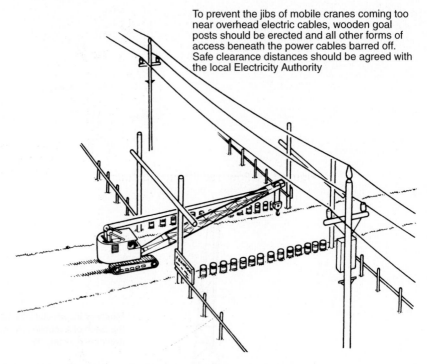

To prevent the jibs of mobile cranes coming too near overhead electric cables, wooden goal posts should be erected and all other forms of access beneath the power cables barred off. Safe clearance distances should be agreed with the local Electricity Authority

Figure 1.26 Care when working near overhead cables

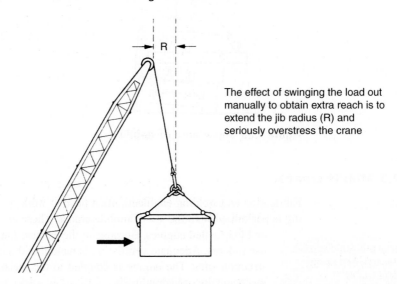

The effect of swinging the load out manually to obtain extra reach is to extend the jib radius (R) and seriously overstress the crane

Figure 1.27 Do not over-reach the jib

for, in doing so, the effect is to extend the effective length of the jib and throw stresses on the crane for which it is not designed as shown in Fig. 1.27. The crane could also become unstable and overturn.

■ A similar problem arises if the load is swung sideways. Again the jib is overstressed and the crane is destabilized and in danger of overturning as shown in Fig. 1.28.

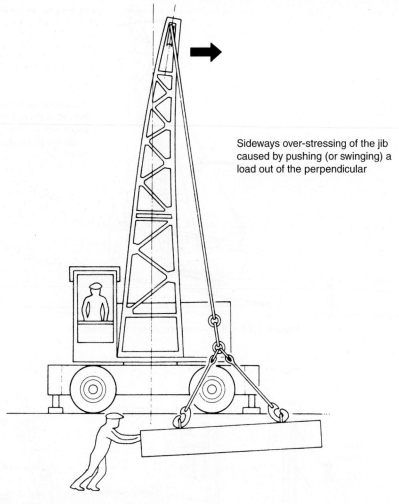

Sideways over-stressing of the jib
caused by pushing (or swinging) a
load out of the perpendicular

Figure 1.28 Do not swing the load

■ Mobile cranes should not be moved with the jib in its minimum radius position,
particularly if moving up an incline as shown in Fig. 1.29. The jib can whip back
and overturn the crane.

1.12.3 Safety

Only fully competent persons (i.e. trained and authorized) are permitted to operate mechanical lifting equipment. Trainees can only use such equipment under the
close supervision of a qualified and authorized instructor. Even after you have been
instructed in the use of lifting equipment, always make the following checks before
attempting to raise a load.

1. Check the lifting equipment is suitable for the load being raised. All lifting equipment should be clearly marked with its SWL.

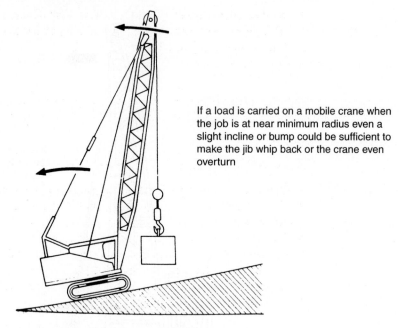

If a load is carried on a mobile crane when the job is at near minimum radius even a slight incline or bump could be sufficient to make the jib whip back or the crane even overturn

Figure 1.29 Care when moving the load

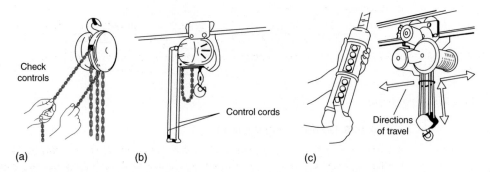

Check controls

Control cords

Directions of travel

(a)   (b)   (c)

Figure 1.30 Check the operating procedure to raise, lower and move the load. (a) Manual chain blocks. Check which way the chain has to be pulled to raise or lower the load; (b) Power hoist. Check which cord raises and which lowers the load; (c) Power hoist with power transverse. Check all the pendant controls to raise, lower and move the load

2. Never leave a load unattended whilst it is supported by lifting equipment, and ensure that before it is released from the lifting equipment, it is resting in a stable condition on a suitable support.
3. Hand chain: check the direction in which the chain must be pulled to raise or lower the load as shown in Fig. 1.30(a).
4. Cord control: when using cord-controlled power-operated hoists, check which cord raises the load and which cord lowers the load as shown in Fig. 1.30(b).

5. Pendant switch control: if the pendant switch controls a travelling hoist, check the direction of travel as well as checking which buttons raise and lowers the load as shown in Fig. 1.30(c).

1.13 Use of lifting equipment

1.13.1 Lifting a load

Before lifting a load using a mechanical lifting device you should:

- Warn everyone near the load and anyone approaching the load to keep clear.
- Check that all slings and ropes are safely and securely attached both to the load and to the hook.
- Take up the slack in the chain, sling or rope gently.
- Raise the load slowly and steadily so that it is just off the ground.
- Check that the load is stable and that the sling has not accidentally caught on a part of the load incapable of sustaining the lifting force.
- Stand well back from the load and lift steadily.

1.13.2 Traversing a load on a travelling crane

Before traversing a load with a travelling crane:

- Make sure the load will not pass over anyone.
- Check that there are no obstacles in the way of the crane and its load.
- Keep well clear of the load and move it steadily.
- Stop the crane immediately if anyone moves across the path of the load.

1.13.3 Lowering a load

Before lowering a load always:

- Check that the ground is clear of obstacles and is capable of supporting the load.
- Place timbers under the load so that the sling(s) will not be trapped and damaged. This will also facilitate the removal of the sling(s) as shown in Fig. 1.31(a).
- Lower the load until it is close to the ground, then gently ease it onto the timbers until the strain is gradually taken off the lifting equipment as shown in Fig. 1.31(b). It may be necessary to manually guide the load into place, in which case safety shoes and protective gloves should always be worn.

Never work under a suspended load

- Always lower the load onto suitable supports as shown in Fig. 1.31(c).

1.13.4 Hand signals

If the hoist or crane operator cannot see the load – as, for example, when working on site with a mobile crane positioning girders on a high building – the assistance of a trained and qualified signaller is required. The signaller should be positioned so as to see the load clearly whilst, at the same time, the hoist or crane operator should be able

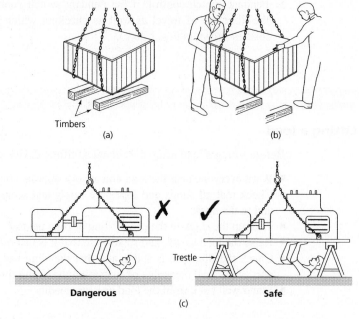

Figure 1.31 Care when lowering a load: (a) lower onto timbers; (b) guide by hand; (c) never work under a suspended load

to see the signaller. Both must be familiar with the Standard Code of Hand Signals as shown in Fig. 1.32.

1.14 Accessories for lifting gear

1.14.1 Hooks

These are made from forged steel and are carefully proportioned so that the load will not slip from them whilst being lifted. The hooks of lifting gear are frequently painted bright yellow to attract attention and to prevent people walking into them.

1.14.2 Slings

These are used to attach the load to the hook of the lifting equipment. There are four types in common use. They must all be marked with tags stating their SWL.

1. *Chain slings* (Fig. 1.33(a)) – as well as general lifting operations, only this type of sling is suitable for lifting loads having sharp edges or for lifting hot materials.
2. *Wire rope slings* (Fig. 1.33(b)) – these are widely used for general lifting. They should not be used for loads with sharp edges or for hot loads; nor should they be allowed to become rusty. Further, they should not be bent round a diameter of less than 20 times the diameter of the wire rope itself.

Raise load   Move load to my right   Travel towards me   Stop

Lower load   Move load to my left   Travel away from me   Emergency stop

Figure 1.32 Standard code of hand signals for crane control

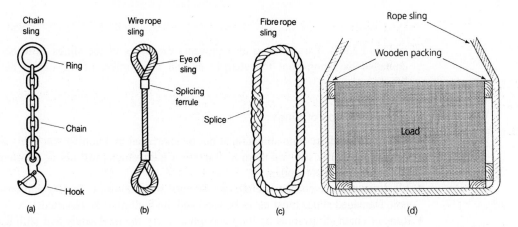

Figure 1.33 Types of sling

3. *Fibre rope slings* (Fig. 1.33(c)) – these may have eyes spliced into them or, more usually, they are endless as shown. They are used for general lifting, and are particularly useful where machined surfaces and paintwork have to be protected from damage.
4. *Belt or strap slings* – because of their breadth these do not tend to bite into the work and cause damage to the surface finish of the work. Rope and belt slings themselves must be protected from being cut or frayed by sharp edges as shown

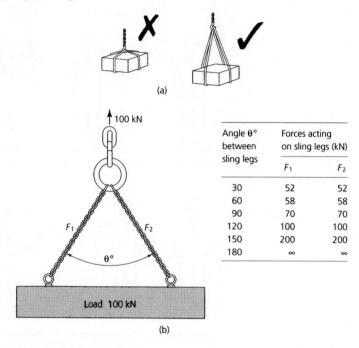

(a)

(b)

Angle θ° between sling legs	Forces acting on sling legs (kN)	
	F_1	F_2
30	52	52
60	58	58
90	70	70
120	100	100
150	200	200
180	∞	∞

Figure 1.34 Length of slings: (a) incorrect and correct length; (b) effect of leg angle on forces acting on the sling

in Fig. 1.33(d). This packing also prevents the fibres of the slings from being damaged by being bent too sharply around the corners of the object being lifted.

Care and condition of slings

- Wire rope and fibre rope slings must not be shortened by knotting since this will damage the fibres, causing them to fracture. Chain slings must not be shortened by bolting the links together.
- All slings must be checked before use for cuts, wear, abrasion, fraying and corrosion. Damaged slings must never be used and the fault must be reported.
- Rope or chain slings must be long enough to carry the load safely and with each leg as nearly vertical as possible as shown in Fig. 1.34(a). The load on a sling increases rapidly as the angle between the legs of the sling becomes greater. This is shown in Fig. 1.34(b).

1.14.3 Rings

These are used for ease of attachment of the sling to the crane hook. They also prevent the sling being sharply bent over the hook. Figure 1.35(a) shows a chain sling fitted with a suitable ring at one end. Figure 1.35(b) shows how a ring is used in conjunction with a rope sling.

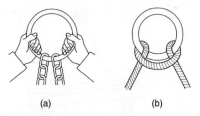

Figure 1.35 Use of rings: (a) with 2 leg chain sling; (b) with a rope sling

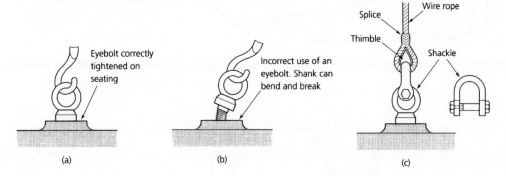

Figure 1.36 Use of eye bolts: (a) correct use; (b) incorrect use; (c) a shackle connects the eye of the sling to an eye bolt

1.14.4 Eyebolts and shackles

Forged steel eyebolts are frequently provided for lifting equipment and assemblies such as electric motors, gearboxes, and small machine tools. An example of the correct use of an eyebolt is shown in Fig. 1.36(a), whilst Fig. 1.36(b) shows how eyebolts must never be used. Forged steel shackles are used to connect lifting accessories together. In the example shown in Fig. 1.36(c), the eye of a wire rope sling is connected to an eyebolt using a shackle.

1.14.5 Special purpose equipment

Figure 1.37(a) shows how plate hooks and a spreader is used. Spreaders are also used when lifting long loads such as bundles of tubing, bundles of reinforcing rod, girders, etc. Figure 1.37(c) shows a plate hook in greater detail.

1.15 Useful knots for fibre ropes

Knots should only be tied in fibre ropes, they must **never** be tied in **wire ropes** as they are not sufficiently flexible and permanent damage will be caused. Some widely used knots are as follows.

- *Reef knot.* This is used for joining ropes of equal thickness (Fig. 1.38(a)).
- *Clove hitch.* This is used for attaching a rope to a pole or bar (Fig. 1.38(b)).

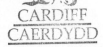

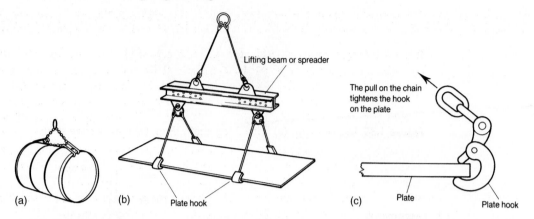

Figure 1.37 Special purpose equipment. (a) Use of barrel claws; (b) use of a spreader; (c) plate hook

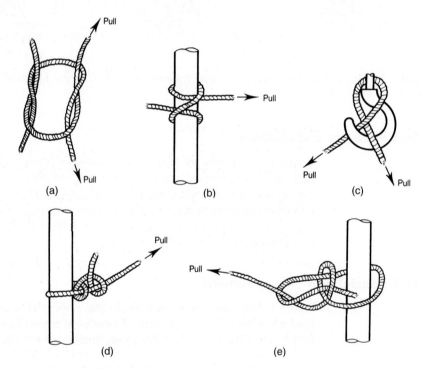

Figure 1.38 Useful knots for fibre ropes. (a) Reef knot; (b) Clove hitch; (c) single loop; (d) two half hitches; (e) bowline

- *Single loop*. This is used to prevent fibre ropes from slipping off crane hooks (Fig. 1.38(c)).
- *Half hitches*. Two half hitches can be used to secure a rope to a solid pole or for securing a rope to a sling (Fig. 1.38(d)).
- *Bowline*. This is used to form a loop which will not tighten under load (Fig. 1.38(e)).

1.16 Inspection (lifting equipment)

It is a *legal requirement* under the Health and Safety at Work etc. Act that all lifting equipment is regularly inspected by qualified engineers specializing in such work and that the results of such inspections are recorded in the register provided. If an inspector condemns any item of equipment it must be taken out of service immediately and either rectified or destroyed. If rectified, it must be re-inspected and approved by a qualified inspector before being taken back into service. The inspector will, on each visit, also confirm the SWL markings for each piece of equipment. No new item of lifting equipment must be taken into service until it has been inspected and certificated. Failure to comply with the appropriate regulations is not only a breach of the legal requirements leading to a court action but, in the event of an accident, it invalidates the company's accident insurance.

1.17 Oxy-acetylene welding

Gas welding involves the use of highly flammable and potentially explosive gases. Since it is a thermal process involving very high temperatures burns and fire are an ever present risk. Therefore, we need to consider some of the basic safety precautions applicable to this process.

1.17.1 Safety in the use of gas cylinders

Cylinders for compressed gases are not in themselves dangerous. They must comply with rigid government standards – as must all pressure vessels – and should be regularly tested. Gas cylinders can be identified by their shape and colour coding. For example *high pressure* cylinders are tall and narrow. If they contain oxygen they are painted black. *Low pressure* cylinders are shorter and are colour-coded according to their contents as described in Table 1.1.

Unlike other flammable gases, acetylene can only be stored safely if dissolved under pressure in acetone and absorbed in type of porous concrete which fills the acetylene cylinder. Some of the main safety precautions that must be observed are:

- All cylinders must be protected from mechanical damage during storage and use.
- Acetylene cylinders must always be stored and used upright.
- Cylinders must be kept cool. On no account should the welding flame, or any other naked light, be allowed to play on the cylinders or regulators. Cylinders must be protected from direct sunlight, wet, and frost on an open site.
- Cylinders must always be stored in well-ventilated surroundings to prevent the build-up of explosive mixtures of gases should leaks occur.

SMOKING MUST NOT BE ALLOWED IN OR NEAR A GAS STORE

- Keep cylinders free from contamination. Although oxygen is not flammable, an oxygen-rich atmosphere in the event of leaks can cause the spontaneous combustion and violent burning of oils and greases together with oily rags and overalls.
- Correct automatic pressure regulators must be fitted to all cylinders prior to use. The main cylinder valve must always be kept closed when the cylinder is not in use or whilst changing cylinders or equipment.

Table 1.1 Colour coding for gas cylinders (based on BS 349)

If a cylinder has no neck band its contents are neither poisonous nor flammable.
If a cylinder has a red neck band or it is wholly coloured red or maroon its contents are flammable.
If a cylinder has a yellow neck band its contents are poisonous.
If a cylinder has both a red and a yellow neck band its contents are both flammable and poisonous.
Full details can be found in BS 349 *Identification of the Contents of Industrial Gas Containers*.

Whilst on site, the cylinders should be kept clean so that they can be easily identified. On no account must the colour of any cylinder be changed by any person either on site or in the workshop. Acetylene cylinders must never be stacked horizontally. Here are some examples of colour coding.

Gas	Ground colour of cylinder	Colour of neck bands
Acetylene	Maroon	None
Air	French grey	None
Ammonia	Black	Signal red and golden yellow
Argon	Peacock blue	None
Carbon monoxide	Signal red*	Golden yellow
Helium	Middle brown	None
Hydrogen	Signal red*	None
Methane	Signal red*	None
Nitrogen	French grey	Black
Oxygen	Black	None

*The name of the contents is also painted clearly on the cylinder in a contrasting colour.

1.17.2 Testing for leaks

Never search for gas leaks with a flame. *Oxygen* is odourless and, whilst it does not burn, it speeds up the combustion process violently. *Acetylene* has an unmistakeable, strong, garlic-like smell and can be readily ignited by a spark or even a piece of red-hot metal. The correct procedure for pressure testing is as follows:

1. Open the control valves on the torch.
2. Release the pressure-adjusting control on the regulators.
3. Open cylinder valves to turn on the gas.
4. Set the working pressure by adjusting the regulator controls.
5. Having established the correct pressure for each gas, close the control valves on the torch.
6. The system is now ready for leak testing as shown in Fig. 1.39.

1.17.3 Fire hazards

To reduce the risk of fire and explosions:

- Do not position gas cylinders and hoses where sparks can fall on them.
- Avoid the use of wooden structures and other combustible materials in the vicinity of welding operations.

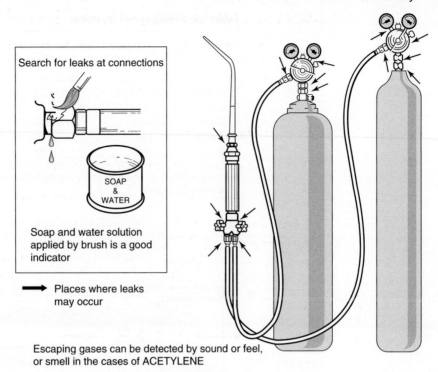

Search for leaks at connections

SOAP
&
WATER

Soap and water solution
applied by brush is a good
indicator

➡ Places where leaks
   may occur

Escaping gases can be detected by sound or feel,
or smell in the cases of ACETYLENE

Figure 1.39 Testing for leaks

■ Keep fire-fighting equipment ready to hand. A responsible person should keep the
site under observation for at least half an hour after completion of the work in
order to watch for, and deal with, any out beak of fire. There is always the risk of
combustible material smouldering for hours before a fire breaks out.

1.17.4 Explosion risks

The heat required for gas welding and cutting operations is produced by the combustion of a suitable fuel gas with oxygen. Usually this is oxygen mixed with acetylene as this give the hottest flame (see Table 1.2). However, other fuel gases may be
used for reasons of convenience and economy where a lower temperature will suf-
fice as when brazing and flame cutting.

Explosions can occur when acetylene gas is present in air in any proportions
between 2% and 82%. You do not need an oxygen-rich atmosphere to cause an
explosion. Acetylene will also explode when under unduly high pressure even when
air or oxygen is not present. The working pressure for acetylene should not exceed
0.62 bars. This is why acetylene is dissolved in acetone and absorbed into porous
concrete in acetylene cylinders.

Another potential cause of explosions in any oxy-fuel gas welding equipment is
a flash-back. This generally occurs because of faulty equipment or incorrect usage,
mainly when lighting up the torch. Always light up with fuel gas only and *gradually* increase the flow of oxygen until the correct flame conditions are attained.

Table 1.2 Welding gas mixtures

Fuel gas	Maximum flame temperature	
	with air (°C)	with oxygen (°C)
Acetylene	1755	3200
Butane	1750	2730
Coal gas	1600	2000
Hydrogen*	1700	2300
Propane	1750	2500

*The oxy-hydrogen flame has an important application in underwater cutting processes.

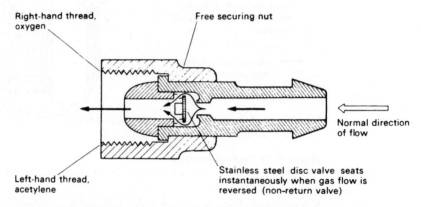

Right-hand thread, oxygen

Free securing nut

Left-hand thread, acetylene

Stainless steel disc valve seats instantaneously when gas flow is reversed (non-return valve)

Normal direction of flow

Note: THESE PROTECTORS WILL PREVENT THE FLOW OF GASES TO THE WELDING OR CUTTING TORCH IF THE HOSES ARE INADVERTANTLY REVERSED

Figure 1.40 Hose protector

Other causes of flash-backs are due to dipping the nozzle tip into the weld pool, or putting the nozzle tip against the work so as to prevent the free flow of gas mixture. Manufacturers' literature should be consulted before using gas welding equipment. Flash-backs can be prevented by the use of 'hose protectors' as shown in Fig. 1.40. Note that all items of oxygen equipment have right-hand threads and that all items of acetylene equipment have left-hand threads to prevent the possibility of incorrect connection.

Copper tube or fittings made from copper must never be used with acetylene. All valves and fittings must contain no more than 70% copper. This is because pure copper reacts with acetylene to form a highly explosive compound called copper acetylide which can be detonated by heat or abrasion. The only exception to this rule is the welding or cutting nozzle itself. This is to prevent overheating.

Explosions can also occur as a result of dust, grit, oil or grease getting into the socket of the cylinder. The outlet sockets of cylinder valves should be examined for cleanliness before fitting the regulators. Not only should the sockets be carefully cleaned, but the cylinder sockets should be blown through by momentarily opening

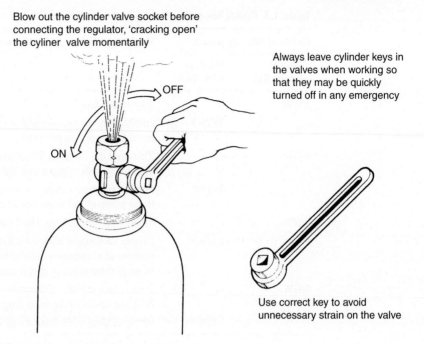

Blow out the cylinder valve socket before connecting the regulator, 'cracking open' the cyliner valve momentarily

Always leave cylinder keys in the valves when working so that they may be quickly turned off in any emergency

OFF

ON

Use correct key to avoid unnecessary strain on the valve

Figure 1.41 Cracking the cylinder

the cylinder valves before fitting the regulators. This is known as *cracking the valves* or *cracking the cylinders* as shown in Fig. 1.41.

1.17.5 Protective equipment (goggles)

Goggles must be worn to protect the welder's eyes from heat, glare and particles of hot metal. These should be fitted with an approved filter lens complying with BSEN 169 and the grade of filter selected should be suitable for the size of the welding flame and the thickness of the metal being welded (see Table 1.3). Figure 1.42 shows typical eye protection suitable for gas welding. Note that the lenses of welding goggles are always of composite construction. The front lens is made of clear shatter-proof glass for safety reasons and ease of replacement if damaged. Behind the shatter-proof glass is the tinted lens which can be changed in accordance with the recommendations found in Table 1.3. This tinted lens is softer, more easily scratched and more expensive than the clear glass that protects it.

1.17.6 Protective equipment (overalls)

Normal workshop overalls are usually made from cotton/polyester mixtures that are not flame resistant materials. However, overalls worn when welding should be made from a flame resistant cloth and should be kept buttoned or zipped up to the neck so that particles of hot metal and slag cannot lodge inside them. Sleeves and trousers should not have cuffs which can trap globules of hot metal and hot slag. In addition, flame resistant head gear and a chrome leather apron should be worn together with

Table 1.3 Filters for gas welding and cutting

Grade of filter required		Recommended for use when welding
Welding without flux	Welding with flux	
3/GW	–	Thin sheet steel
–	3/GWF	Aluminium, magnesium and aluminium alloys. Lead burning, oxyacetylene cutting
4/GW	–	Zinc-base die castings. Silver soldering. Braze welding light gauge copper pipes and light gauge steel sheet
–	4/GWF	Oxygen machine cutting – medium sections. Hand cutting, flame gouging and flame descaling
5/GW	–	Small steel fabrications. Hard surfacing
–	5/GWF	Copper and copper alloys. Nickel and nickel alloys. Heavier sections of aluminium and aluminium alloys. Braze welding of un-preheated heavy gauge steel and cast iron
5/GW	–	Heavy steel sections. Preheated cast iron and cast steel. Building up and reclaiming large areas
–	6/GWF	Braze welding of preheated cast iron and cast steel

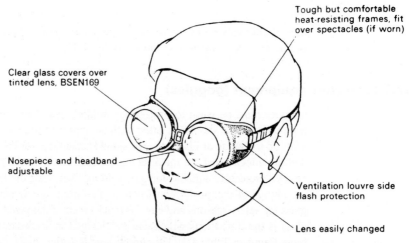

Tough but comfortable heat-resisting frames, fit over spectacles (if worn)

Clear glass covers over tinted lens, BSEN169

Nosepiece and headband adjustable

Ventilation louvre side flash protection

Lens easily changed

Note: GOGGLES WITH LENSES SPECIFIED FOR USE WHEN GAS WELDING OR CUTTING MUST NOT BE USED FOR ARC WELDING OPERATIONS

Figure 1.42 The essential features of good quality welding goggles

leather gloves. Metal tongs should be used for holding or moving hot metal. Chrome leather gauntlet gloves and leather spats and a safety helmet should also be worn when welding overhead or on vertical surfaces. This full protective clothing should always be worn when flame cutting.

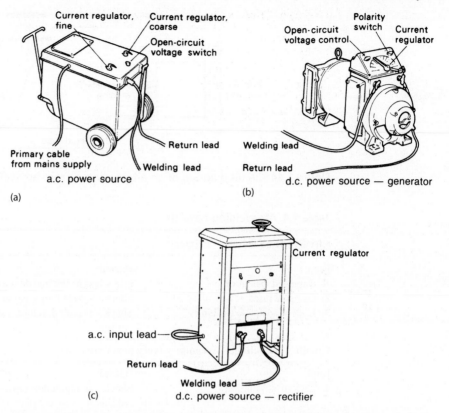

Figure 1.43 Mains-operated arc welding equipment. (a) Single operator mobile a.c. welding set; (b) motor generator d.c. welding set; (c) transformer-rectifier, d.c. welding set

1.18 Arc-welding equipment (mains operated)

This equipment is designed to change the high-voltage alternating current mains supply into safe, low-voltage, heavy current supply suitable for arc welding. Figure 1.43 shows some examples of some typical arc welding sets.

It can be seen that the output of arc-welding sets can have an alternating current (a.c.) wave-form or be a direct current (d.c.) that flows continuously in one direction. The output potential should be between 50 volts and 100 volts for safety but the current may be as high as 500 amperes. Figure 1.44 shows a typical alternating current arc-welding circuit. It can be seen that, in addition to the transformer which reduces the voltage, there is a tapped choke to control the current flow to suit the gauge of electrode used.

1.18.1 Hazards associated with mains-operated arc-welding equipment

The hazards associated with mains-operated arc-welding equipment are listed in Table 1.4 which should be read in conjunction with the schematic circuit diagram

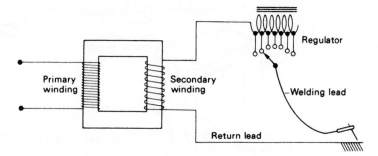

Figure 1.44 Circuit diagram of an alternating-current arc-welding set

Table 1.4 Arc-welding hazards

Circuit – High voltage – Primary

Fault:	Hazard:
1. Damaged insulation	Fire – loss of life and damage to property
2. Oversize fuses	Shock – severe burns and loss of life
3. Lack of adequate earthing	Shock – if fault develops – severe burns and loss of life

Circuit – low voltage – Secondary (very heavy current)

Fault:	Hazard:
1. Lack of welding earth	Shock – if a fault develops – severe burns and loss of life
2. Welding cable – damaged insulation	Local arcing between cable and any adjacent metalwork at earth potential causing fire
3. Welding cable – inadequate capacity	Overheating leading to damaged insulation and fire
4. Inadequate connections	Overheating – severe burns – fire
5. Inadequate return path	Current leakage through surrounding metalwork – overheating – fire

shown in Fig. 1.45. The circuit divides into two parts:

1. The primary (mains-voltage) circuit which should be installed and maintained by a qualified electrician.
2. The secondary (low-voltage) external welding circuit which is normally set up by a welder to suit the job in hand.

To eliminate these hazards as far as possible, the following precautions should be taken. These are only basic precautions and you should always check whether the equipment and working conditions require special, additional precautions.

■ Ensure that the equipment is fed from the mains via a switch-fuse of the correct rating to protect the equipment and incoming cable and also so that the welding set can be isolated from the supply. Easy access to this switch must be provided at all times. The isolating switch should also incorporate a residual current detector (also known as an earth leakage isolator – see section 1.4).

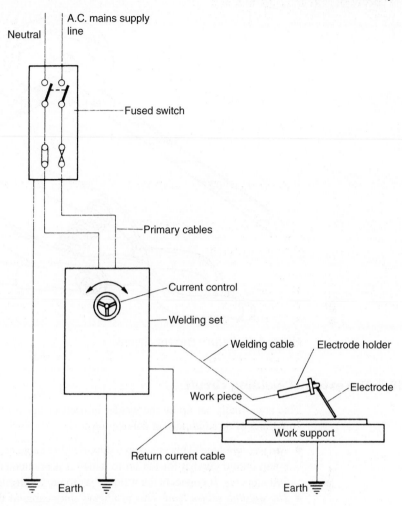

Figure 1.45 Manual metal-arc welding circuit diagram

- Ensure that the incoming (primary cable) is armoured against mechanical damage and that the insulation is suitable for a 415 volt supply.
- Ensure that all cable insulation is undamaged and all terminations are secure and undamaged. *If in doubt do not use the equipment until it has been checked by a qualified electrician.*
- Ensure that all welding equipment is adequately earthed with conductors capable of carrying the heavy currents used in welding.
- Ensure that the current regulator has an 'off' position so that in the event of an accident the welding current can be stopped without having to trace the primary cables back to the isolating switch.
- Ensure that all cables and fittings for the 'external welding circuit' are adequate for the heavy currents involved.

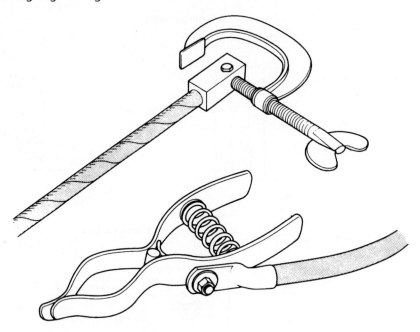

Figure 1.46 Return current clamps

1.18.2 The external welding circuit

This is normally set up by the welder himself or herself to suit the job in hand and comprises of three important connections.

- *The welding lead.* This is a heavy duty flexible cable capable of carrying the maximum output current for the set to which it is connected. This may be as much as 500 amperes. It connects the welding set to the electrode holder.
- *The welding return lead.* This is a heavy duty cable of the same type and capacity as the welding lead. It carries the return current from the work-piece back to the welding set, thus completing the external welding circuit. It is connected to the work-piece by means of screw or spring clamps as shown in Fig. 1.46.
- *The welding earth.* This is necessary on all welding circuits to maintain the work-piece and any other associated equipment and structures connected to it at earth potential.
- *Electrode holders.* These should be soundly connected to the welding lead and have adequate rating for the maximum welding current in order to prevent heating up and becoming to hot to handle. Electrode holders may be partially insulated or fully insulated as shown in Fig. 1.47. The protective guard (disc) of the partially insulated type of holder not only protects the welder's hands but also prevents short circuits should the holder be laid down on the work-piece.

The proper selection and care of welding cables is essential for safety and efficiency. The most common faults are:

- Poor connections.
- Overlong cables leading to excessive volt-drop.

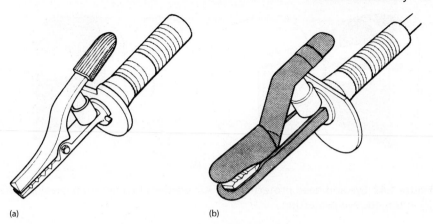

(a)   (b)

Figure 1.47 Electrode holders: (a) Partially insulated type; (b) Fully insulated type

- Defective insulation leading to short circuits.
- The use of cables of incorrect current carrying capacity.

1.18.3 Protective clothing

Protective clothing for arc-welding has to be more comprehensive that that used for oxy-acetylene welding since the welder needs to be protected from the harmful radiations produced by the 'arc'.

Eye and head protection

For all arc-welding operations it is essential to protect the welder's head, face and eyes from ultra-violet radiation, heat (infra-red radiation), spatter and hot slag. Remember that arc-welding generates greater heat than gas welding and also generates much higher temperatures. For these reasons, either a helmet (visor) must be worn or a hand-shield must be used. The helmet-type head shield is pivoted so that it can be raised when not welding and lowered when welding is taking place. Examples of these protective items are shown in Fig. 1.48(a) and 1.48(b). Increasingly, automatic welding screens are being used and an example is shown in Fig. 1.48(c). This hands-free welding helmet allows the wearer to see clearly when no welding is taking place. However, the moment the arc is struck the filter glass darkens almost instantaneously and its shade can be adjusted to suit the work being done and the welding current required as listed in Table 1.5. The power to change the filter from clear to dark when welding commences is provided by a small battery or by solar cells.

Never wear gas welding goggles when arc-welding

The filter glasses used when arc-welding are much denser that those used for gas welding.

As in gas welding goggles, there is an outer layer of toughened plain glass that can be easily and cheaply replaced protecting the softer and more expensive filter glass. The filter glass must be suitable for the welding current being used as listed in Table 1.5.

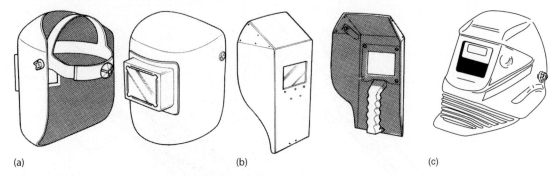

(a)   (b)   (c)

Figure 1.48 Eye and head protection. (a) Arc welder's helmet; (b) arc welder's hand-shield; (c) automatic eye protection

Table 1.5 Filters for manual arc welding

Grade of filter required	Approximate range of welding current (amps)
8/EW	Up to 100
9/EW	100 to 200
10/EW	Up to 200
11/EW	Up to 300
12/EW	For use with currents over 300
13/EW	For use with currents over 300
14/EW	For use with currents over 300

Each filter purporting to comply with BS 679 should be permanently marked as follows:

1. BS 679
2. The certification mark of the British Standards Institute
3. The manufacturer's name, symbol or licence number
4. The figures and letters denoting its shade and type of welding process (gas or electric)

Filter glasses are expensive, therefore they should be used with a clear plain cover glass on the outside in order to protect them from damage by spatter and fumes.
These cover glasses are relatively cheap and easily replaceable.

Arc-eye

Exposure to welding arc radiation can cause burns similar to sunburn and also cause severe irritation to your eyes. This condition is known as 'arc-eye'. The symptoms, which include watering eyes, headache and partial loss of vision, become apparent four to eight hours after exposure. If you suffer these symptoms you should:

- Wash your eyes using a sterile eye bath and an approved eye lotion.
- Repeat the treatment every four hours.
- Wear dark glasses in brightly lit areas until the symptoms subside.
- Seek medical advice if the symptoms have not cleared up within 36 to 48 hours.

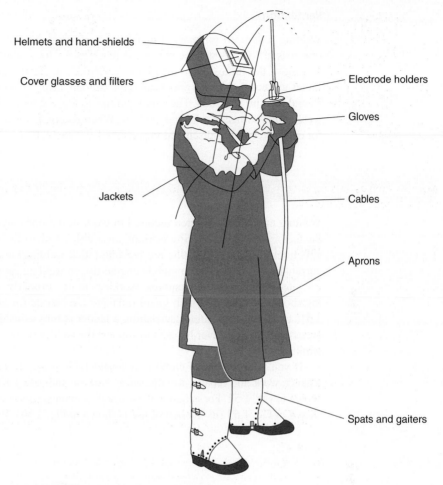

Helmets and hand-shields

Cover glasses and filters

Electrode holders

Gloves

Jackets

Cables

Aprons

Spats and gaiters

Figure 1.49 Fully protected arc welder

Prevention is better than cure. Always wear suitable eye-protection when arc weld-ing. Always ensure that arc-welding processes are screened off so that persons work-ing in the vicinity of arc-welding operations are protected from welding flashes. These screens can be opaque (solid) or transparent, if transparent they should com-ply with BSEN 1598.

Body protection

The welder must protect his or her body and clothing from radiation and burns caused by flying globules of molten metal (spatter). The welder should wear flame resistant overalls and a protective leather apron and leather gauntlet gloves. If work-ing on vertical surfaces or working overhead, or if deep gouging or cutting, full protective clothing should be worn as shown in Fig. 1.49.

Ventilation

Welding rods used for manual metallic-arc-welding are usually coated with a flux (the reason for this will be explained in Chapter 10). Fume extraction is essential to remove the fumes given off by the hot flux. This should be designed so that the fumes are not drawn upwards passed the welders face. Adequate general ventilation should also be provided. The flux solidifies to leave a hard coating of slag over the weld zone that needs to be chipped off. When chipping, clear goggles must be worn for eye protection as discussed in sections 1.9.2 and 1.10.4.

1.19 Working on site

Welding engineers are often required to work in the open and on site, as well as in the factory environment. Site working invariably involves the welding of heavy steel structures so manual metallic-arc-welding (stick welding) is used with the welding current being supplied by a mobile engine-driven welding set as shown in Fig. 1.50.

Site working often requires working aloft. Properly constructed scaffolding should be provided with guard rails and toe boards for safety as shown in Fig. 1.51. Under current safety legislation, a ladder is only considered to be a means of access and should never be used to support the welder or any other employee whilst working.

If you are required to climb to a considerable height to carry out regular maintenance work and repairs, a static safety harness and safety block should be used as shown in Fig. 1.52. For occasional use a self-contained safety block should be used as shown in Fig. 1.53 (its method of use is shown in Fig. 1.54). With normal movement,

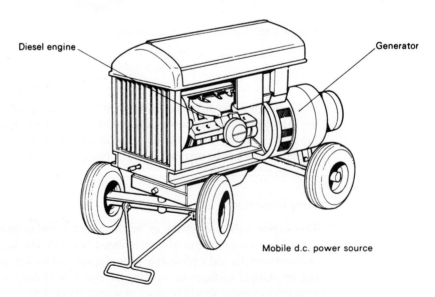

Diesel engine

Generator

Mobile d.c. power source

Figure 1.50 A mobile engine-driven direct-current welding set

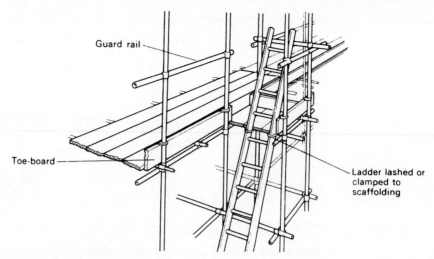

Figure 1.51 Overhead platforms

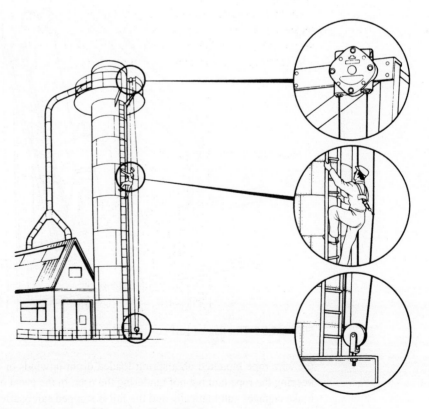

Figure 1.52 Use of a static safety block

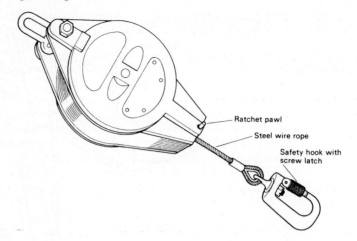

Ratchet pawl

Steel wire rope

Safety hook with
screw latch

Figure 1.53 The self-contained safety block

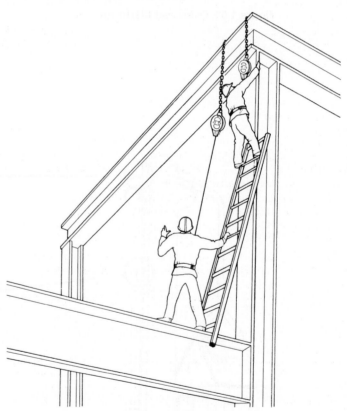

Figure 1.54 The use of self-contained safety blocks

the wire rope mounted on a spring-loaded drum unwinds or winds up automatically, keeping the rope taut but not hindering the user. In the event of a slip or fall, a friction brake engages automatically and the fall is stopped safely after about 300 mm with no jolt or jerk.

Exercises

1.1 Health and Safety at Work Act and other important industrial legislation

   a) **i)** What do the initials HSE stand for?

       ii) What is a prohibition notice?

      iii) What is an improvement notice?

      iv) Which person is responsible for issuing the notices in questions i) and ii) above?

   b) As an employee you also have duties under the Act. Copy out and complete Table 1.6 by writing brief comments regarding your duties in the following circumstances.

   c) Copy out and complete Table 1.7 by adding the name of the most appropriate industrial regulation(s) for each of the situations given.

   d) Copy out and complete Fig. 1.55 by stating:

       i) The category of the sign (e.g. warning sign, mandatory sign, etc.).

      ii) The meaning of the sign.

     iii) Where the sign should be used.

1.2 Electrical Hazards

   a) Explain why portable electrical equipment should be:

       i) Earthed unless it is 'double-insulated'.

      ii) Operated from a low-voltage supply.

     iii) Protected by an earth leakage isolator incorporating a residual current detector.

   b) When you are issued with portable electrical equipment from the stores you should make a number of visual checks before accepting and using the equipment. Describe these checks.

   c) If the checks you made in b) above showed the equipment to be faulty, what action should you take?

   d) In the event of a workmate receiving a severe electric shock that renders him/her unconscious, what emergency action should you take?

Table 1.6 Exercise 1.1(b)

Circumstances	Duties
You are uncertain how to operate a machine needed to complete your task	
You need to carry some sheet metal with very sharp edges	
You are working on site and you have mislaid your safety helmet	
You find that the belt guard has been removed from a machine you have been told to use	
You have spilt hydraulic oil on the floor whilst servicing a machine	
The earth wire has come disconnected from a portable power tool you are using	
Your supervisor has told you to clear up the rubbish left by another worker	
You find someone smoking in a prohibited area	

Table 1.7 Exercise 1.1(c)

Situation	Appropriate industrial regulations
Use of grinding machines and abrasive wheels	
Eye protection	
Electrical control equipment, use and maintenance	
Use of substances that can be harmful to health (solvents, etc.)	
Safe use of power presses	
Protection against high noise levels	
Safe use of milling machines	
Use of protective clothing	

Sign	Meaning	Category	Where used

Figure 1.55 Exercise 1.1(d)

1.3 Fire hazards

a) List THREE main causes of fire on industrial premises.

b) If you detect a fire in a storeroom at work, what action should you take?

c) Figure 1.56 shows some various types of fire extinguisher.

   i) State the types of fire upon which each extinguisher should be used and any precautions that should be taken.

   ii) State the colour coding that identifies each type of fire extinguisher.

1.4 Accidents

a) Why should all cuts, bruises and burns be treated by a qualified first-aider, and what risks does a well-meaning but unqualified person run in attempting first aid?

b) Apart from rendering first aid what other action must be taken in the event of an accident?

c) State how you would identify a first aid post.

d) What action should you take if you come across someone who has received a serious accident (broken bones, severe bleeding, partial or complete loss of consciousness, etc.)?

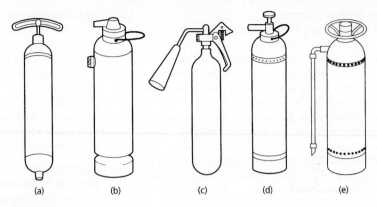

(a)   (b)   (c)   (d)   (e)

Figure 1.56 Exercise 1.3(c)

Table 1.8 Exercise 1.5(a)

Clothing/equipment	Situation/environment
Ear protectors	
Overalls	
PVC apron	
Leather apron	
Leather gloves	
PVC/rubber gloves	
Safety helmet	
Clear goggles	
Visor	
Barrier cream	
Safety boots	
Goggles with filter lenses	

e) Briefly describe the accident reporting procedures for your place of work or training workshop.

1.5 Working environment

a) Copy out and complete Table 1.8 by stating the type of working environment in which you would need to use the following items of safety clothing and equipment listed in the table.

b) With the aid of a sketch explain what is meant by:
  i) A transmission guard.
  ii) A cutter guard.

c) For each of the examples listed below, sketch an appropriate item of safety equipment and state how it works.
  i) A drill chuck guard.
  ii) A chipping screen.
  iii) An interlocked transmission guard.

1.6 Lifting and carrying

a) State the maximum recommended weight that may be lifted without the aid of mechanical lifting equipment.

b) With the aid of sketches show the correct and incorrect way to lift a load.

c) What precautions should be taken when carrying loads?

d) What precautions should be taken when moving a heavy load with a lifting team?

e) Lifting equipment should be marked with its SWL.
  i) State what these initials stand for.
  ii) State how often lifting equipment needs to be tested and examined.
  iii) State what records need to be kept.

1.7 Oxy-acetylene welding

   a) State how you would distinguish between oxygen and acetylene cylinders.

   b) List THREE important precautions that should be taken when handling welding gas cylinders.

   c) Explain briefly how to test for leaks in oxy-acetylene welding equipment.

   d) Explain briefly the fire and explosion hazards associated with oxy-acetylene welding equipment and how such hazards can be avoided.

   e) Describe what essential protective clothing and equipment should be worn when oxy-acetylene welding and flame-cutting.

1.8 Electric arc-welding

   a) What is the most dangerous hazard associated with the primary (mains) circuit of electric arc-welding equipment? Explain what steps can be taken to minimize this hazard.

   b) Briefly describe the precautions that should be taken to minimize any hazards associated with the external (secondary) electric arc-welding circuit.

   c) Describe what essential protective clothing and equipment should be worn when electric-arc welding.

   d) Explain what is meant by 'arc-eye' and explain how this condition can be treated.

   e) Explain why good ventilation is essential when arc-welding with coated electrodes.

1.9 Working on site

Describe the hazards likely to be faced when working on site compared with working in a factory environment. Pay particular attention to climbing, working at heights, craning and lifting, protective clothing and personal safety equipment.

2 Personal development

When you have read this chapter, you should understand how to:

- Create and maintain effective working relationships with supervisory staff
- Create and maintain working relationships with other people, members of the same working groups and other employees in the same organization

2.1 Basic relationships

Even the smallest businesses have to communicate with, and relate to, a surprising number of people either through necessity or because it is the law. This is shown in Fig. 2.1.

In the first group (*necessity*) are your suppliers of raw materials and tools and equipment used in production. Also you have to deal with the customers who buy your products and the transportation companies who deliver your products to your customers. You also need a bank and since, from time to time, you may need an overdraft, it's as well to maintain good relationships with your bank-manager.

There is no law that says you need a solicitor or an accountant. However, you will require a solicitor to draw up all the documents required when setting up a business and when problems arise with customers, suppliers and the Local Authority (e.g. noise complaints from the neighbours). You will require an accountant to audit your accounts, advise on financial matters, sort out your tax returns, make sure that you avoid over-payment of tax, and to deal with the Customs and Excise officials over your VAT payments and returns. Therefore, it is a *necessity* that you make every effort to maintain *good working relationships* with them.

> There is no law that says you have to have suppliers, customers and bankers but you would not get far in business without them

In the second group (*law*) you have to communicate with such persons as Local Authority inspectors (planning officers, etc.), tax inspectors, VAT inspectors, and Health and Safety inspectors. These people have the power of the legal system behind them so it pays to maintain *good working relationships* with them.

In our working lives we also have constantly to relate to and communicate with other people so that we can exchange technical data, implement management decisions

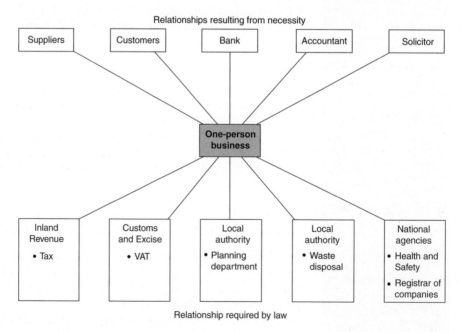

Figure 2.1 Structure of relationships

and safety policy, and relate to other personnel within the company as well as customers and buyers who work outside the company. In this section we are concerned mainly with the people with whom you will work on a daily basis; not only your workmates but also your immediate supervisors and managers.

Having made the point that no one can work in isolation, even if they are the sole proprietor of a one person firm, let's consider the situation if you are an employer or an employee in a small, medium or large company. Like it or not, you are going to be one of a team. Like it or not, you are going to have to communicate, participate and co-operate. You are going to have to maintain *good working relationships*. When dealing with other people, you can adopt one of two possible attitudes: you can either *confront* them or you can *cooperate* with them.

2.1.1 Confrontation

Confrontation is how the aggressive, bullying person works. A confrontational person demands and threatens to get his or her own way. It may work in the short-term as long as the aggressor has the whip-hand. However, such aggressive bullies never win the respect of the people with whom they work. They can never rely upon the loyalty of the people they have continually confronted when a favour is required – it would be no good expecting 'good will' co-operation where extra effort is required to complete an urgent order on time.

2.1.2 Cooperation

This is how sensible, civilized people work. They collaborate and help each other. In this way they gain respect for each other. This results in the development of efficient working relationships and efficient working practices. In an emergency everyone can be relied upon to make maximum effort and to help each other.

2.1.3 'Reading' people

As you become more experienced in dealing with people, you will realise that the most important skill is learning to 'read' their moods. You must be able to realise with whom you can have a joke and with whom you can't. You need to know who only wants a 'yes or no' answer and who prefers to discuss a problem. You need to know when to be friendly and when to be aloof, when to offer a word of sympathy or advice, and when to leave somebody alone to get over a bad mood.

2.2 Relationship with managers, supervisors and instructors

You are *employed* by the firm for whom you work, but you are *responsible* to your immediate superior. Depending on the structure of the company your immediate superior may be an instructor, a charge-hand, a foreman or forewoman, a supervisor, a manager or, in a very small firm, the 'Boss'.

Figure 2.2 shows the structure of a training department for a large company. The structure will vary from firm to firm but whatever the size of the company and the

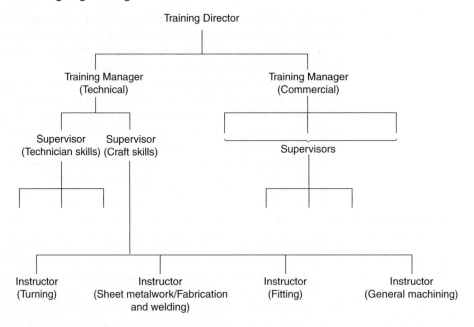

Figure 2.2 Training personnel structure

The change from the school environment to the adult working environment often poses unforeseen problems, it is essential to know who you should turn to when you need advice

structure of its training facility; it is always a good idea to find out what the structure is. You need to know who influences your training package, who actually trains you and who is responsible for your, welfare, discipline and assessment.

First and foremost, it is most important that you get on well with your instructor, your supervisor and your training manager. Each of them will require a different approach. This is not only because they are different people, but also because they have a different status and a different level of importance in the company. Let's now see how you can make a 'good impression' on these people and establish good working relationships with them. For example:

- Develop a habit of good time-keeping and regular attendance, even under difficult conditions.
- Be neat and tidy in your appearance.
- Keep your work area neat and tidy and your tools and instruments in good condition.
- Keep your paperwork up to date, fill it in neatly and keep it clean in a plastic folder.
- File your paperwork systematically so that you can produce it for your instructor or your training manager on demand. 'Attention to detail' always makes a good impression.
- Be reliable so that people quickly find that they can depend upon you.
- Be conscientious: always try your hardest and do your best.
- Reasonable requests for information should be dealt with promptly, accurately and in a co-operative manner providing they do not unduly interfere with or interrupt your work.

- If responding to any request is going to take time and interrupt your work, or if it requires you to leave your working area, always seek permission from your supervisor or instructor before carrying out the request. Always turn off your equipment before leaving it.
- If you are in the middle of an intricate piece of work that requires your full concentration, don't just down tools, but ask politely if you may complete your task before responding to the request.
- No matter how tired you are or how inconvenient, trivial and unnecessary the request may seem to you, always try to be cheerful, helpful and efficient. NEVER answer in a surly, uncooperative, couldn't care less, any old time will do, manner.

Your relationships with other people, particularly your instructor, must be a dialogue of instruction and advice. If you are in doubt you must always discuss your problem with your instructor until you are certain that you fully understand what you have to do.

> Your instructor is also there to help you with any personal problems you may have or problems with other people with whom you have to work

He or she wants to get to know you as a person so that they can get the best from you and help you to make to a success of your training. Should your instructor be talking to another trainee or a supervisor or manager, don't just barge in, either get on with another job and go back later or wait to one side, respectfully, until it is your turn. Be patient, on no account should you try and start work on a job or on a machine without instruction just because your instructor is busy and you are tired of waiting for him or her.

2.3 Attitude and behaviour

2.3.1 Attitude

When you enter the world of industry you are a very new, very unimportant and very expendable member of the work force. You know little or nothing about the skills of engineering so, if you are going to complete your training successfully and become a useful member of the company and of society as a whole, you will have a great deal to learn.

> Your training is a major investment for your employer

Employers need to train and employ reliable people who they can rely upon and who will give them a reasonable return for their investment in time and money. Those who demonstrate good attitudes are the most likely to succeed. It is no good being the most skilful apprentice or trainee if you are also the most temperamental. Whilst high levels of skill are important, so is consistency, reliability, loyalty and the ability to work in a team.

The greatest incentive to learning a trade is the earning power it gives you. To learn a trade you need the skilled help and advice of a lot of people. You must respect their skill and experience if you are to get their help and advice in return.

Apart from the advice already given, here are some further suggestions.

- Dress in the way recommended by your company. Many firms provide smart overalls bearing the company logo. Do not turn up to work looking scruffy. For example, a long hairstyle not only gives a bad impression – it can also be very dangerous (as discussed in Chapter 1).
- For hygiene reasons change into clean overalls daily if possible. Dirty, oily overalls can cause serious hygiene and health problems. A tidy person has a tidy and receptive mind.

- Listen carefully to the instructions your instructor gives to you, particularly safety instructions. Never operate a machine or carry out a process if you are in doubt. Always confirm the correct procedure with your instructor.
- Keep a log of the operations you are taught and the work you do because your practical skill training has to be assessed in order for you to obtain your certification. Since you may have to present your logbook at a future job interview it is worthwhile spending some time on it. Keep it neat and clean in a plastic folder.
- Show consistency, commitment and dedication in carrying out the tasks set you. Work to as high a standard as you can and always be trying to improve your standards. Have pride in your work, you never know who is going to look at it. This applies not only to the production of components and assemblies but equally to organizational tasks.

2.3.2 Behaviour

In an industrial environment horseplay and fooling around – such as pushing, shouting, throwing things and practical joking by a person or a group of persons – infers reckless and boisterous behaviour. Engineering equipment is potentially very dangerous and this sort of behaviour cannot be tolerated in an industrial environment.

As well as the negative attitude to behaviour just described, there are positive attitudes to be taken as well. For example, keep your workstation clean and tidy and clean up any spillages immediately top prevent accidents due to slipping. Accumulated rubbish can lead to fires when welding or flame-cutting.

2.4 Implementing company policy

Company policy may be dictated by the 'Boss' in a small company or it may be set by the board of directors in a large company. These people are not free agents and they have to abide by national and international laws and guidelines in setting out a strategy for the company. They have to consider the demands of the shareholders, and they are also responsible for the success, profitability and growth of the company upon which the job security and rewards of all who work for the company depend. For these reasons company policy should be understood and obeyed. In successful companies this is not an entirely autocratic process and there are various committee structures through which ideas from the shop floor can be fed back up the command chain to the senior management. This is particularly true for safety issues – see Chapter 1.

2.4.1 Communication

No company can exist without lines of communications both internal and external. Without internal lines of communication the company policy could not be communicated to the work force, nor would the senior management know if their policies were being carried out. Figure 2.3 shows a typical management structure for a company.

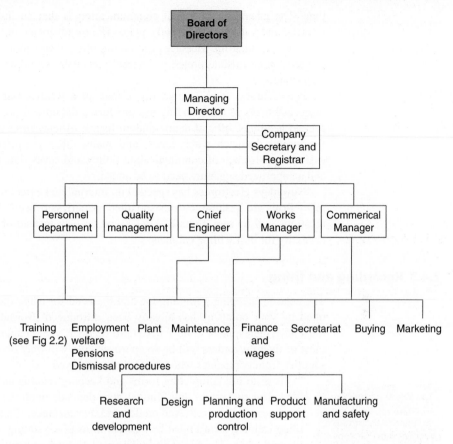

Figure 2.3 Management structure

This structure not only represents the lines of communication by which the senior management can ensure that decisions are passed down the line, it represents a route by which messages and requests are sent back up to the various levels of management. These channels of communication are part of company policy and, to bypass them, can lead to confusion and friction between the parties concerned.

Wherever possible, always use the standard forms provided when communicating within your company. This will result in your requests being treated seriously. Such forms may range from the stores requisition forms that you fill in daily, to job application forms and internal promotion application forms. Always follow laid down procedures.

External lines of communication are equally important so that the company can communicate with its customers and suppliers. Market research, public relations and advertising are essential to the success of a company and depend upon the use of suitable means of communication. This is why many companies employ firms of consultants specializing in these fields.

Verbal communications can take place via the telephone on a 'one to one' basis or via meetings when information has to be given to a number of people at the same time. The advantage of verbal communication is that an instant response can be received and a discussion can take place. The disadvantage of verbal communication is its lack of integrity. Messages can be forgotten, they can be repeated inaccurately, or they can be misinterpreted. All verbal messages should be backed up by written confirmation.

For accuracy, send a letter, fax, e-mail or a written memorandum in the first place. All firms of any standing use pro-forma documents such as letter paper with printed headings, official memorandum forms, official order forms, official invoice forms and despatch patch notes, and many other pre-printed documents. This ensures consistency of communication policy and saves time since only the details and an approved signature need to be added.

Nowadays electronics has speeded up internal and external communications and many firms are heading towards so-called 'paperless offices'. Increasingly, communications will be sent digitally. Files saved on disk instead of on paper remove the necessity for bulky filing cabinets.

2.4.3 Recording and filing

The need for keeping a training log and the need for using the standard forms supplied by your company has already been introduced. Nowadays, most companies must have their quality control system to be BS EN 9000 approved. This is because most of their customers will be so approved, and will only be able to purchase their supplies from companies who are similarly approved.

To trace the progress of all goods from supplier to customer, records must be kept and filed

It is no use completing forms and keeping records unless they are properly filed. The success of any filing system depends upon the ease with which any documents can be retrieved on demand. For instance, if a file is removed from a filing cabinet, a card must be inserted in its place stating who has borrowed the file and when. The file must be returned as soon as possible so that it does not become lost. These days, most files are now kept on computers.

2.5 Creating and maintaining effective working relationships with other people

As has been stated previously, you cannot work in isolation. Sooner or later you have to relate to other people. In fact, most working situations rely upon teamwork.

2.5.1 Positive attitudes

At work you should always try to adopt a positive and constructive attitude to other people. This can be difficult when you are tired or the person you are relating to is off-hand, aggressive, demanding, and asking for the near impossible. However, they are often under pressure themselves and allowances have to be made. Sometimes people are just out to annoy and provoke a confrontation. Try not to become involved. It is better to walk away from a quarrel than let it get out of hand. Always try to cool down the aggressor.

Sooner or later you are bound to come up against someone with whom you cannot get on. This may be a workmate, or an instructor. Often there is no apparent reason for this; it is simply a clash of incompatible personalities. If you cannot resolve the matter amicably yourself, don't leave the situation to deteriorate, but seek advice from the appropriate member of staff such as your supervisor or manager. He or she may be able to solve the problem even if it may involve you being moved to another section. Remember that, during your training, your personal attitudes and your ability to work as a team member is as much under scrutiny as the products that you produce.

2.5.2 Teamwork

Quite often you will have to work as a member of a team. This requires quite different skills in interpersonal relationships than when you are working on your own or under the guidance of your instructor. For example, consider the lifting of a large and heavy packing case when mechanical lifting gear is not available. Like any team, the lifting party has to have a team leader (captain). That person must have the respect and confidence of all the other members of the team because of his or her experience and expertise. The team should be picked from people who it is known can work together amicably and constructively. One 'oddball' going his or her own way at a crucial moment could cause an accident and injury to other members of the team.

Although the team leader is solely responsible for the safe and satisfactory completion of the task, he or she should be sensible enough to consider comments and contributions from other members of the team. If you are a member of such a team and you think you have spotted a potential hazard in the job to be done, then it is your duty to draw it to the attention of the team leader. Eventually, however, discussion has to cease and the job has to be done. At this point the team leader has to make up his or her mind about how the job is to be done.

The team leader should not take an active part in the exercise, but should stand back where he or she can see everything that is going on. So, in the event of a potentially hazardous situation developing, the team leader is free to step in and correct the situation in order to prevent an accident.

2.5.3 Personal property

During a working lifetime most workers acquire an extensive set of personal tools. Some may be bought and some may be made personally. You will be mightily unpopular in any workshop if you borrow any of these tools without the owner's consent. The same applies to overalls or any other personal belongings. Although we have considered company policy, each and every workshop has a code of conduct all of its own. This is not written down, it is not company policy, it is a code of behaviour that has grown up over the years amongst the people working in that shop.

Woebetide anyone who disregards this code of conduct

Respect it, obey it, and you will find that your relationships with your workmates and supervisors will be much improved. You will receive more useful help and wise advice and will establish worthwhile friendships that can stand you in good stead throughout your working life.

2.1 Effective working relationships
  a) You are engaged in an intricate welding operation when a colleague asks for your assistance. Explain how you would deal with this situation.
  b) You are having difficulty in understanding an engineering drawing and you want advice. Your instructor is engaged in conversation with the training manager. Explain what you should do in this situation.
  c) Your supervisor has directed you to help with a team activity in another department. Explain how you would introduce yourself to the team leader and how you would try to relate to the other members of the team.

2.2 Dress, presentation and behaviour
  a) Describe the dress code at your place of work or your training centre and explain why the dress code should be adhered to.
  b) Explain **three** possible consequences of 'fooling about' in an engineering workshop.
  c) Explain why you, as an engineering trainee, should:
    i) Adopt a short, neat hair style.
    ii) Not wear dirty overalls.
    iii) Write up your logbook carefully and neatly, keep it in a plastic folder, and make sure it is available on demand for examination by your supervisor.

2.3 Instructions
  a) Draw an organization chart to show the chain of command in your training centre or in a company with which you are familiar.
  b) Upon receiving a verbal instruction, describe what you would do to ensure that you have understood it correctly.
  c) If a written instruction is unclear or badly printed, describe what you would do to avoid making a mistake in carrying out the instruction.

2.4 How to ask for help
  a) Describe a situation where your instructor might have sent you to another person, such as a more senior colleague, for advice. Explain who that person might be in your training centre or company.
  b) To avoid bothering your instructor when he or she is busy, describe:
    i) The sort of practical assistance you might seek from a colleague.
    ii) The sort of information you might seek from a colleague.

  c) State whom you would approach for advice, and why you have chosen that person, in the following circumstances:
    i) Clarification of instructions or unclear advice from a colleague.
    ii) Safe working practice concerning a new material that has been introduced into the workshop.
    iii) Assistance in completing forms.
    iv) Reporting personal injuries and accidents.
    v) Discussing personal problems.
  d) Give **one** example of the *correct approach* to another person when seeking that person's help or advice, and **one** *inappropriate approach* to another person when seeking that person's help or advice.

2.5 How to give help when asked
  a) List **five** important criteria that you must remember when giving help or advise to another person.
  b) Describe **three** situations when you should refuse to offer help or advice.
  c) Explain how you would try to make such a refusal without giving offence.

2.6 Reporting deficiencies in tools, equipment and materials
  a) Give five reasons why it is necessary to report deficiencies in tools, equipment and materials.
  b) Briefly describe the procedures used in your training centre or company for reporting defective tools, equipment and materials.

2.7 Respect for other people's opinions and property
  a) You may have to work with people whose values on work and life in general disagree with your own. Should you:
    i) Argue aggressively with them OR
    ii) Respect their views despite your personal reservations.
  b) You are in a hurry and a long way from the stores. You know that your workmate has the equipment you need in his or her personal toolkit. Describe the correct procedure for borrowing and returning such equipment.
  c) You are in a hurry to get home at the end of your shift. You are returning the tools you have been using to the stores. Should you clean them and check them or leave that to the stores personnel to save yourself time? Give reasons for your answer.

2.8 Teamwork and co-operation

  a) Why is it necessary to take the time and trouble to gain some knowledge and understanding of what other people do in your training centre or company, both within your department and in other departments? How could this lead to improved co-operation and teamwork?

  b) How do some companies expand their trainees' and apprentices' insight into the work of other departments in the organization?

  c) Give reasons for your answers to the following. When working as a team:

    i) should you take part in discussions concerning the work to be done?

    ii) should you ask for clarification of matters you do not understand?

    iii) from whom should you take instructions?

2.9 Difficulties in working relationships

  a) State **five** possible *causes* of difficulty that may arise in your relationships with your work-mates and more senior staff.

  b) With whom should you discuss such problems in the first place?

  c) Describe the procedures that exist for formally reporting such difficulties in your training centre or company if you can get no satisfaction from b) above?

3 Engineering materials and heat treatment

When you have read this chapter, you should understand:

- How to define the basic properties of engineering materials
- How to correctly identify and select a range of engineering metals and alloys
- How to correctly identify and select a range of non-metallic materials suitable for engineering applications
- Safe working practices as applicable to heat treatment processes
- The principles and purposes of heat treatment
- The through hardening of plain carbon steels
- How to temper hardened steels
- How to anneal and normalize steels
- The basic heat treatment of non-ferrous metals and alloys
- The principles, advantages and limitations of heat treatment furnaces
- The temperature control of heat treatment furnaces
- The advantages, limitations and applications of quenching media

3.1 States of matter

Almost all matter can exist in three physical states by changing its temperature in appropriate conditions. These states are solids, liquids and gases.

- Ice is *solid* water and exists below 0°C.
- Water is a *liquid* above 0°C and below 100°C.
- Steam is water *vapour* above 100°C and becomes a *gas* as its temperature is raised further (superheated).

Metals such as brass, copper or steel are solid (frozen) at room temperatures but become liquid (molten) if heated to a sufficiently high temperature. If they are heated to a high enough temperature they will turn into a gas. On cooling, they will first turn back into a liquid and then back into a solid at room temperature. Providing no chemical change takes place (e.g. the oxidation of the metal through contact with air at high temperatures) we can change substances backwards and forwards through the three states by heating and cooling as often as we like.

There are exceptions: for example, when a thermosetting plastic has been heated during the moulding process, it undergoes a chemical change called 'curing'. Once 'cured' it can never again be softened nor turned into a liquid by heating. It can, however, be destroyed by over-heating. Another example is the non-metallic element iodine. When heated this *sublimes* directly from a solid to a vapour without becoming a liquid.

3.2 Properties of materials

To compare and identify engineering materials, it is important to understand the meaning of their more common properties. For example, it is no good saying one material is stronger or harder than another material unless we know what is meant by the terms 'strength' and 'hardness'.

3.2.1 Strength properties

Tensile strength

This is the ability of a material to withstand a stretching load without breaking. This is shown in Fig. 3.1(a). The load is trying to stretch the rod. Therefore, the rod is said to be in a *state of tension* – it is being subjected to a tensile load. To resist this load without breaking, the material from which the rod is made needs to have sufficient *tensile strength*.

Compressive strength

This is the ability of a material to withstand a squeezing load without breaking. This is shown in Fig. 3.1(b). The load is trying to squash (crush or compress) the material from which the component is made. Therefore, the component is said to be in a *state of compression* – it is being subjected to a compressive load. To resist this load without breaking, the material from which it is made needs to have sufficient *compressive strength*.

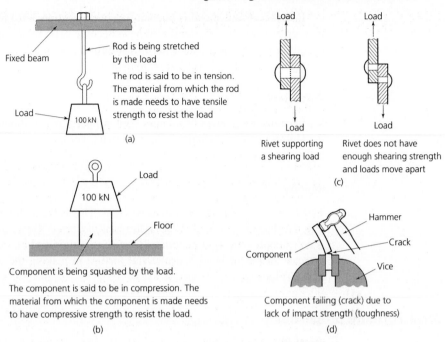

Figure 3.1 Mechanical properties – strength: (a) tensile strength; (b) compression strength; (c) shear strength; (d) impact strength

Shear strength

This is the ability of a material to withstand an *offset* load without breaking (shearing). This is shown in Fig. 3.1(c). The loads are trying to pull the joint apart and the rivet is trying to resist them. The loads are not in line, but are *offset*. The rivet is subjected to a shear load. The material from which it is made must have sufficient *shear strength* or the rivet will fail as shown, and the loads will move apart. The rivet is then said to have *sheared*. The same effect would have occurred if the loads had been pushing instead of pulling.

Note: riveted joints should be designed so that the load always acts in shear across the shank of the rivet as shown. It must never pull on the heads of the rivet. The heads are only intended to keep the rivet in place.

Toughness

This is the ability of a material to withstand an impact load. This is shown in Fig. 3.1(d). The impact loading is causing the metal to crack. To resist this impact loading without breaking, the material from which it is made needs to have sufficient *toughness*. Strength and toughness must not be confused. Strength refers to tensile strength – the ability to withstand an axial pulling load. For example, when you buy a rod of high-carbon steel (e.g. silver steel) it is in the soft condition and it is strong and tough. It has a relatively high tensile strength and its toughness will enable it to withstand relatively high impact loading before it cracks.

If, however, this metal is quench-hardened (section 3.16) its *tensile strength* will have greatly *increased*, it will also have become very *brittle*. In this hard and brittle condition it will now break with only a light tap with a hammer – it can *no longer resist impact loads* – it has *lost its toughness*.

Brittleness

We have just mentioned brittleness. This property is the opposite of toughness. It is the ability to shatter when subject to impact. For example, the way a glass window behaves when struck by a stone.

Rigidity

This property is also referred to as stiffness. This is the ability of a metal to retain its original shape under load. That is, to resist *plastic or elastic deformation*. Cast iron is an example of a rigid material. Because it is rigid and because it can be cast into intricate shapes, it is a good material for use in making the beds and columns for machine tools.

3.2.2 Forming properties

Elasticity

This property enables a material to change shape under load and to return to its original size and shape when the load is removed. Components such as springs are made from elastic materials as shown in Fig. 3.2(a). Note that springs will only return to their original length providing they are not overloaded.

Plasticity

This property enables a material to deform under load and to retain its new shape when the load is removed. This is shown in Fig. 3.2(b). The coin is made from a copper alloy that is relatively soft and plastic. It takes the impression of the dies when compressed between them, and retains that impression when the dies are opened. When deformed by a tensile force, as when wire drawing, the property of plasticity is given the special name *ductility*. When the deforming force is compressive, for example when coining, the property of plasticity is given the special name *malleability*.

Ductility

As stated above, this property enables a material to deform in a plastic manner when subjected to a tensile (stretching) force. For example, when wire drawing as shown in Fig. 3.2(c). Also, the bracket shown in Fig. 3.2(d) requires a ductile material. The outer surface of the material stretches as it bends and is in a state of tension. At the same time the material must remain bent when the bending force is removed so it must be plastic. A *ductile* material combines *tensile strength* and *plasticity*. Note that even the most ductile of metals still show a degree of elasticity. Therefore, when bending the bracket to shape, it must be bent through slightly more than a right-angle. This 'overbend' allows for any slight 'springback'.

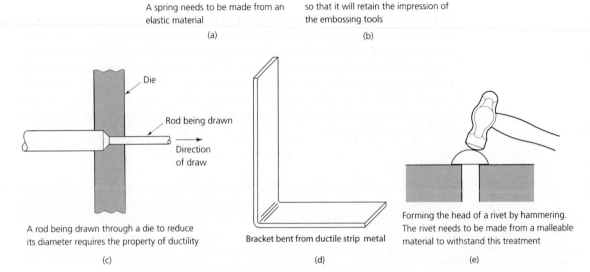

A spring needs to be made from an elastic material

A coin is made from a plastic material so that it will retain the impression of the embossing tools

(a)

(b)

Die

Rod being drawn

Direction of draw

A rod being drawn through a die to reduce its diameter requires the property of ductility

(c)

Bracket bent from ductile strip metal

(d)

Forming the head of a rivet by hammering. The rivet needs to be made from a malleable material to withstand this treatment

(e)

Figure 3.2 Mechanical Properties – flow: (a) elasticity; (b) plasticity; (c) ductility; (d) bracket bent from ductile strip metal; (e) malleability

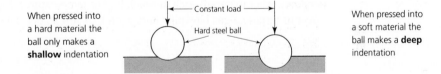

When pressed into a hard material the ball only makes a **shallow** indentation

Constant load

Hard steel ball

When pressed into a soft material the ball makes a **deep** indentation

Figure 3.3 Hardness

Malleability

As stated above, this property enables a material to deform in a plastic manner when subjected to a compressive (squeezing) force. For example, when forging or as when rivet heading as shown in Fig. 3.2(e). The material must retain its shape when the compressive force is removed so it must be plastic. A *malleable* material combines *compressive strength* and *plasticity*.

Hardness

This is the ability of a material to resist scratching and indentation. Figure 3.3 shows a hard steel ball being pressed into the surface of two pieces of material using the same standard load. When pressed into a hard material, the ball only makes a shallow indentation. When pressed into a soft material, under the same test conditions, the ball sinks into the material further and makes a deeper impression. This is the basic principle of all standard hardness tests.

3.2.3 Heat properties

Heat conductivity

This is the ability of a material to conduct heat. Metals are good conductors of heat and non-metals are poor conductors of heat. Figure 3.4(a) shows an electrically heated soldering iron. The bit is made of copper because this is the best of the common metals for conducting heat. It conducts the heat from the heating element to the joint to be soldered. Copper also has an affinity for soft-solder so it can be easily 'tinned'. The handle is made of wood or plastic as these materials are easily shaped and are poor conductors of heat. They are heat-insulating materials. They keep cool and are pleasant to handle.

Refractoriness

Refractory materials are largely unaffected by heat. The fire-bricks used in furnaces are *refractory materials*. They do not burn or melt at the operating temperature of the furnace. They are also good heat-insulating materials. However, the plastic or wooden handle of the soldering iron shown in Fig. 3.4(a) had good heat insulating properties but these are not refractory materials since plastics and wood are destroyed by high temperatures.

> You must not mix up poor heat conductivity with good refractory properties

Fusibility

This is the ease with which materials melt. Soft solders melt at relatively low temperatures other materials melt at much higher temperatures. Figure 3.4(b) shows the effect of turning a gas blowpipe onto a stick of soft solder. The solder quickly melts.

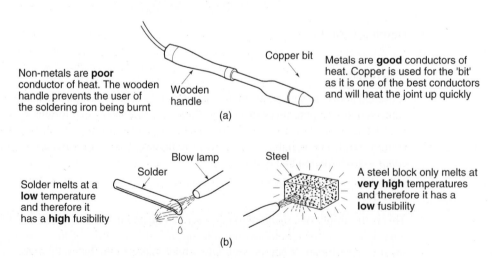

Figure 3.4 Heat processes: (a) heat conductivity; (b) fusibility

The same flame turned onto a block of steel makes the steel hot (possibly red-hot) but does not melt it.

- The soft solder melts at a relatively low temperature because it has high fusibility.
- The steel will not melt in the flame of the blowpipe because it has low fusibility and will only melt at a very high temperature.

3.2.4 Corrosion resistance

This is the ability of a material to withstand chemical or electro-chemical attacks. A combination of such every day things as air and water will chemically attack plain carbon steels and form a layer of rust on its exposed surfaces. Stainless steels are alloys of iron together with carbon, nickel and chromium, they will resist corrosion. Many non-ferrous metals are corrosion resistant which is why we use copper for water pipes, and zinc or lead for roofing sheets and flashings. The sheet metal worker will often meet with such materials as *tin plate* and *galvanized steel* (galvanized iron). The former is thin sheet steel coated with the metal tin whilst the latter is thin sheet steel coated with zinc. These coatings resist corrosion of the steel substrate.

3.2.5 Hot and cold-working

Metals can be cast to shape by melting them and pouring them into moulds. Metals can also be cut to shape by using hand tools and machine tools. However, there is the alternative of *working* them to shape. This is how a Blacksmith shapes metal by hammering it to shape on the anvil. Metals that have been shaped by working (also called 'flow forming') are said to be *wrought*. Metals may be worked hot or cold. In either case the crystalline structure of the metal is distorted by the working process and the metal becomes harder, stronger and less ductile.

- In *cold-working* metals the crystalline structure (grain) becomes distorted due to the processing and remains distorted when the processing has finished. This leaves the metal harder, stronger and less ductile. Unfortunately, further cold working could cause the metal to crack.
- In *hot-working* metals the grain also becomes distorted but the metal is sufficiently hot for the grains to reform as fast as the distortion occurs, leaving the metal soft and ductile. Some grain refinement will occur so that the metal should be stronger than at the start of the process. The metal is easier to work at high temperatures and less force is required to form it. This is why the Blacksmith gets the metal red-hot before forging it to shape with a hammer.
- *Recrystallization* is the term used when the distorted grains reform when they are heated. The temperature at which this happens depends upon the type of metal or alloy and how severely it has been previously processed by cold working.
- *Cold-working* is the flow forming of metal *below* the temperature of recrystallization (for example, cold heading rivets).
- *Hot-working* is the flow forming of metals *above* the temperature of recrystallization (for example, rolling red-hot ingots into girders in a steel mill).

Figure 3.5 shows examples of hot- and cold-working, and Tables 3.1 and 3.2 summarize the advantages and limitations of hot- and cold-working processes.

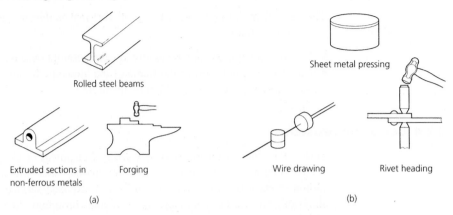

Figure 3.5 Examples of (a) hot-working and (b) cold-working

Table 3.1 Hot-working processes

Advantages	Limitations
1. Low cost 2. Grain refinement from cast structure 3. Materials are left in the fully annealed condition and are suitable for cold working (heading, bending, etc.) 4. Scale gives some protection against corrosion during storage 5. Availability as sections (girders) and forgings as well as the more usual bars, rods, sheets, strip and butt-welded tube	1. Poor surface finish – rough and scaly 2. Due to shrinkage on cooling the dimensional accuracy of hot-worked components is of a low order 3. Due to distortion on cooling and to the processes involved, hot working generally leads to geometrical inaccuracy 4. Fully annealed condition of the material coupled with a relatively coarse grain leads to a poor finish when machined 5. Low strength and rigidity for metal considered 6. Damage to tooling from abrasive scale on metal surface

Table 3.2 Cold-working processes

Advantages	Limitations

1. Good surface finish
2. Relatively high dimensional accuracy
3. Relatively high geometrical accuracy
4. Work hardening caused during the cold-working processes:
  (a) increases strength and rigidity
  (b) improves the machining characteristics of the metal so that a good finish is more easily achieved
	1. Higher cost than for hot-worked materials. It is only a finishing process for material previously hot worked. Therefore, the processing cost is added to the hot-worked cost
	2. Materials lack ductility due to work hardening and are less suitable for bending, etc.
	3. Clean surface is easily corroded
	4. Availability limited to rods and bars, also sheets and strip, solid drawn tubes

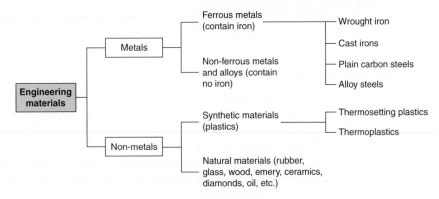

Figure 3.6 Classification of engineering materials

To keep things simple, we will group materials of similar types together and then consider the properties and uses of some examples from each group. These main groups are shown in Fig. 3.6.

3.3.1 Metals

For the purposes of this book, we can consider metals as substances that have a lustrous sheen when cut, are good conductors of heat, and are good conductors of electricity. Some examples are aluminium, copper and iron. Sometimes metals are mixed with non-metals. For example, cast irons and plain carbon steels are mixtures of iron and carbon with traces of other elements. Sometimes metals are mixed with other metals to materially alter their properties such mixtures of metals are called *alloys*. For example, brass is an alloy of copper and zinc.

3.3.2 Non-metals

These can be elements, compounds of elements and mixtures of compounds. They include wood, rubber, plastics, ceramics and glass. Some materials are compounds of metals and non-metals. For example, naturally occurring abrasive grits, such as emery and corundum contain between 70% and 90% of aluminium oxide (a compound of aluminium and oxygen). Aluminium oxide (also known as alumina) is used in firebricks for furnace linings.

Organic compounds are based on the element *carbon* chemically combined with other substances. Some examples of organic materials can be *natural* materials such as wood and some rubbers, or *synthetic* materials such as plastics.

Ferrous metals and alloys are based on the metal *iron*. They are called ferrous metals because the Latin name for iron is *Ferrum*. Iron is a soft grey metal and is rarely found in the pure state outside a laboratory. For engineering purposes the metal iron is usually associated with the non-metal carbon.

3.4.1 Plain carbon steels

Plain carbon steels consist, as their name implies, mainly of iron with small quantities of carbon. There will also be traces of impurities left over from when the metallic iron was extracted for its mineral ore. A small amount of the metal manganese is added to counteract the effects of the impurities. However, the amount of manganese present is insufficient to change the properties of the steel and it is, therefore, not considered to be an alloying element. Plain carbon steels may contain:

- 0.1% to 1.4% carbon.
- up to 1.0% manganese (not to be confused with magnesium).
- up to 0.3% silicon.
- up to 0.05% sulphur.
- up to 0.05% phosphorus.

Figure 3.7 shows how the carbon content of a plain carbon steel affects the properties of the steel. For convenience we can group plain carbon steels into three categories:

- low carbon steels (below 0.3% carbon).
- medium carbon steels (0.3% to 0.8% carbon).
- high carbon steels (between 0.8% and 1.4% carbon).

3.4.2 Low carbon steels

These are also referred to as *mild steels*. If the carbon content is kept between 0.1% and 0.15%, the steel is often referred to as 'dead mild' steel. This steel is very ductile and very soft, and can be pressed into complicated shapes for car body panels at room

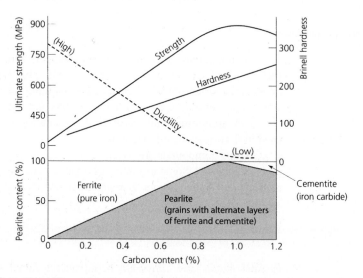

Figure 3.7 The effect of carbon content on the properties of plain carbon steels

temperature without cracking. It is used for the sub-strate of tin-plate and galvanized steel. It is slightly weaker than the next group of low carbon steels to be considered.

If the carbon content is between 0.15% and 0.3%, the steel is stronger, but slightly less soft and ductile. It is often referred to as 'mild' steel. It can be forged, rolled and drawn both in the hot and in the cold condition. It is easily machined with high-speed steel cutting tools. Because of its ease of manufacture and the very large quantities produced, mild steels are the cheapest and most plentiful of the steel products. Hot worked mild steel is available as structural sections for fabrication engineering (e.g. I-section beams, rolled steel joists, reinforcing rods and mesh for concrete, etc.), forgings, sheet, strip, plate, rods, bars and seam-welded tubes. Cold worked mild steel – also known as bright drawn mild steel (BDMS) – is available as bright drawn bars and rods; solid drawn tubes; sheet and strip; and formed wire goods.

3.4.3 Medium carbon steels

There are two groups of medium carbon steels:

1. From 0.3% to 0.5% carbon. These can be heat treated to make them tough and strong.
2. From 0.5% to 0.8% carbon. These can be heat treated to make them fairly hard yet remain a degree of toughness (impact resistance).

Medium carbon steels are harder, stronger and tougher than low carbon steels. They are also more expensive. They cannot be bent or formed in the cold condition to the same extent as low carbon steels without cracking. However, medium carbon steels hot forge well, but close temperature control is required to prevent:

- 'Burning' at high temperatures over 1150°C, as this leads to embrittlement. The metal cannot be reclaimed and the forging has to be scrapped.
- Cracking when forging is continued below 700°C. Cracking is due to work hardening as the steel is in the 'cold' condition from a forging point of view.

Medium carbon steels with a carbon content in the 0.3% to 0.5% range are used for such products as drop-hammer die blocks, laminated springs, wire ropes, screwdriver blades, spanners, hammer heads and heavy duty forgings.

Medium carbon steels with a carbon content in the 0.5% to 0.8% range are used for such products as wood saws, cold chisels, forged blanks for connecting rods, crankshafts, gears and other stressed components such as high-tensile pipes and tubes.

3.4.4 High carbon steels

These are harder, less ductile and more expensive than both mild and medium carbon steels. They are also less tough and are mostly used for springs, cutting tools and forming dies. High carbon steels work-harden readily and, for this reason, they are not recommended for cold working. However, they forge well providing the temperature is controlled at between 700°C and 900°C. There are three groups of high carbon steels:

1. From 0.8% to 1.0% carbon where both toughness and hardness are required. For example, sheet metal pressing tools where the number of components to be produced

does not warrant the cost of expensive alloy steels. Also, cold chisels for fine work, some hand tools, shear blades, coil springs and high-tensile wire (piano wire).

2. From 1.0% to 1.2% carbon for sufficient hardness for most metal cutting tools for example wood-drills, screw-cutting taps and screw-cutting dies.

3. From 1.2% to 1.4% carbon where extreme hardness is required for wood-working tools and knives where a very keen cutting edge is required.

For all of these applications the steel has to be heat-treated to enhance its properties. Heat treatment processes are considered later in this chapter.

3.5 Ferrous metals (alloy steels)

These are essentially plain carbon steels to which other metals (*alloying elements*) have been added in sufficient quantities to materially alter the properties of the steel. The most common alloying elements are:

- *Nickel*, to refine the grain and strengthen the steel.
- *Chromium*, to improve the response of the steel to heat treatment; also to improve the corrosion resistance of the steel.
- *Molybdenum*, to reduce temper brittleness during heat treatment, welding, and operation at sustained high temperatures.
- *Manganese*, improves the strength and wear resistance of steels. Steels containing a high percentage of manganese (14%) are highly wear resistant and these steels are used for such applications as bulldozer blades and plough blades.
- *Tungsten* and *cobalt*, improve the ability of steels to remain hard at high temperatures and are used extensively in cutting tool materials.

Alloy steels are used where great strength is required, corrosion resistance is required, or where the ability to remain hard at high temperatures is required. In this book we only need to consider high-speed steels for cutting tools, and stainless steels where corrosion resistance is required.

3.5.1 High-speed steels

High-speed steels are a group of alloy steels containing metallic elements such as tungsten and cobalt. They are used to make tools suitable for cutting metals. These cutting tools are for use with machine tools where the heat generated by the cutting process would soon soften high carbon steel tools. High-speed steels can operate continuously at 700°C whereas high carbon steel starts to soften at 220°C. Table 3.3 lists the composition and uses of some typical high speed steels.

3.5.2 Stainless steels

These are also alloy steels. They contain a high proportion of chromium to provide corrosion resistance. Various grades of stainless steel are available to suit various applications. For example:

- *Ferritic* stainless steel (BS 403S17), which has 14% chromium but only 0.04% carbon and 0.5% nickel, is easily formed by pressing and spinning and is used

Table 3.3 Typical high-speed steels

Type of steel	Composition* (%)						Hardness (VNP)	Uses
	C	Cr	W	V	Mo	Co		
10% tungsten	0.68	4.0	19.0	1.5	–	–	800–850	Low-quality alloy, not much used
30% tungsten	0.75	4.7	22.0	1.4	–	–	850–950	General-purpose cutting tools for jobbing work shops
6% cobalt	0.8	5.0	19.0	1.5	0.5	6.0	800–900	Heavy-duty cutting tools
Super HSS 12% cobalt	0.8	5.0	21.0	1.5	0.5	11.5	850–950	Heavy-duty cutting tools for machining high-tensile materials

*C = Carbon, Cr = chromium, W = tungsten, V = vanadium, Mo = molybdenum, Co = cobalt.

widely for low cost domestic utensils. Unfortunately, it cannot be hardened and has the lowest strength and corrosion resistance of the stainless steels.

■ *Martensitic* stainless steel (BS 420S45) has 13% chromium, 0.3% carbon and 1.0% nickel. It can be quench hardened and is used for cutlery.
■ *Austenitic* stainless steel (BS 302S25) has 18% chromium, 0.1% carbon and 8% nickel; hence it is widely known as 18/8 stainless steel. It is widely used for fabrications, domestic, architectural and decorative purposes. It is the most corrosion resistant of all the stainless steels.

3.6 Ferrous metal (cast irons)

These are also ferrous metals containing iron and as much as 3% carbon. They do not require the expensive refinement processes of steel making, and they provide a relatively low cost engineering material that can be easily cast into complex shapes at much lower temperatures that those associated with cast steel. The cast irons are not likely to be met with by fabrication and welding engineers.

3.7 Abbreviations

Table 3.4 lists some of the abbreviations used for ferrous metals. They may be found on storage racks in the stores and on engineering drawings. Such abbreviations are very imprecise and refer mainly to groups of materials that can vary widely in composition and properties within the group. It is better to specify a material precisely using a British Standard coding.

3.8 Non-ferrous metals and alloys

Non-ferrous metals and alloys refer to all the multitude of metals and alloys that do not contain iron or, if any iron is present, it is only a minute trace. The most

Table 3.4 Abbreviations for ferrous metals

Abbreviation	Metal
CI	Cast iron (usually 'grey' cast iron)
SGCI	Spheroidal graphite cast iron
MS	Mild steel (low-carbon steel)
BDMS	Bright drawn mild steel
HRPO	Hot-rolled pickled and oiled mild steel
CRCA	Cold-rolled close annealed mild steel
GFS	Ground flat stock (gauge plate)
LCS	Low-carbon steel
SS	Silter steel (centreless ground high-carbon steel)
HSS	High-speed steel
Bright bar	Same as BDMS
Black bar	Hot-rolled steel still coated with scale

widely used non-ferrous metals and alloys are:

- Aluminium and its alloys.
- Copper and its alloys.
- Zinc based die-casting alloys.
- Titanium and its alloys used in the aerospace engineering including airframe and engine components.

In this book we are only interested in the first two groups.

3.8.1 Aluminium and its alloys

Aluminium is the lightest of the most commonly used metal. Its electrical and thermal conductivity properties are very good, being second only to copper. It also has good corrosion resistance and is cheaper than copper. Unfortunately, it is relatively weak in the pure state and is difficult to solder and weld. Special techniques and materials are required for these processes. Pure aluminium is available as foil, sheet, rod, wire, sections (both drawn and extruded). It is also the basis of a wide range of alloys. These can be classified as:

- Wrought alloys (not heat-treatable).
- Wrought alloys (heat-treatable).
- Casting alloys (not heat-treatable).
- Casting alloys (heat-treatable).

The composition and uses of some typical examples of the wrought alloys are listed in Table 3.5. These are the most useful alloys for sheet metal and fabrication engineers.

3.8.2 Copper and its alloys

Copper has already been introduced as a corrosion resistant metal with excellent electrical and thermal conductivity properties. It is also relatively strong compared

Table 3.5 Typical aluminium alloys

Composition						Category	Applications
Copper	Silicon	Iron	Manganese	Magnesium	Other elements		
0.1 max	0.5 max	0.7 max	0.1 max	–	–	Wrought; Not heat-treatable	Fabricated assemblies. Electrical conductors. Food and brewing processing plant. Architectural decoration
0.15 max	0.6 max	0.75 max	1.0 max	4.5–5.5	0.5 chromium	Wrought; Not heat-treatable	High-strength shipbuilding and engineering products. Good corrosion resistance
4.2	0.7	0.7	0.7	0.7	0.3 titanium (optional)	Wrought; Heat-treatable	Traditional 'Duralumin' general purpose alloy. Widely used for stresses components in aircraft and elsewhere
–	0.5	–	–	0.6	–	Wrought; Heat-treatable	Corrosion-resistant alloy for lightly stressed components such as glazing bars, window sections and automotive body components

to aluminium and very easy to join by soldering or brazing. It is much heavier than aluminium and also more costly. Copper is available as cold drawn rods, bars, wire and tubes. Cold rolled sheet and strip, extruded sections, castings and powder for sintered components. So-called 'pure copper' is available in various grades: for example:

■ *Cathode copper*. This is used for the production of copper alloys. As its name implies, cathode copper is manufacture by an electrolytic refining process.
■ *High conductivity copper*. This is better than 99.9% pure and is used for electrical conductors and heat exchangers.
■ *Tough pitch copper*. This is a general purpose, commercial grade copper containing some residual copper oxide from the refining process. It is this copper oxide content that increases its strength and toughness but reduces its electrical conductivity and ductility. It is suitable for roofing sheets, chemical plant, general presswork, decorative metalwork and applications where special properties are not required.

- *Phosphorus deoxidized, non-arsenical copper*. This is a welding quality copper. Removal of the dissolved oxygen content prevents gassing and porosity. Also the lack of dissolved and combined oxygen improves the ductility and malleability of the metal. It is used in fabrications, castings, cold impact extrusion and severe presswork.
- *Arsenical tough pitch and phosphorous deoxidized copper*. The addition to traces of the metal, arsenic improves the strength of the metal at high temperatures. Arsenical coppers are used for boiler and firebox plates, flue tubes and general plumbing.

The main groups of copper-based alloys are the:

- High copper content alloys (e.g. cadmium copper, silver copper, etc.)
- Brass alloys (copper and zinc)
- Tin bronze alloys (copper and tin)
- Aluminium bronze alloys (copper and aluminium)
- Cupro-nickel alloys (copper and nickel)

In this book we are only interested in the brass alloys and the tin-bronze alloys.

Brass alloys

Brass alloys of copper and zinc tend to give rather weak and porous castings. The brasses depend largely upon hot and/or cold working to consolidate the metal and improve its mechanical properties. The brass alloys can only be hardened by cold-working (work hardening). They can be softened by heat treatment (the annealing process). The composition and uses of the more commonly available brass alloys are given in Table 3.6.

Table 3.6 Typical brass alloys

Name	Composition			Applications
	Copper	Zinc	Other elements	
Cartridge brass	70	30	–	Most ductile of the copper–zinc alloys. Widely used in sheet metal pressing for severe deep drawing operations. Originally developed for making cartridge cases, hence its name
Standard brass	65	35	–	Cheaper than cartridge brass and rather less ductile. Suitable for most engineering processes
Basis brass	63	37	–	The cheapest of the cold-working brasses. It lacks ductility and is only capable of withstanding simple forming operations
Muntz metal	60	40	–	Not suitable for cold working, but hot works well. Relatively cheap due to its high zinc content, it is widely used for extrusion and hot-stamping processes
Free-cutting brass	58	39	3% lead	Not suitable for cold working, but excellent for hot working and high-speed machining of low-strength components
Admiralty brass	70	29	1% tin	This is virtually cartridge brass plus a little tin to prevent corrosion in the presence of salt water
Naval brass	62	36	1% tin	This is virtually Muntz metal plus a little tin to prevent corrosion in the presence of salt water

Tin bronze alloys

These are alloys of copper and tin together with a *de-oxidizer*. The de-oxidizer is essential to prevent the tin content from oxidizing at high temperatures during casting and hot working. Oxidation is the chemical combination of the tin content with the oxygen in the atmosphere, and it results in the bronze being weakened and becoming hard, brittle and 'scratchy'. Two de-oxidizers are commonly used:

- A small amount of phosphorus in the *phosphor bronze* alloys
- A small amount of zinc in the *gun metal* alloys

The composition and uses of some typical tin bronze alloys are listed in Table 3.7. Unlike the brasses, which are largely used in the wrought condition (rod, sheet, etc.), only low tin content bronzes can be worked and most bronze components are in the form of castings. The tin bronze alloys are more expensive than the brass alloys but they are stronger and give sound, pressure-tight castings that are widely used for steam and hydraulic valve bodies and mechanisms. They are highly resistant to corrosion.

Table 3.7 Typical tin–bronze alloys

Name	Composition					Applications
	Copper	**Zinc**	**Phosphorus**	**Tin**	**Lead**	
Low-tin bronze	96	–	0.1–0.25	3.9–3.75	–	This alloy can be severely cold worked to harden it so that it can be used for springs where good elastic properties must be combined with corrosion resistance, fatigues resistance and electrical conductivity, e.g. contact blades
Drawn phosphor-bronze	94	–	0.1–0.5	5.9–5.5	–	This alloy is used in the work-hardened condition for turned components requiring strength and corrosion resistance, such as valve spindles
Cast phosphor-bronze	rem.	–	0.03–0.25	10	–	Usually cast into rods and tubes for making bearing bushes and worm wheels. It has excellent anti-friction properties
Admiralty gunmetal	88	2	–	10	–	This alloy is suitable for sand casting where fine-grained, pressure-tight components such as pump and valve bodies are required
Leaded-gunmetal (free-cutting)	85	5	–	5	5	Also known as 'red brass', this alloy is used for the same purposes as standard, Admiralty gunmetal. It is rather less strong but has improved pressure tightness and machine properties
Leaded (plastic) bronze	74	–	–	2	24	This alloy is used for lightly loaded bearings where alignment is difficult. Due to its softness, bearings made from this alloy 'bed in' easily

3.9 The identification of metals

Materials represent a substantial investment in any manufacturing company. It is essential that all materials are carefully stored so that they are not damaged or allowed to deteriorate before use. Ferrous metals must be stored in a warm, dry environment so that rusting cannot occur. This is particularly the case when storing bright drawn sections and bright rolled sheets. Rusting would quickly destroy the finish and cause such materials to be unfit for use. The similarity in appearance between many metals of different physical properties makes it essential that some form of permanent identification should be marked on them (e.g. colour coding), so that they can be quickly and accurately identified. Mistakes, resulting in the use of an incorrect material, can be very costly through waste. It can also cause serious accidents if, for example, a weak metal is used instead of a strong one for highly stressed components.

3.10 Non-metals (natural)

Non-metals are widely used in engineering today. Some of the materials occur naturally. For example:

1. *Rubber* is used for anti-vibration mountings, coolant and compressed air hoses, transmission belts, truck wheel tyres.
2. *Glass* is used for spirit level vials (the tube that contains the bubble), lenses for optical instruments.
3. *Emery and corundum* (aluminium oxides) is used abrasive wheels belts and sheets, and as grinding pastes. Nowadays it is usually produced artificially to control the quality.
4. *Wood* for making casting patterns.
5. *Ceramics* for cutting tool tips and electrical insulators.

3.11 Non-metals (synthetic)

These are popularly known as *plastics*. A plastic material is said to be one that deforms to a new shape under an applied load and retains its new shape when the load is removed. Yet, the range of synthetic materials we call *plastics* are often tough and leathery, or hard and brittle, or even elastic. They are called *plastics* because, during the moulding operation by which they are formed, they are reduced to a plastic condition by heating them to about twice the temperature of boiling water. There are many families of 'plastic' materials with widely differing properties. However, they all have certain properties in common.

1. *Electrical insulation.* All plastic materials are, to a greater or lesser extent, good electrical insulators (they are also good heat insulators). However, their usefulness as insulators is limited by their inability to withstand high temperatures and their relative softness compared with ceramics. They are mainly used for insulating wires and cables and for moulded switch gear and instrument components and cases.

2. *Strength/weight ratio*. Plastic materials vary considerably in strength. All plastics are much less dense than metals, resulting in a favourable strength/weight ratio. The high strength plastics and reinforced plastics compare favourably with the aluminium alloys and are often used for stressed components in aircraft construction.

3. *Degradation*. Plastics do not corrode like metals. They are all inert to most inorganic chemicals. They can be used in environments that are chemically hostile to even the most corrosion resistant metals. They are superior to natural rubber in their resistance to attack by oils and greases. However, all plastics degrade at high temperatures and many are degraded by the ultraviolet content of sunlight. Plastics that are exposed to sunlight (window frames and roof guttering) usually contain a pigment that filters out the ultraviolet rays. Some thermoplastics can be dissolved by suitable solvents.

4. *SAFETY*. Solvents used in the processing of plastics are often highly toxic and should not be inhaled but used in well-ventilated surroundings. Make sure you know the likely dangers before starting work on plastic materials and always follow the safe working practices laid down by the safety management.

DANGER! Plastics can give off very dangerous toxic fumes when heated. For example, note the number of people who have died from inhaling the smoke from plastic furniture padding in house fires!

Plastic materials can be grouped into two distinct families. These are the *thermosetting plastics* and the *thermoplastics*. Typical examples of each of these families will now be considered. Thermosetting plastics are often referred to simply as 'thermosets'.

3.11.1 Thermosetting plastics

These undergo a chemical change called 'curing' during hot moulding process. Once this chemical change has taken place, the plastic material from which the moulding is made can never again be softened and reduced to a plastic condition by re-heating.

Thermosetting resins are usually mixed with other substances (additives) to improve their mechanical properties, improve their moulding properties, make them more economical to use, and provide the required colour for the finished product. A typical moulding material could consist of:

- Resin   38% by weight
- Filler   58% by weight
- Pigment   3% by weight
- Mould release agent 0.5% by weight
- Catalyst   0.3% by weight
- Accelerator   0.2% by weight

The pigment gives colour to the finished product. The mould release agent stops the moulding sticking to the mould. It also acts as an internal lubricant and helps the plasticized material to flow to the shape of the mould. The catalyst promotes the curing process and the accelerator speeds up the curing process and reduces the time the moulds have to be kept closed, thus improving productivity.

Fillers are much cheaper than the resin itself and this is important in keeping down the cost of the moulding. Fillers also have a considerable influence on the properties of the mouldings produced from a given thermosetting resin. They improve

Table 3.8 Some typical thermosetting plastic materials

Material	Characteristics
Phenolic resins and powders	These are used for dark-coloured parts because the basic resin tends to become discoloured. These are heat-curing materials
Amino (containing nitrogen) resins and powders	These are colourless and can be coloured if required; they can be strengthened by using paper-pulp filters, and used in thin sections
Polyester resins	Polyester chains can be cross-linked by using a monomer such as styrene; these resins are used in the production of glass-fibre laminates
Epoxy resins	These are also used in the production of glass-fibre laminates

the impact strength (toughness) and reduce shrinkage during moulding. Typical fillers are:

1. *Shredded paper* and *shredded cloth* give good strength and reasonable electrical insulation properties at a low cost.
2. *mica granules* give good strength and heat resistance (asbestos is no longer used).
3. *aluminium powder* gives good mechanical strength and wear resistance.
4. *Wood flour (fine sawdust)* and *calcium carbonate (ground limestone)* provide high bulk at a very low cost but with relatively low strength.
5. *Glass fibre (chopped)* good strength and excellent electrical insulation properties.

Some typical examples of thermosetting plastics and their uses are given in Table 3.8.

Thermoplastics

These can be softened as often as they are reheated. They are not as rigid as the thermosetting plastics but they tend to be tougher. Additives (other than a colourant and an internal lubricant) are not normally used with thermoplastics. Some typical examples of thermoplastics and their uses are given in Table 3.9.

3.12 Forms of supply

There is an almost unlimited range to the forms of supply in which engineering materials can be made available to a manufacturer or to a fabricator. Figure 3.8 shows some of these forms of supply.

- *Sections* such as steel angles, channel sections, I-section beams, joists and T-sections in a wide range of sizes and lengths. They are usually hot-rolled with a heavily scaled finish. Such sections are mostly used in the steel fabrication and civil engineering and construction industries. Bright-drawn steel angle sections are available in the smallest sizes. Non-ferrous metal sections are normally extruded to close tolerances and

Table 3.9 Some typical thermoplastic materials

Type	Material	Characteristics
Cellulose plastics	Nitrocellulose	Materials of the 'celluloid' type are tough and water resistant. They are available in all forms except moulding powders. They cannot be moulded because of their flammability
	Cellulose acetate	This is much less flammable than the above. It is used for tool handles and electrical goods
Vinyl plastics	Polythene	This is a simple material that is weak, easy to mould, and has good electrical properties. It is used for insulation and for packaging
	Polypropylene	This is rather more complicated than polythene and has better strength
	Polystyrene	This is rather cheap, and can be easily moulded. It has a good strength but it is rigid and brittle and crazes and yellows with age
	Polyvinyl chloride (PVC)	This is tough, rubbery, and almost non-inflammable. It is cheap and can be easily manipulated: it has good electrical properties
Acrylics (made from an acrylic acid)	Polymethyl methacrylate	Materials of the 'Perspex' type have excellent light transmission, are tough and non-splintering, and can be easily bent and shaped
Polyamides (short carbon chains that are connected by amide groups – NHCO)	Nylon	This is used as a fibre or as a wax-like moulding material. It is fluid at moulding temperature, tough, and has a low coefficient of friction
Fluorine plastics	Polytetrafluoroethylene (PTFE)	This is a wax-like moulding material; it has an extremely low coefficient of friction. It is very expensive
Polyesters (when an alcohol combines with an acid, an 'ester' is produced)	Polyethylene terephthalate	This is available as a film or as 'Terylene'. The film is an excellent electrical insulator

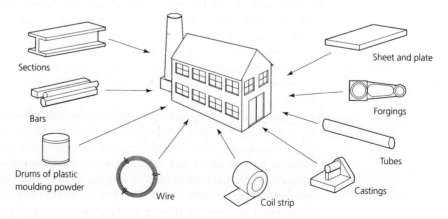

Figure 3.8 Forms of supply

have a bright finish. Both standard sections and sections to customers' own requirements are made this way. The sizes available are very much smaller than for steel sections.

- *Bars* may be 'flats' which are available in rectangular sections, they may be 'squares', or they may be 'rounds' which have a cylindrical section and this term applies to the larger sizes. In the smaller size they are usually referred to as 'rods'. They are available in hot-rolled (black) or cold-drawn (bright) finishes. Hexagon section bars for making nuts and bolts are always cold-drawn.
- *Plastics* may be supplied in powder or granular form for moulding into various shapes; or in semi-finished forms in rounds, flats, squares, tubes, and sheet. They may also be ordered as sections to manufacturers' own requirements, such as curtain rails and insulation blocks ready for cutting off to length. Reinforced plastics such as 'Tufnol' are also available in standard sections and mouldings.
- *Wire* is available either bright-drawn or hot rolled depending upon its size and the use to which it is going to be put. Bright-drawn, high-carbon steel wire (piano-wire) is used for making springs; copper wire is used for electrical conductors. The smaller the diameter of any wire, the longer the strand length that can be supplied in.
- *Coil strip* is used for cold stamping and pressing where continuous automatic feed to the presses is required. It is available in a range of thicknesses and finishes. A steel strip is available in such finishes as bright-rolled (BR), hot-rolled, pickled and oiled (HRPO) and cold-rolled, closed-annealed (CRCA). It can be sheared on continuous rotary shears to the customer's specification where accurate control of the width is required. Alternatively, it can be left with a 'mill-edge' where the flat surfaces are bright rolled and the edges are left rounded and in the hot rolled state. A non-ferrous strip is usually bright rolled and sheared to width. The rolling process tends to harden the strip which is sold in various 'tempers' according to the amount of cold-working it has received since annealing. For example, dead-soft, soft, quarter-hard, half-hard, etc.
- *Castings* can be made in most metals and alloys but the moulding process will vary depending upon the type of metal, the size of the casting, the accuracy of the casting and the quantities involved. There are no 'standard' castings. They are made to the customer's own patterns or, in the case of die-casting, in the customer's own dies.
- *Tubes* and *pipes* can be made in ferrous and non-ferrous metals and alloys and in plastic. Tubes refer to the smaller sizes and pipes to the larger sizes. Steel pipes may be cold-drawn or hot-drawn without seams for the highest pressures. They may also be rolled from strip with a butt-welded seam running along the length of the pipe or tube where lower pressures are involved, or they are only used for sheathing (electrical conduit is made like this). Plastic tube may be rigid or flexible.
- *Forgings* may be manufactured by forming the red-hot metal with standard tools by a 'black-smith' when only small quantities are required. The size may range from horse-shoes, farm implements, and decorative wrought-iron work made by hand, to turbine shafts and ships' propeller shafts forged under huge hydraulic presses. Where large quantities of forging of the same type are required, these are drop-forged or press forged. The red-hot steel is forged in dies which impart the finished shape to the forging. Light alloy (aluminium alloy) forgings are used by the aircraft industry. Because of the low melting temperature of such alloys

compared with steel, they have to be forged at a lower temperature. This requires a greater forging pressure and care must be taken that cracking does not occur.

- *Sheet and plate* starts off as a very wide coiled strip and is then passed through a series of rollers to flatten it. It is sheared to length by 'flying shears' that cut it whilst the flattened strip is moving. The terminology is somewhat vague and depends upon the metal thickness. Generally, 'sheet' can be worked with hand tools and ranges from foil (very thin sheet) to about 1.5 mm thick. Then 'thin plate' up to about 6 mm thick. After that it becomes thick plate. Both thin and thick plates have to be cut and formed using power driven tools. Sheet is available in both cold- and hot-rolled finishes, whereas plate is only available as hot-rolled. For the sizes of any standard metallic products you should consult the appropriate British Standard Specifications.

3.13 Heat treatment processes (introduction)

Heat treatment processes as a means of modifying the properties of metals, have already been mentioned earlier in this chapter. Table 3.10 summarises and defines the more common heat treatment processes. Because of the wide range of non-ferrous metals and alloys that exist, the heat treatment processes for non-ferrous metals vary widely and all such processes are quite different to the processes used for the heat treatment of plain carbon steels. However, some of the more important processes for the heat treatment of copper based and aluminium based alloys will be included in this chapter.

Table 3.10 Heat treatment definitions

Term	Meaning
Annealed	The condition of a metal that has been heated above a specified temperature, depending upon its composition, and then cooled down in the furnace itself or by burying it in ashes or lime. This annealing process makes the metal very soft and ductile. Annealing usually precedes flow-forming operations such as sheet metal pressing and wire and tube drawing
Normalized	The condition of a metal that has been heated above a specified temperature, depending upon its composition, and then cooled down in free air. Although the cooling is slow, it is not as slow as for annealing so the metal is less soft and ductile. This condition is not suitable for flow forming but more suitable for machining. Normalizing is often used to stress relieve castings and forgins after rough machining
Quench hardened	The condition of a metal that has been heated above a specified temperature, depending upon its composition, and then cooled down very rapidly by immersing it in cold water or cold oil. Rapid cooling is called **quenching** and the water or oil is called the **quenching bath**. This rapid cooling from elevated temperatures makes the metal very hard. Only medium- and high-carbon steels can be hardened in this way
Tempered	Quench-hardened steels are brittle as well as hard. To make them suitable for cutting tools they have to be reheated to a specified temperature between 200 and 300°C and again quenched. This makes them slightly less hard but very much tougher. Metals in this condition are said to be *tempered*

3.14 Heat treatment processes (safety)

General safety was introduced in Chapter 1. However, heat treatment can involve large pieces of metal at high temperatures and powerful furnaces. Therefore, it is now necessary to consider some safety practices relating specifically to heat treatment processes, before discussing the processes involved.

3.14.1 Protective clothing

Overalls used in heat treatment shops should be made from a flame resistant or a flame retardant material and be labelled accordingly. In addition, a leather apron should be worn to prevent your overalls coming into contact with hot work-pieces and hot equipment.

Gloves should be worn to protect your hands. These should be made from leather or other heat resistant materials and should have gauntlets to protect your wrists and the ends of the sleeves of your overalls. Leather gloves offer protection up to 350°C. Headwear, goggles and visors, such as those described in Chapter 1, should be worn when there is any chance of danger to your eyes, the skin of your face and your hair and scalp. Such dangers can come from sources such as:

1. Splashes from the molten salts when using salt bath furnaces.
2. Splashes from hot liquids when quenching.
3. The accidental ignition of oil quenching baths due to overheating.
4. The radiated heat from large furnaces when their doors are opened.
5. Accidents such as a 'flash-back' when lighting up furnaces.

The safety shoes and boots as recommended for wearing in workshops were introduced in Chapter 1. They are also the most suitable for use in heat treatment shops. They not only protect you from cuts and bruising but, being made of strong leather, they also protect against burns from hot objects.

Remember, your first instinctive reaction to accidentally picking up anything hot, is to let go quickly and drop it. This is when the toe protection of industrial safety shoes really earns its keep

In addition, it is advisable that leather spats are worn. These protect your lower legs and ankles from splashes of molten salts or spillage of hot quenching fluids. Spats are particularly important if you wear safety shoes rather than safety boots.

3.14.2 Safety notices

Whilst metal is red hot it is obviously in a dangerous condition. However, most accidents occur when the metal has cooled down to just below red heat. Although no longer glowing, it is still hot enough to cause serious burns and to start fires if flammable substances come into contact with it. Hot work-pieces must never be stored in gangways and warning notices must be used as shown in Fig. 3.9. Such notices must satisfy the legal requirements of the Health and Safety Executive.

3.14.3 Fire

Quenching baths using a quenching-oil must have an airtight lid. In the event of the oil overheating and igniting, the lid can be closed which puts the fire out. Quenching

Figure 3.9 Safety notices must be placed by hot objects

tanks should always have sufficient reserve capacity so that the oil does not over-heat. If the oil is allowed to overheat, it:

- Will not cool the work quickly enough to harden it.
- May catch fire.

Only quenching-oil, with a high flash point and freedom from fuming, should be used. *Lubricating oil must never be used*. A suitable fire extinguisher, or several fire extinguishers if a large quenching tank is used, should be positioned conveniently near to the bath in case of an emergency.

The type of extinguisher used should be suitable for use on oil fires

Furnaces and blowpipes must not be lit or closed down without proper instruction and permission. Incorrect setting of the controls and incorrect light-ing up procedures can lead to serious explosions. All personnel working in heat treatment shops must be alert to the possibility of fires, and be conversant with, and trained in, the correct fire drill.

3.15 The heat treatment of plain carbon steels

Plain carbon steels are subjected to heat treatment processes for the following reasons:

- To improve the properties of the material as a whole; for example, by imparting hardness to prevent wear or softness and grain refinement to improve its machin-ing properties.
- To remove undesirable properties acquired during previous processing; for example, hardness and brittleness imparted by cold working.

The heat-treatment processes we are now going to consider are used to modify the properties of plain carbon steels. They are:

- Through (quench) hardening.
- Tempering.
- Annealing.
- Normalizing.
- Case hardening.

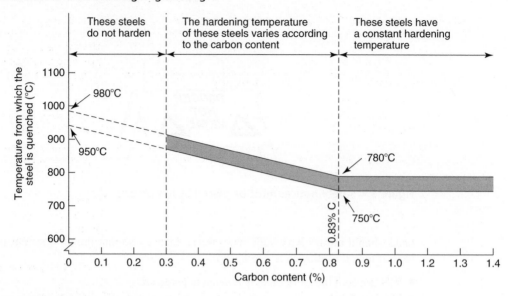

Figure 3.10 Hardening temperatures for carbon steels

3.15.1 Through-hardening

Basically the process by which we 'through-harden' (also referred to as *quench hardening*) consists of heating a suitable steel to a critical temperature that is dependent upon the carbon content of the steel and cooling it quickly (quenching it) in water or quenching oil. The hardness attained will depend upon:

- The carbon content of the steel (the higher the carbon content the harder and more brittle the steel).
- The rate of cooling (the faster the cooling the harder the steel and the more likely it is to crack).

For example, to harden a chisel made from 0.8% carbon steel, the temperature to which the steel is heated should be between 800°C and 830°C. That is, the steel should glow 'cherry red' in colour. It is then quenched by cooling it in cold water. Any increase in temperature will make no difference to the hardness of the chisel. It will only result in 'grain growth' and weakening of the steel. Grain growth means that the hotter the steel becomes and the longer it is kept at excessively high temperatures, the bigger the grains in the metal will grow by merging together. As previously stated, the steel has to be heated to within a critical temperature range if it is to be correctly hardened. The temperature range for hardening plain carbon steels is shown in Fig. 3.10.

Figure 3.10 shows that the hardening temperatures for plain carbon steels lie in a narrow band depending upon the carbon content of the steel. If the hardening temperature for any given steel is not achieved, it will not harden. If it is exceeded, no increase in hardness is achieved but grain growth will occur and the steel will be weakened. It has already been stated that the more quickly a component is cooled the harder it becomes for any given carbon content. However, some care is required

Table 3.11 Effect of carbon content and rate of cooling on hardness

Type of steel	Carbon content (%)	Effect of heating and quenching (rapid cooling)
Low carbon	Below 0.25	Negligible
Medium carbon	0.3–0.5	Becomes tougher
	0.5–0.9	Becomes hard
High carbon	0.9–1.3	Becomes very hard

Carbon content (%)	Quenching bath	Required treatment
0.30–0.50	Oil	Toughening
0.50–0.90	Oil	Toughening
0.50–0.90	Water	Hardening
0.90–0.13	Oil	Hardening

Notes:
1. Below 0.5% carbon content, steels are not hardened as cutting tools, so water hardening has not been included.
2. Above 0.9% carbon content, any attempt to harden the steel in water could lead to cracking.

because the faster you cool the work-piece, the more likely it is to crack and distort. Therefore, the work-piece should never be cooled more quickly than is required to give the desired degree of hardness (*critical cooling rate*). The most common substances used for quenching (*quenching media)* are:

- *Brine* (a solution of sodium chloride and water) is the most rapid quenching bath – it will give the greatest hardness and is the most likely to cause cracking.
- *Water* is less severe and is the most widely used quenching bath for plain carbon steels.
- *Oil* is the least severe of the liquid quenching media. Only plain carbon steels of the highest carbon content will harden in oil, and then only in relatively small sections. Oil quenching is mostly used with alloy die steels and tool steels.
- *Air blast* is the least severe of any of the quenching media used. It can only be applied to heavily alloyed steels of small section such as high-speed steel tool bits.

Table 3.11 summarises the effect of carbon content and rate of cooling for a range of plain carbon steels.

3.15.2 Quenching, distortion and cracking

Quenching and distortion

When quenching hot metal, some thought must be given to the way the work is lowered into the bath to avoid distortion and to get the most effective quenching. For example, when hardening a chisel, the shank of the chisel is held with tongs and that the chisel is dipped vertically into the quenching bath.

- This results in the cutting end of the steel entering the bath first and attaining maximum hardness whilst the quenching medium is at its minimum temperature.

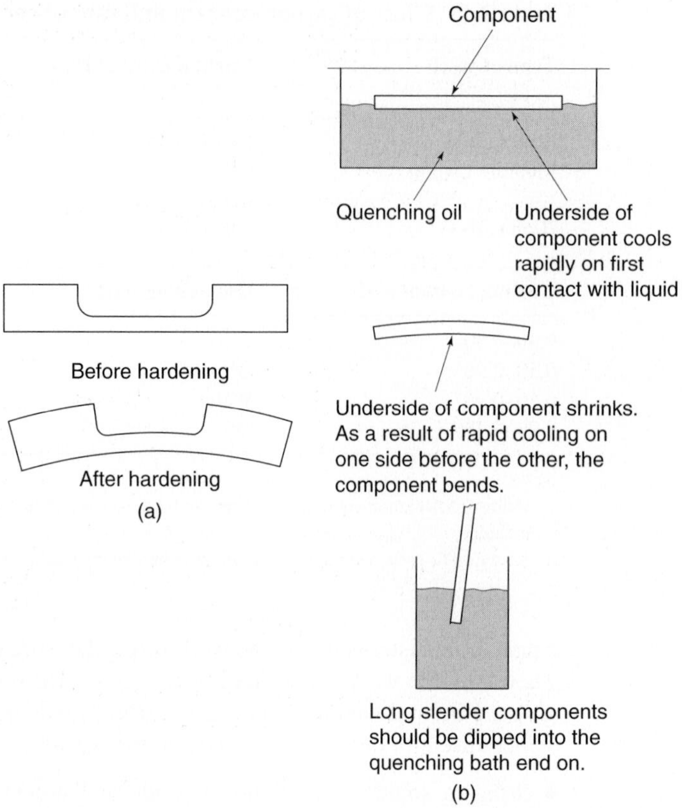

Component

Quenching oil   Underside of
component cools
rapidly on first
contact with liquid

Before hardening

After hardening

(a)

Underside of component shrinks.
As a result of rapid cooling on
one side before the other, the
component bends.

Long slender components
should be dipped into the
quenching bath end on.

(b)

Figure 3.11 Causes of distortion: (a) distortion caused by an unbalanced shape being hardened; (b) how to quench long slender components

- The shank is masked to some extent by the tongs and the result of this is reduced hardness. This does not matter in this example as we want the shank to be tough rather than hard so that it does not shatter when struck with a hammer.
- The chisel should be stirred around in the bath so that it is constantly coming into contact with fresh and cold water. It also prevents steam pockets being generated round the chisel that would slow up the cooling rate and prevent maximum hardness being achieved.
- Dipping a long slender work-piece like a chisel vertically into the bath prevents distortion. This is shown in Fig. 3.11. This figure also shows what happens if the component is quenched flat and also how the shape of the component itself can cause uneven cooling and distortion.

Cracking

Figure 3.12 shows some typical causes of cracking occurring during and as a result of heat treatment. Careful design and the correct selection of materials can result in fewer problems in the hardening shop.

- Avoid sharp corners and sudden changes of section.
- Do not position holes, slots and other features near the edge of the work-piece.

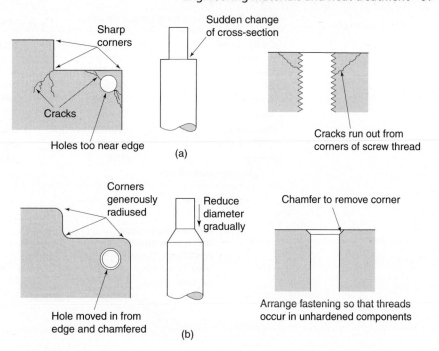

Figure 3.12 Causes of cracking: (a) incorrect engineering that promotes cracking; (b) correct engineering to reduce cracking

- Do not include screw threads in a hardened component. Apart from the chance of cracking occurring, once hardened you cannot run a die down the thread to ease it if it has become distorted during the hardening process.
- For complex shapes (which are always liable to cracking and distortion during hardening) always use an alloy steel that has been formulated so that minimum distortion and shrinkage (movement) will occur during heat treatment. Such steels are oil or air hardening and this also reduces the chance of cracking.

3.15.3 Tempering

When you heat and quench a plain carbon steel as described previously, you not only harden the steel, but also make it *very brittle*. In this condition it is unsuitable for immediate use. For instance, a chisel would shatter if you hit it with a hammer. After hardening we have to carry out another process known as *tempering*. This greatly reduces the brittleness and increases the toughness. However, the tempering process also reduces the hardness to some extent.

Tempering consists of reheating the hardened steel work-piece to a suitable temperature and again quenching it in oil or water. The tempering temperature to which the work-piece is reheated depends only upon the use to which the work-piece is going to be put. Table 3.12 lists some suitable temperatures for tempering components made from plain carbon steels.

Table 3.12 Tempering temperatures

Component	Temper colour	Temperature (°C)
Edge tools	Pale straw	220
Turnign tools	Medium straw	230
Twist drills	Dark straw	240
Taps	Brown	250
Press tools	Brownish-purple	260
Cold chisels	Purple	280
Springs	Blue	300
Toughening (crankshafts)	–	450–600

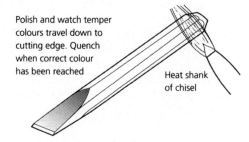

Polish and watch temper colours travel down to cutting edge. Quench when correct colour has been reached

Heat shank of chisel

Figure 3.13 Tempering a cold chisel

In a workshop, the tempering temperature is usually judged by the 'temper colour' of the oxide film that forms on the surface of the work-piece. After hardening, the surface of the work-piece is polished so that the colour of the oxide film can be clearly seen. Figure 3.13 shows a chisel being tempered. The chisel is not uniformly heated. As shown, the shank is heated in the flame and the temper colours are allowed to 'run down' the chisel until the cutting edge reaches the required colour. When the cutting edge is the required temper colour, the chisel is immediately 'dipped' vertically into the quenching bath again. This gives the cutting edge the correct temper but leaves the shank softer and tougher so that it can withstand being struck with a hammer. Complex and large components and batches of components should be tempered in a furnace with atmosphere and temperature control to ensure consistent results.

3.15.4 Annealing

Annealing processes are used to soften steels that are already hard. This is the heat-treatment process most applicable to fabrication and welding engineering Hardness may be imparted in two ways.

- *Quench hardening*. This has as previously described.
- *Work-hardening*. This occurs when the metal has been cold-worked (see section 3.2.5). It becomes hard and brittle at the point where cold-working occurs as this causes the grain structure to deform. For example, if a strip of metal is held in a vice, bending the metal back and forth causes it to work-harden at the point of bending. It will eventually become sufficiently hard and brittle to break off at that point.

Full annealing

The temperatures for full annealing are the same as for hardening. To *anneal* (soften) the work-piece, the hot metal is allowed to cool down as slowly as possible. Small components can be buried in crushed limestone or in ashes. Larger components and batches of smaller components will have been heated in furnaces. When the correct temperature has been reached, the component is 'soaked' at this temperature so that the temperature becomes uniform throughout its mass. The furnace is then shut down, the flue dampers are closed and the furnace is sealed so that it cools down as slowly as possible with the work inside it. Although such slow cooling results in some grain growth and weakening of the metal, it will impart maximum ductility. This results in the metal being in the correct condition for cold forming.

Stress-relief annealing

This process is reserved for steels with a carbon-content below 0.4%. Such steels will not satisfactorily quench harden but, as they are relatively ductile, they will be frequently cold-worked and become work-hardened. Since the grain structure will have become severely distorted by the cold-working, the crystals will begin to reform and the metal will begin to soften (theoretically) at 500°C. In practice, the metal is rarely so severely stressed as to trigger *recrystallization* at such a low temperature. Stress relief annealing is usually carried out between 630°C and 700°C to speed up the process and prevent excessive grain growth. Stress relief annealing is also known as:

- *Process annealing* since the work-hardening of the metal results from cold-working (forming) processes.
- *Inter-stage annealing* since the process is often carried out between the stages of a process when extensive cold-working is required. For example, when deep-drawing sheet metal in a press.

The degree of stress relief annealing and the rate of cooling will depend not only upon the previous processing the steel received before annealing, but also upon the processing it is to receive *after* annealing. If further cold-working is to take place, the maximum softness and ductility is required. This is achieved by prolonged heating and very slow cooling to encourage grain growth. However, if grain refinement, strength and toughness are more important, then heating and cooling should be more rapid.

3.15.5 Normalizing

Plain carbon steels are normalized by heating them to the temperatures shown in Fig. 3.14. This time we want a finer grain structure in the steel. To achieve this, we have to heat the metal up more quickly and cool it more quickly. The work-piece is taken out of the furnace when its normalizing temperature has been achieved and allowed to cool down in the free air of the heat-treatment shop. The air should be able to circulate freely round the work-piece. However, the work-piece must be sited so that it is free from cold draughts. Warning notices that the steel is dangerously hot must be placed around it.

Plain carbon steel components that have been normalized will not be as ductile as components that have been annealed. Therefore, normalizing is not used prior to further cold working. However, normalized components will be stronger and

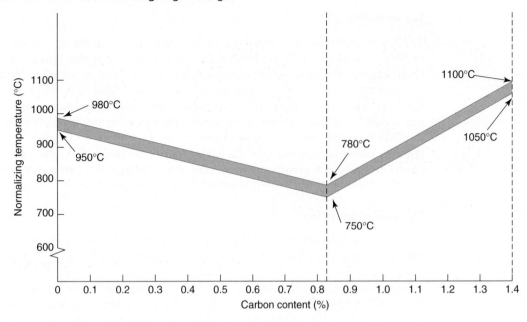

Figure 3.14 Normalizing temperatures for plain carbon steels

machine to a better finish. Normalizing will also remove any residual stresses in the steel produced by previous processing and prevent distortion during and after finish machining.

Figure 3.15 shows the stages in case-hardening a low carbon steel component. This gives it a hard, wear resistant skin and a tough, shock-resistant core.

3.16 The heat treatment of non-ferrous metals and alloys

None of the non-ferrous metals, and only a very few non-ferrous alloys, can be quench hardened like plain carbon steels. The majority of non-ferrous metals can only be hardened by cold-working processes. Alternatively, they can be manufactured from cold-rolled (spring temper) sheet or strip, or they can be manufactured from cold drawn wire. Work-hardened non-ferrous metals can be annealed by a re-crystallization process that is similar to the process annealing for plain carbon steels. The main difference is that non-ferrous metals do not have to be cooled slowly. They can be quenched after heating and this has the advantage that the rapid cooling causes the metal to shrink suddenly – and this removes the oxide film. In the case of copper and its alloys, this is even more effective if the metal is pickled in a weak solution of sulphuric acid whilst still warm.

Suitable annealing temperatures are:

- Aluminium   500°C to 550°C   (pure metal)
- Copper   650°C to 750°C   (pure metal)
- cold-working brass   600°C to 650°C   (simple alloy of copper and zinc)

SAFETY   If an acid bath is used, protective clothing and eye protection such as goggles or, better still, a visor **must** be worn

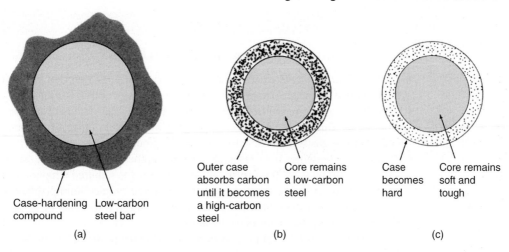

Figure 3.15 Case hardening: (a) carburising; (b) after carburizing; (c) after quenching a component from a temperature above 780°C

Heat-treatable aluminium alloys ('duralumin' is such an alloy) require somewhat different treatment. They can be softened by *solution treatment* and hardened by *natural ageing* or they can be hardened artificially by *precipitation treatment.*

The alloy 'duralumin' contains traces of copper, magnesium, manganese and zinc aluminium makes up the remainder of the alloy.

3.16.1 Solution treatment

To soften duralumin-type aluminium alloys, they are raised to a temperature of about 500°C (depending upon the alloy). At this temperature the alloying elements can form a solid solution in the aluminium. The alloy is quenched from this temperature to preserve the solution at room temperature. Gradually, the solid solution will break down with age and the alloy will become harder and more brittle. Therefore, solution treatment must be carried out immediately before the alloy is to be processed. The breakdown of the solution can be delayed by refrigeration at between –6°C and –10°C. Conversely, it can be speeded up by raising the temperature.

3.16.2 Precipitation treatment

The natural hardening mentioned above is called *age-hardening.* This is the result of hard particles of aluminium-copper compounds precipitating out of the solid solution. This hardens and strengthens the alloy but makes it less ductile and more brittle. Precipitation hardening can be accelerated by heating the alloy to about 150°C to 170°C for several hours. This process is referred to as *artificial ageing* or *precipitation hardening.* The times and temperatures vary for each alloy and the alloy manufacture's heat treatment specifications must be carefully observed, especially for critical components such as those used in the aircraft industry.

3.17 Heat-treatment furnaces

The requirements of heat-treatment furnaces are as follows:

- *Uniform heating of the work.* This is necessary in order to prevent distortion of the work due to unequal expansion, and also to ensure uniform hardness.
- *Accurate temperature control.* The critical nature of heat-treatment temperatures requires the furnace be capable of operating over a wide range of temperatures, but it must be easily adjustable to the required process temperature.
- *Temperature stability.* It is essential that the temperature is not only accurately adjustable but once set the furnace must remain at the required temperature. This is achieved by ensuring that the mass of the heated furnace lining (refractory) is very much greater than the mass of the work (charge). It can also be achieved by automatic temperature control, or by both.
- *Atmosphere control.* Should the work be heated in the presence of air, the oxygen in the air attacks the surface of the metal to form metal oxides (scale). This not only disfigures the surface of the metal, it can also change the composition of the metal at its surface. For example, in the case of steels, the oxygen can also combine with the carbon at the surface of the metal. Reducing the carbon content results in the metal surface becoming less hard and/or tough. To provide atmosphere control, the air in the furnace is replaced with some form of inert gas which will not react with the work-piece material. Alternatively the work may be heat treated by totally immersing it in hot, molten salts.
- *Economical use of fuels.* It is essential – if heat-treatment costs are to be kept to a minimum – for the furnaces to be run continuously and economically on a shift work basis since the fuel required to keep firing up furnaces from cold is much greater than that required for continuous running. Thus it is more economical for small workshops to contract their heat treatment out to specialist firms who have sufficient volume of work to keep their furnaces in continuous use.
- *Low maintenance costs.* Furnaces are lined with a heat-resistant material such as firebrick. Since a furnace must be taken out of commission each time this lining is renewed, it should be designed to last as long as possible.

3.17.1 Semi-muffle furnace

Figure 3.16 shows a semi-muffle furnace. The flame from the burner does not play directly onto the charge, but passes under the hearth to provide 'bottom heat'. This results in fairly uniform heating. The advantages and limitations of this type of furnace are as follows:

Advantages

- Relatively low initial cost.
- Simplicity in use and maintenance.
- Fuel economy.
- Fairly rapid heating.
- Heating is fairly uniform.
- Limited atmosphere control can be achieved by varying the gas-air mixture through a system of dampers. The flue outlets are situated just inside the furnace

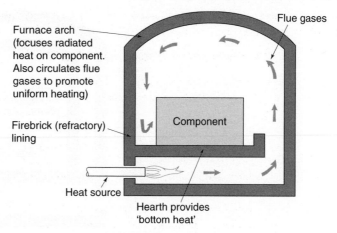

Furnace arch
(focuses radiated
heat on component.
Also circulates flue
gases to promote
uniform heating)

Flue gases

Firebrick (refractory)
lining

Component

Heat source

Hearth provides
'bottom heat'

Figure 3.16 Gas-heated semi-muffle furnace

door so that any atmospheric oxygen that may leak past the door is swept up the flue before it can add to the scaling of the work.

- Reasonable temperature control.
- Reasonable temperature stability due to the mass of the furnace lining.

Limitations

- Heating is still relatively uneven compared with more sophisticated furnace types.
- Atmosphere control is somewhat limited. Although oxidation can be reduced by careful control of the gas-air mixture, some scaling will still take place and there will be still be flue gas contamination of the work.

3.17.2 Muffle furnace (gas heated)

Figure 3.17 shows a full muffle furnace. You can see from the figure that the work is heated in a separate compartment called a *muffle*. The work is completely isolated from the heat source (flame) and the products of combustion. The advantages and limitations for this type of furnace are as follows:

Advantages

- Uniform heating.
- Reasonable temperature control.
- Good temperature stability due to the mass of refractory material forming the muffle and the furnace lining compared with the mass of the work.
- Full atmosphere control is possible. Any sort of atmosphere can be maintained within the muffle since no combustion air is required in the muffle chamber.

Limitations

- Higher initial cost.
- Maintenance more complex and costly.

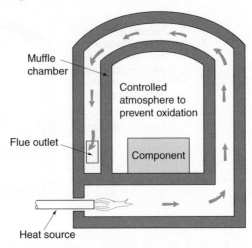

Figure 3.17 Gas-heated muffle furnace

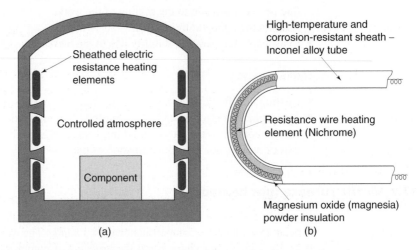

Figure 3.18 Electrically-heated muffle furnace: (a) the electric resistance furnace; (b) heating element

- Greater heat losses and slow initial heating results in lower fuel economy unless the furnace can be operated continuously.

3.17.3 Muffle furnace (electric resistance)

Figure 3.18 shows a typical electric resistance muffle furnace. The electric heating elements are similar to those found in domestic electric ovens. They are independent of the atmosphere in which they operate. Therefore they can be placed within the muffle chamber itself, resulting in a higher operating efficiency compared with the gas heated muffle furnace and more than offsets the higher energy cost for electricity compared with gas. The advantages and limitations of this type of furnace are as follows:

Advantages

- Uniform heating of the work.
- Accurate temperature control.
- Ease of fitting automatic control instrumentation.
- High temperature stability.
- Full atmosphere control.
- Comparatively easy maintenance.

Limitations

- Higher energy source costs.
- Lower maximum operating temperatures, as above 950°C to 1000°C the life of the resistance elements is low.

3.18 Temperature measurement

The importance of temperature measurement and control during heat-treatment processes has already been discussed in this chapter. For the high temperatures met with in heat-treatment furnaces one or other of the high-temperature measuring devices known as *pyrometers* are required.

3.18.1 Thermocouple pyrometer

This is the most widely used temperature measuring device for heat-treatment purposes. Figure 3.19(a) shows the principle of the thermocouple pyrometer. If the junction of two wires made from dissimilar metals (such as a copper wire and an iron wire) form part of a closed electric circuit and the junction is heated, a small electric current will flow. The presence of this current can be indicated by a sensitive galvanometer. Increasing the temperature difference between the hot- and cold-junctions increases the current in the circuit. If the galvanometer is calibrated in degrees of temperature, we have a temperature measuring device called a pyrometer.

Figure 3.19(b) shows how these principles can be applied to a practical thermocouple pyrometer. The component parts of this instrument are:

- The thermocouple probe (hot junction).
- The indicating instrument (milli-ammeter).
- The 'ballast' or 'swamp' resistor.
- The compensating leads.

The thermocouple probe consists of a junction of two wires of dissimilar metals contained within a tube of refractory metal or of porcelain. Porcelain beads are used to insulate the two wires and locate them in the sheath as shown in Fig. 3.19(c).

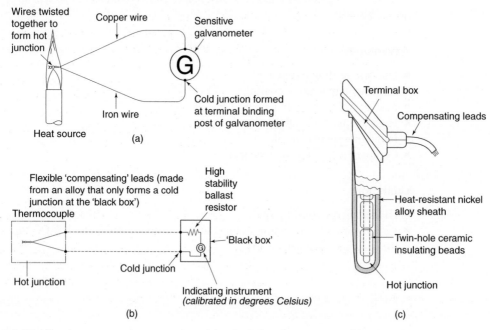

Figure 3.19 The thermocouple pyrometer: (a) principle of operation; (b) pyrometer circuit; (c) thermocouple probe

Table 3.13 lists the more usual hot junction material combinations, together with their temperature ranges and sensitivities.

The indicating instrument

This is a sensitive milli-ammeter calibrated in degrees Celsius (°C) so that a direct reading of temperature can be made. A common error is to set this instrument to read *zero* when the system is cold. In fact, should be set to read the *atmospheric temperature* at the point of installation. The terminals of this instrument form the cold junction and should be placed in a cool position where they are screened from the heat of the furnace.

Table 3.13 Thermocouple combinations

Thermocouple	Sensitivity (millivolts/°C)	Temperature range (°C)
Copper-constantan	0.054	−220 to +300
Iron-constantan	0.054	−220 to +750
Chromel-alumel	0.041	−200 to +1200
Platinum-platinum/rhodium	0.0095	0 to +1450

Notes:
Constantan = 60% copper, 40% nickel.
Chromel = 90% nickel, 10% chromium.
Alumel = 95% nickel, 2% aluminium, 3% manganese.
Platinum/rhodium = 90% platinum, 10% rhodium.

The 'ballast' or 'swamp' resistor

This is contained within the case of the indicating instrument. Its purpose is to improve the stability to the system. The resistance of electrical conductors increases as their temperature increases and the conductors that make up a pyrometer circuit are no exception. The variation in resistance with temperature would seriously affect the calibration of the instrument if the ballast resistor were not present. This resistor is made from manganin wire. Manganin is an alloy whose resistance is virtually unaffected by changes in temperature. By making the resistance of this ballast resistor very large compared with the resistance of the rest of the circuit, it *swamps* the effects of any changes in resistance that may occur in the rest of the circuit and renders them unimportant.

Compensating leads

These are used to connect the thermocouple probe to the indicating instrument. They are made of a special alloy so that they form a cold junction with the terminals of the indicating instrument but not with the terminals of the probe. To avoid changes in calibration, the compensating leads must not be changed in length, nor must alternative conductors be used. The thermocouple, compensating leads and the indicating instrument must always be kept together as a set.

3.18.2 The radiation pyrometer

This device is used to measure the temperature:

- Of large hot components that have been removed from the furnace.
- Where the furnace temperature so high it would damage the thermocouple probe.
- Where the hot component is inaccessible.
- Where the temperature of the component in the furnace needs to be measured rather that the temperature of the furnace atmosphere itself.

The principle of this type of pyrometer is shown in Fig. 3.20. Instead of the thermocouple probe being inserted into the furnace atmosphere, the radiant heat from the

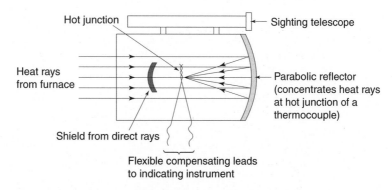

Figure 3.20 The radiation pyrometer

furnace or the component being heated in the furnace is focused onto the thermo-couple by a parabolic mirror. Remember, that as the temperature of a work reaches the furnace temperature, the rate at which the temperature of the work increases slows down. It is difficult to assess just when, if ever, the component reaches furnace temperature. Certainly, the soaking time involved would give rise to excessive grain growth. Furnaces are frequently operated above the required process temperature, and the work is withdrawn from the furnace when it has reached its correct temperature as measured by a radiation pyrometer.

3.19 Atmosphere control

When natural gas, liquid petroleum gas or oil, is burnt in a furnace, excess air is usually present to ensure complete and efficient combustion. The resulting *products of combustion* (flue gases) contain oxygen, carbon dioxide, sulphur, nitrogen and water vapour. These all react to a greater or lesser degree with the surface of the work-piece whilst it is in the furnace. They will produce heavy scaling and, in the case of steel, surface decarburization and softening. The situation is not quite as serious in the case of a muffle furnace since the fuel is burnt in a separate chamber and cannot come into contact with the work. However, the oxygen and water vapour in the air are still present in the muffle chamber and will cause some scaling and decarburization of the work. Little can be done to offset this effect in simple furnaces. However, in muffle furnaces air in the muffle chamber can be replaced by alternative atmospheres, depending upon the process being performed and the metal being treated. This is known as *atmosphere control*. These controlled atmospheres can be based upon natural gas (methane) and LPG gases such as propane and butane. For special applications, ammonia gas and 'cracked' ammonia gas are used.

Exercises

3.1 Material properties

a) Name the properties required by the materials used in the following applications:

    i) A metal-cutting tool

    ii) A forged crane hook

    iii) A motor car body panel

    iv) The conductors in an electric cable

    v) A crane sling

    vi) The sheathing of an electric cable

    vii) A kitchen sink

    viii) A garden hose pipe

    ix) Concrete for a machine foundation

    x) The body of a hand press.

b) Giving reasons for your choice, name a suitable plain carbon steel and state its heat treatment condition for each of the following applications.

    i) Cold chisel

    ii) Engineer's file

    iii) Wire for a coil spring

    iv) Sheet steel for pressing out car body panels

    v) Rod for making snap head rivets.

3.2 Material applications and classification

a) Copy out and complete Table 3.14.

Table 3.14 Exercise 3.2(a)

Material	Typical application
Cast iron	
High-speed steel	
Duralumin	
Stainless steel (austenitic)	
Gunmetal	
Phosphor bronze	
70/30 Brass	
60/40 Brass	
Free-cutting brass	
Tufnol	
Nylon	
PTFE	
Perspex	
Polystyrene	
PVC	
Glass fibre-reinforced polyester	
Epoxy resin	
Urea-formaldehyde	

b) Copy out and complete Table 3.15 by explaining briefly the meaning of the following terms:

Table 3.15 Exercise 3.2(b)

Term	Meaning	Example
Ferrous metal		
Non-ferrous metal		
Thermoplastic		
Thermosetting plastic		
Synthetic material		
Natural material		
Metallic		
Non-metallic		
Alloy		

3.3 Forms of supply, identification, and specification

a) Table 3.16 lists a number of material applications. Copy out and complete the table by naming the 'form of supply' in which you would expect to receive the material for each application.

b) i) State the meaning of the following abbreviations as applied to plain carbon steels: MS, BDMS, HRPO, CRCA.

    ii) What do the terms 'quarter-hard', 'half-hard', etc., refer to when ordering non-ferrous sheet metal and rolled strip?

c) Describe the methods of material identification used in the raw material stores at your place of work or your training workshop.

3.4 Heat-treatment safety

a) Briefly describe the type of clothing and protective equipment you should wear when carrying out heat treatment processes.

Table 3.16 Exercise 3.3(a)

Applications	Form of supply
Car body panels	
Tin plate funnel	
Steel rivets	
Copper boiler	
Structural steel work	
The two main raw materials for GRP boat hull mouldings	
Plastic window frames	
Fabricated machine frames	

b) Sketch THREE warning signs you would expect to find in a heat treatment shop.

3.5 Reasons for heat treatment

a) State the main TWO purposes for the heat treatment of metallic materials.

b) Explain why a coppersmith would anneal a blank cut from a sheet of copper before beating it to shape, and why would he/she need to re-anneal the metal from time to time as forming proceeds.

3.6 Hardening plain carbon steels

a) What two factors does the hardness of a plain carbon steel depend upon when through-hardening?

b) Explain why steels have to be tempered after hardening and how the degree of temper is controlled when this is done over a brazing hearth in the workshop.

c) When through-hardening, what is the effect of:
   i) overheating the steel
   ii) underheating the steel.

d) When through hardening, explain how the hot metal should be quenched and what precautions must be taken to avoid cracking and distortion.

3.7 Annealing and normalizing

a) Describe the essential differences between annealing and normalizing plain carbon steels.

b) Describe the essential differences between full-annealing and sub-critical annealing as applied to plain carbon steels.

c) Describe the essential differences between the annealing of plain carbon steels and the annealing of non-ferrous metals (other than the heat treatable aluminium alloys).

d) Described how 'duralumin' is softened. What is the name of the process used, and what is the name of natural process by which this aluminium alloy gradually becomes hard again?

3.8 Heat treatment equipment

a) List the main requirements of a heat treatment furnace.

b) With the aid of sketches describe any heat treatment furnace with which you are familiar. Draw particular attention to its main features. List the main advantages and limitations for the furnace type chosen.

c) Describe the precautions that must be taken when starting up and shutting down furnaces.

d) Describe the need for, and a method of, atmosphere control in heat treatment furnaces.

e) Describe a method of temperature measurement suitable for a furnace used for the occasional hardening of high carbon steel components.

4

Using and communicating technical information

When you have read this chapter you should understand how to:

- Select information sources to undertake work tasks
- Extract, interpret and evaluate engineering information
- Record and process engineering information
- Interpret (read) drawings in first and third angle projection
- Interpret welding symbols and their uses in fabrication and structural components
- Develop patterns for fabricated components using geometrical constructions
- Develop patterns for sheet-metal intersections using geometrical constructions

4.1 Selection of information sources

It is necessary, for clear communications that cannot be misinterpreted, to select means of communication which ensure that the correct information is provided and used. Therefore, wherever possible, engineering drawings are used to transmit and receive information concerning components to be manufactured and assembled. However, some information has to be given in writing. For example:

- Manufacturing instructions such as the name of the parts to be made, the number required, any special finishes required and the date by which they are required.
- Technical data such as screw thread sizes, and manufacturers' recommended cutting speeds and feeds.
- Stock lists such as material sizes, standard 'bought-in parts', and standard cutting tools.
- Training logbooks.

Verbal instructions and telephone messages should be confirmed in writing or by fax. The latter is particularly useful if illustrations are involved. In industry and commerce all information must be produced in a way that is:

- Easy to understand with no risk of errors.
- Complete, with no essential details missing.
- Quick and easy to complete.

These goals are best achieved by the use of standardized forms. By providing much of the information in the form of boxes which require a tick, the interpretation of hand-writing which may be difficult to read is removed.

Manufacturing organizations are concerned with making the goods available to their customers at a price their customers are prepared to pay, and in delivering those goods in the correct quantities at the correct time. This involves teamwork within the organizations and close liaison with their customers and suppliers, and can only be achieved by the selection of efficient communication and the efficient handling of engineering information.

4.2 Interpretation of information (graphical)

There are many ways in which information can be presented and it is essential to select the most appropriate method. This will depend upon such factors as:

- The information itself.
- The accuracy of interpretation required.
- The expertise of the audience to whom the information is to be presented.

Much of the information required for the manufacture of engineering products is numerical. This can be presented in the form of tables where precise information concerning an individual item can be outlined. Sometimes, all that is needed is a general overview of a situation that can be seen at a glance. In this case the numerical data is most clearly presented by means of graphs and diagrams. There are many different types of graphs available depending upon the relationship between the

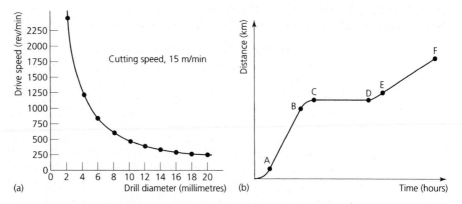

Figure 4.1 Line graphs: (a) points connected by a smooth curve (points related mathematically); (b) points connected by straight lines

quantities involved and the numerical skills of the user for the graph. Let's look at some graphs in common use.

4.2.1 Line graphs

Figure 4.1(a) shows a graph giving the relationship of drill speed and drill diameter for a cutting speed of 15 m/min. In this instance, a continuous curve flowing through the points is plotted. This is because the points plotted on the graph are related by a mathematical expression and any value of drill speed or drill diameter calculated from that expression will lie on the curve.

This is not true in every instance, as shown by Fig. 4.1(b). This graph connects time and distance that a vehicle might travel.

- From A to B the distance that the vehicle travels is proportional to the time taken. That is, the straight line indicates that the vehicle is travelling at a constant speed.
- The curved bit at the beginning of the line shows that the vehicle accelerated from a standing start. The curved bit at the end shows that the vehicle slowed down smoothly to a stop.
- From C to D there is no increase in distance with time. The vehicle is stationary.
- From E to F the vehicle recommences its journey at a reduced speed since the line slopes less steeply.

In this graph it is correct for the points to be connected by separate lines since each stage of the journey is unrelated to the previous stage or to the next stage. It would have been totally incorrect to draw a flowing curve through the points in this instance.

4.2.2 Histograms

Figure 4.2 shows the number of notifiable accidents which occur each year in a factory over a number of years. The points cannot be connected by a smooth, continuous curve as this would imply that the statistics follow some mathematical equation.

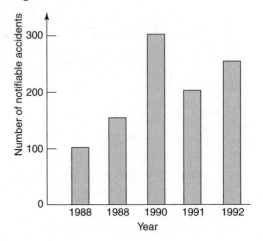

Figure 4.2 Histograms

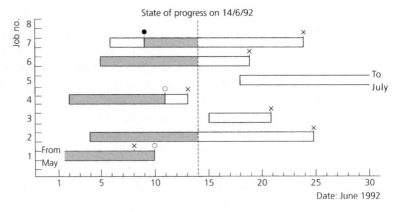

Figure 4.3 Bar chart: x = scheduled completion date; ○ = completion date;
● = start delayed; shaded area = work completed to date

Neither can they be connected by a series of straight lines; this would imply that, although the graph does not represent a mathematical equation, nevertheless the number of accidents increased or decreased continuously and at a steady rate from one year to the next. In reality, the number of accidents are scattered throughout the year in a random manner and the total for one year is independent of the total for the previous year or the next year. The correct way to present this information is by a histogram as shown.

4.2.3 Bar charts

These are frequently used for indicating the work in progress and they are used in production planning. An example is shown in Fig. 4.3.

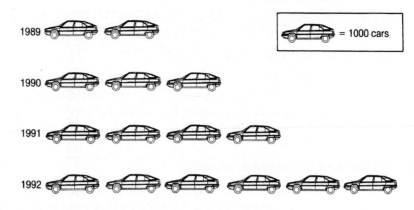

Number of cars using a car park each month

Figure 4.4 Ideograph (pictogram): number of cars using a car park each month

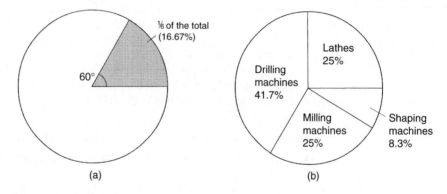

Figure 4.5 Pie chart

4.2.4 Ideographs (pictograms)

These are frequently used for presenting statistical information to the general public. In Fig. 4.4 each symbol represents 1000 cars. Therefore, in 1990 the number of cars using the visitors' car park at a company was 3000 (1000 cars for each of 3 symbols). Similarly, in 1991 the number of cars using the car park was 4000 and in 1992 the number had risen to 6000.

4.2.5 Pie charts

These are used to show how a total quantity is divided into its individual parts. Since a complete circle is 360°, and this represents the total, then a 60° sector would represent $60/360 = 1/6$ of the total. This is shown in Fig. 4.5(a). The total number of fabricated products produced by various methods of assembly in a company can be represented by a pie chart, as shown in Fig. 4.5(b).

4.3 Interpretation of information (tables, charts and schedules)

4.3.1 Manufacturers' catalogues

Manufacturers' catalogues and technical manuals are essential for keeping up to date with suppliers' product lines. Also such catalogues and technical manuals usually include performance data and instructions for the correct and most efficient use for the products shown.

4.3.2 British and European Standards

At the start of the industrial revolution there was no standardization of components. Every nut and bolt was made as a fitted pair and they were not interchangeable with any other nut and bolt. It's not surprising, therefore, that screwed fasteners were the first manufactured goods to be standardized although, initially, only on a national basis.

> Imagine finding a box full of nuts and bolts of seemingly the same size and having to try every nut on every bolt until you found which nuts fitted which bolts

Modern industry dictates that a vast range of standardized materials and components must provide the interchangeability required for international trading and uniformity of quality. Initially this work was carried out by such organ-izations as the British Standards Institute (BSI) in the UK, by DIN in Germany, and by ANSI in America. Since 1947, the International Standards Organization (ISO) has been steadily harmonizing *national standards* and changing them into *international standards* in order to promote international trading in manufactured goods. The aims of standardization as defined by the BSI are:

- The provision of efficient communication amongst all interested parties. The promotion of economy in human effort, materials and energy in the production and exchange of goods through the mass production of standardized components and assemblies.
- The protection of consumer interests through adequate and consistent high quality of goods and consumer services.
- The promotion of international trade by the removal of barriers caused by differences in national practices.

4.3.3 Production schedules

These are usually in the form of bar charts or computer listings. The former will show the planned start and finish dates for various jobs and the machines onto which they are to be loaded. The actual progress of the jobs is superimposed on the ideal schedule so that any 'slippage' in production and the reason can be seen at a glance so that remedial action can be taken and, if necessary, the customer advised of possible delay. An example was shown in Fig. 4.3. Computer listings of production schedules and stock balances are updated regularly (on a daily basis) so that the sales staff of as company know what components and assemblies are in stock, and how soon new stock should be available if a particular item has sold out.

4.3.4 Product specifications

In addition to scheduling the work that is to be done and when it is to be done, it is also necessary to issue full instructions to the works concerning the product to be made. That is, a *production specification* must be issued. The production specification, for example, might be a fabricated road bridge needing a complex document covering such things as the materials to be used, the method of production of the various components and sub-assemblies, the order of manufacture, the scheduled dates and times of delivery on site to maintain smooth progress and the methods of assembly and testing.

On a simpler basis is the *works order* issued in a batch production or in a jobbing-workshop. The works order provides information needed to manufacture a batch of simple fabrications. An example of such a works order form is shown in Fig. 4.6.

The example shown provides the following information as it:

- Identifies the component to be made.
- Identifies the drawings to be used.
- States the quantity of the product to be made.
- Specifies the material that is to be used.
- Specifies any special jigs, fixtures, tools and cutters that will be needed and their location in the stores.

ABC Engineering Co. Ltd		Job No.	
Date issued		Date required	
Component			
Drawing numbers			
Quantity			
Material size		Type	Quantity
Tooling			
Finish/Colour			
Date commenced		Date finished	Operator
Inspection report			Inspector
Special requirements			
Destination			Authorised by

Figure 4.6 Typical works order form

- Specifies any heat-treatment and finishing process that may be required.
- Specifies issue date for the order and the date by which it is required.
- Specifies the destination of the job (stores, inspection department, etc.).
- Includes any special variations required by a particular customer.
- Identifies the personnel employed in the manufacture and the inspection of the job.
- Carries the signature that gives the managerial authority for the work to be done.
- Provides room for the actual dates to be inserted when the job was commenced, and when it finished.

You will notice that all of this information is entered on a standard form. This saves time in issuing the information. It is much easier to fill in the blanks than have to write out all the information from scratch. It is also easy to see if a 'box' is blank. This would indicate that a vital piece of information is missing. It is also easier for the person doing the job to see exactly what is required since the same sort of information always appears in the same place on the form every time.

4.3.5 Reference tables and charts

There are a number of 'pocket books' published for the different branches of engineering. A typical 'pocket book' for use in manufacturing workshops would contain tables of information such as:

- Conversion tables for fractional to decimal dimensions in inch units, and conversion tables for inch to metric dimensions.
- Conversion tables for fractional (inch), letter, number and metric twist drill sizes.
- Standard screw-thread and threaded fastener data tables.
- Tables for spacing holes around pitch circles as an aid to marking out.
- Speeds and feeds for typical cutting-tool and work-piece material combinations for different processes.

This list is by no means exhaustive but just a brief indication of the sort of useful data provided. In addition, many manufacturers produce wall charts of similar data as it affects their particular products. These are not only more convenient for the user than having to open and thumb through a book with oily hands, but they are also good publicity for the manufacturers who issue them.

4.3.6 Drawings and diagrams

Engineers use drawings and diagrams to communicate with each other and with the public at large. The type of drawing or diagram will depend upon the audience it is aimed at and their ability to correctly interpret such information. The creation and interpretation of engineering drawings is considered later in this chapter.

4.4 Evaluating engineering information

Keep alert for errors in the information given. Suppose you have made several batches of a component from stainless steel and suddenly the works order form specifies silver steel. Is this a genuine change or a clerical error? Although the manager will have

It is surprising how long an out of date copy of regulations can keep circulating before someone spots it and destroys it

signed it, he or she is a very busy person and may have missed the error. Therefore, check with your supervisor before starting the job. Better to be sure than sorry.

If standards are referred to, check that the issue on the shop floor is up to date. Standard specifications and EU regulations change rapidly these days. Out of date editions should be withdrawn immediately and the latest edition issued.

4.5 Recording and processing engineering information

The need for, and importance of, accurate record keeping is increasing all the time in nearly all the areas of company activity. Let's now look at some of the more important aspects that affect all employees.

4.5.1 Quality control

Quality control now affects nearly all manufacturing companies, both large and small. This is because a firm that wants to sell its goods to a BS EN 9000 approved firm, must itself be approved and, in turn, obtain its supplies from approved sources (in the UK the British Standard for Quality Assurance is BS EN 9000). The definition of quality upon which this standard is based is in the sense of '*fitness for purpose*' and '*safe in use*'; also that the product or service has been designed to meet the needs of the customer.

A detailed study of quality control and total quality management is beyond the scope of this book. However if you are employed in any branch of the engineering industry it is almost inevitable that you are employed in a company that is BS EN 9000 approved and that this will influence your working practices. A key factor in this respect is '*traceability*'. Therefore, BS EN 9000 is largely concerned with documentation procedures. All the products needed to fulfil a customer's requirements must be clearly identifiable throughout the organization. This is necessary in order that any part delivered to a customer can have its history traced from the source from which the raw material was purchased, through all the stages of manufacture and testing, until it is finally delivered to the customer. This is shown diagrammatically in Fig. 4.7.

The need for this *traceability* might arise in the case of a dispute with a customer due to nonconformity with the product specification, for safety reasons if an accident occurs due to failure of the product, and for statutory and legal reasons. For these reasons, like everyone else involved, your contribution to the above chain of events has to be accurately recorded and the records kept indefinitely. Otherwise the company's goods may not be certified and acceptable to the customer.

4.5.2 Health and safety

This is discussed in detail in Chapter 1. Here, also, record keeping is essential. Some of the documents and data that need to be maintained are:

- The accident register.
- The regular inspection and certification of lifting tackle and pressure vessels (boilers and compressed air receivers).

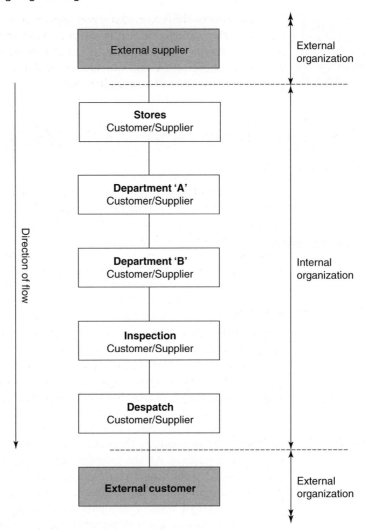

Figure 4.7 Quality control chain (each stage is a customer of the previous stage and a supplier to the next stage, i.e. department 'A' is a customer of the stores and a supplier to department 'B')

- The dates of fire drills and the time taken to evacuate the premises: records of visits to the premises by Local Authority fire and safety officers and their reports.

4.5.3 Legal and financial reasons

Registered companies need to keep legal records to comply with the Company's Act. They must publish their Memorandum and Articles of Association and lodge a copy at Company's House when the company is set up. They must keep accurate minutes of all meetings of the Board of Directors and make annual returns including a current list of directors and other information immediately following each Annual General Meeting (AGM) of the company.

Similarly, it is important that a company keeps accurate and complete financial records so that it can keep its costs under control and ensure that a profit is made. It also needs these records to satisfy the accountants when they make their annual visit to audit the accounts and draw up the balance sheet as required by the directors and shareholders as well as the tax authorities.

4.6 Methods of record keeping

4.6.1 Computer files

These days, most records are kept on computer files in the form of magnetic disks, magnetic tapes and optical discs (compact discs). However, these can be easily destroyed by fire, theft and computer viruses (bugs). For this reason such data should be regularly backed up so that, in an emergency, the data can be reinstated with the minimum of down time and loss of business. Such back-up copies should be kept in a burglar-proof and fire-proof safe.

4.6.2 Micro-film and microfiche

Paper records are bulky and easily lost or destroyed. Forms and technical drawings can be easily copied onto microfilm or microfiche systems. These can store large quantities of information photographically in a small space. Such material can be conveniently catalogued and, when required, it can be read through a suitable viewer or enlarged to its original size in the form of a photographic print.

4.6.3 Registers and logbooks

Registers are used for various purposes; for example, for recording the maintenance history of machine tools, the testing and inspection of equipment such as lifting gear, pressure vessels and fire extinguishers, accidents to employees, and fire drills. Lorry drivers and sales representatives keep logbooks to maintain records of their journeys. As a young trainee, one of the most important documents to be kept is your *training logbook*. The format of logbooks varies from one training establishment to another. No matter what format is chosen, your log book should:

- Record the training you have undergone.
- Show details of the exercises you have undertaken and how you carried them out.
- Show how successfully you have completed each exercise.
- Show your instructor's comments on your performance and his or her signature verifying the entry.

4.7 Communications (miscellaneous)

4.7.1 Safety and hazard notices

There is a saying that '*in an emergency people panic in their own language*'. Therefore, all safety notices and operating instructions for potentially hazardous plant and processes should be printed in as many languages as there are employees from different ethnic backgrounds. Wherever possible, internationally recognized hazard signs should be used.

4.7.2 Safety and hazard signs

All signs must comply with the Safety Signs Regulations 1980. These are recognized internationally and combine geometrical shape, colour and a pictorial symbol to put across the message. Some examples can be found in Section 1.7.2.

4.7.3 Colour coding

This is another means of communication that overcomes language barriers. Table 4.1 shows the colour codes for the contents of gas cylinders. A cylinder that is

Table 4.1 Colour codes for cylinder contents

Gas	Ground colour of cylinder	Colour of bands
Acetylene	Maroon	None
Air	Grey	None
Ammonia	Black	Red and yellow
Argon	Blue	None
Carbon monoxide	Red (+ name)	Yellow
Coal gas	Red (+ name)	None
Helium	Medium brown	None
Hydrogen	Red (+ name)	None
Methane	Red (+ name)	None
Nitrogen	Dark grey	Black
Oxygen	Black	None

Table 4.2 Colour codes for electrical cables

Service	Cable		Colour
Single phase Flexible	Live		Brown
	Neutral		Blue
	Earth		Green/yellow
Single phase Non-flexible	Live		Red
	Neutral		Black
	Earth		Green/yellow
Three phase Non-flexible	Line (live)	⎰ Colour ⎱ denotes ⎱ phase	Red White Blue
	Neutral		Black
	Earth		Green/yellow

coloured wholly red or maroon or has a red band round it near the top, contains a flammable gas. In the case of red cylinders the name of the gas should also be stated on the cylinder. Maroon-coloured cylinders only contain acetylene gas for welding. A cylinder having a yellow band round the top contains a poisonous gas.

The identification colours for electric cables are shown in Table 4.2. The earth conductor is nowadays likely to be referred to as a *circuit protective conductor.*

Finally, pipe runs and electrical conduits are colour-coded according to their contents as listed in Table 4.3. The method of application of the colour code is shown in Fig. 4.8.

Table 4.3 Colour codes for pipe contents

Colour	Contents
White	Compressed air
Black	Drainage
Dark grey	Refrigeration and chemicals
Signal red	Fire (hydrant and sprinkler supplies)
Crimson or aluminium	Steam and central heating
French blue	Water
Georgian green	Sea, river and untreated water
Brilliant green	Cold water services from storage tanks
Light orange	Electricity
Eau-de-Nil	Domestic hot water
Light brown	Oil
Canary yellow	Gas

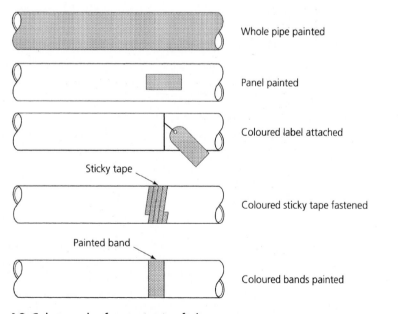

Figure 4.8 Colour codes for contents of pipes

4.7.4 Posters

Posters are also used to put over safety messages. They may be humorous or dramatic. The picture reinforces the caption so that the message is clear even for people who cannot read the words for some reason. Such posters should be displayed at strategic points adjacent to the hazard they represent. They should be changed frequently so as to attract attention.

4.8 Engineering drawing (introduction)

Figure 4.9(a) shows a drawing of a simple clamp. This is a pictorial drawing. It is very easy to see what has been drawn, even to people who have not been taught how to read an engineering drawing. Unfortunately, such drawings have only a limited use in engineering. If you try to put all the information that is required to make the clamp onto this drawing it would become very cluttered and difficult to interpret. Therefore, we use a system called *orthographic drawing* when we make engineering drawings.

An example of an orthographic drawing of our clamp is shown Fig. 4.9(b). We now have a collection of drawings, each one looking at the clamp from a different direction. This enables us to show every feature of the clamp that can be seen and

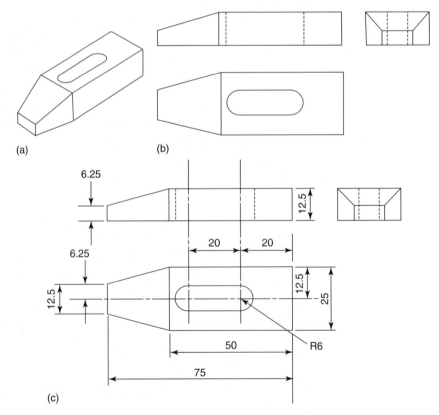

Figure 4.9 Clamp: (a) pictorial drawing; (b) orthographic drawing; (c) fully dimensioned in millimetres

also some things that cannot be seen (hidden details). Features that cannot be seen are indicated by broken lines. Finally, we can add the sizes (dimensions) that we need in order to make the clamp. These are shown in Fig. 4.9(c). A drawing that has all the information required to make a component part, such as Fig. 4.9(c), is called a *detail drawing*. A drawing showing and listing all the components assembled together is called a *general arrangement drawing*.

4.9 First angle orthographic drawing

There are two systems of orthographic drawing used by engineers:

- First angle or English projection.
- Third angle or American projection.

In this section we are going to look at *first angle* projection. We are again going to use the clamp you first met in Fig. 4.10(a). We look at the clamp from various directions.

- Look down on the top of the clamp and draw what you see as shown in Fig. 4.10(a). This is called a *plan view.*
- Look at the end of the clamp and draw what you see as shown in Fig. 4.10(a). This is called an *end view.*
- Look at the side of the clamp and draw what you see is shown in Fig. 4.10(a). Although this is a side view, it is given a special name. It is called an *elevation.*
- You can now assemble these views together in the correct order as shown in Fig. 4.10(b) to produce a *first angle orthographic drawing* of the clamp.

As well as the things that can be seen from the outside of the clamp, we also included some 'hidden detail' in the end view and elevation. Hidden detail indicates a slot through the clamp in this example. The slot is shown by using *broken lines*. We did this because if we had shown the slot as an oval in the plan view it could have meant one of two things:

- A slot passing right through the clamp.
- A slot recessed part way into the clamp.

It *could not* have been an oval shaped lump on top of the clamp as this would have shown up in the end view and in the elevation.

Sometimes only two views are used when the plan and elevation are the same; for example, a cylindrical component such as a shaft. Figure 4.11 shows that an elevation and an end view provide all the information we require.

Finally, let's see how a *first angle orthographic drawing* is constructed.

- First draw the ground lines and a plane at 45° as shown in Fig. 4.12(a).
- Then start to draw in the construction lines faintly using lines that are half the thickness of the final outline. Figure 4.12(b) shows the construction lines in place.
- Then follow each construction line round all the views in order to avoid confusion.
- Finally, we 'line in' the outline so that it stands out boldly as shown in Fig. 4.12(c).

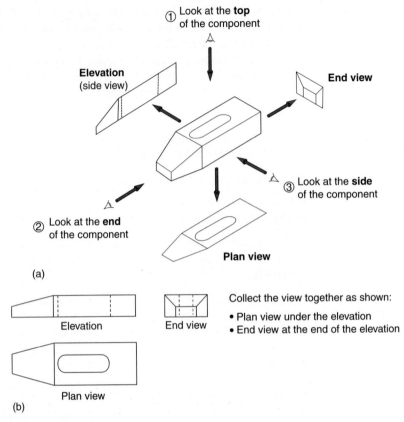

(a)

(b)

Figure 4.10 Principles of drawing in first angle projection: (a) plan view, end view and elevation (side view); (b) collected views together make up an orthographic drawing

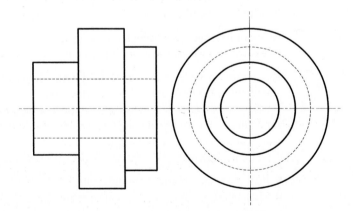

Figure 4.11 First angle drawing of a cylindrical component: the elevation and the plan view are the same and need only be drawn once

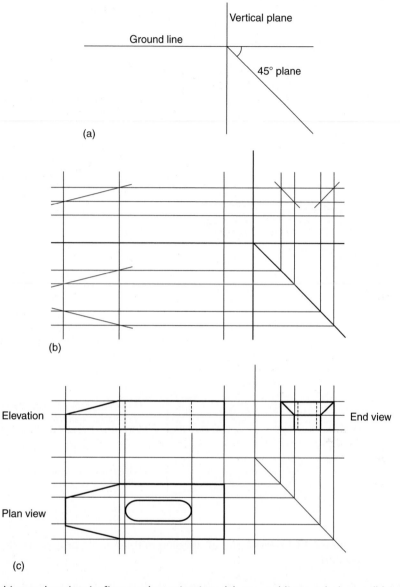

Figure 4.12 Making a drawing in first angle projection: (a) ground line and planes; (b) initial Construction lines; (c) line in the outline (outline is twice the thickness of the construction lines)

4.10 Third angle orthographic drawing

To draw the clamp in third angle (American) orthographic projection, you merely have to rearrange the relative positions of the views. Each view now appears at the same side or end of the component from which you are looking at it. This is shown in Fig. 4.13(a). That is:

Look down on the clamp and draw the plan view above the side view or elevation

Look at the left-hand end of the clamp and draw the end view at the same end.

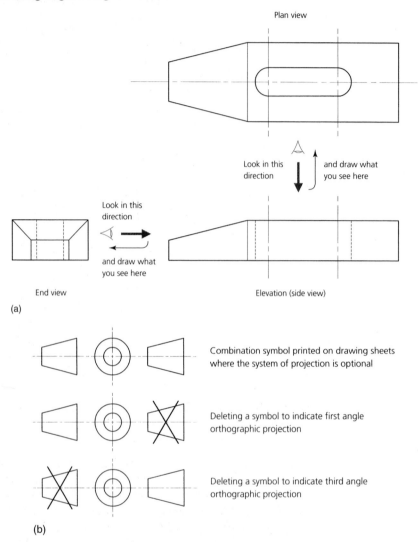

Figure 4.13 (a) Principles of drawing in third angle projection; (b) projection symbol

So what is the advantage of third angle projection? Consider the general arrangement drawing for an airliner drawn to a fairly large scale so that fine detail can be shown. In first angle projection, the end view looking at the nose of the aircraft would be drawn somewhere beyond the tail. An end view looking at the tail of the aircraft would be drawn somewhere beyond the nose. It is much more convenient to draw the end view of the nose of the aircraft at the nose end of the elevation. Also, it is more convenient to draw an end view of the tail of an aircraft next to the tail of the elevation.

Sometimes the projection used is stated in words, more usually it is indicated by the use of a standard symbol. Figure 4.13(b) shows the combined projection symbol and how it is used.

To avoid confusion, always state the projection used on the drawing

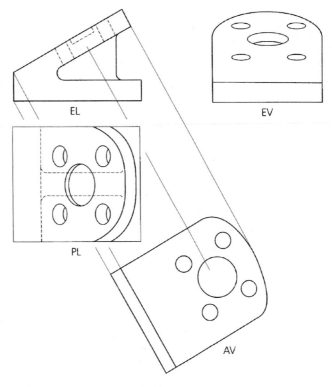

Figure 4.14 Auxiliary view: EL = elevation; EV = end view; PL = plan; AV = auxiliary view

So far we have only considered features that are conveniently arranged at right-angles to each other so that their true shape is shown in the plan, elevation or the end view. This is not always the case and sometimes we have to include an *auxiliary view*. This technique is important in the production of working drawings so that the positions of features on the surface that is inclined not only appear undistorted but can also be dimensioned. Figure 4.14 shows a bracket with an inclined face. When it is drawn in first-angle projection, it can be seen that the end view showing the inclined surface and its features are heavily distorted. However, they appear correct in size and in shape in the auxiliary view (AV) which is projected at right-angles (perpendicular) to the inclined face.

4.11 Conventions

An engineering drawing is only a means of recording the intentions of the designer and communicating those intentions to the manufacturer. It is not a work of art and, apart from the time spent in its preparation, it has no intrinsic value. If a better and cheaper method of communication could be discovered, then the engineering drawing would no longer be used. We are already part way along this road with CAD where the drawings are stored digitally on magnetic or optical disks and can be transmitted between

In the UK we use the British Standard for Engineering Drawing Practice as published by the British Standards Institute (BSI)

companies by the internet. However, hard copy in the form of a printed drawing still has to be produced for the craftsperson or the technician to work to.

As an aid to producing engineering sketches and drawings quickly and cheaply we use *standard conventions*. These are recognized internationally and are used as a form of drawing 'shorthand' for the more frequently used details.

This standard is based upon the recommendations of the International Standards Organization (ISO) and, therefore, its conventions and guidelines, and drawings produced using such conventions and guidelines are accepted internationally.

4.11.1 Types of line

Figure 4.15 shows the types of line recommended by the British Standards Institute, together with some typical applications. The following points should be noted in the use of these lines.

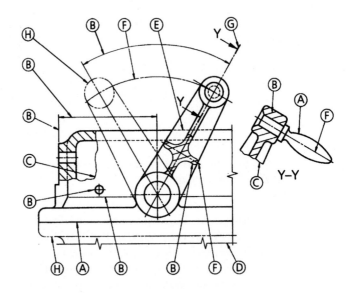

Line		Description	Application
A	——————	Continuous thick	Visible outlines and edges
B	——————	Continuous thin	Dimension, projection and leader lines, hatching, outlines of revolved sections, short centre lines, imaginary intersections
C	⌇⌇⌇	Continuous thin irregular	Limits of partial or interrupted views and sections, if the limit is not an axis
D	─⋎─⋎─	Continuous thin straight with zigzags	
E	- - - - - - -	Dashed thin	Hidden outlines and edges
F	— — — —	Chain thin	Centre lines, lines of symmetry, trajectories and loci, pitch lines and pitch circles
G	┌─ · — / — · ┘	Chain thin, thick at ends and changes of direction	Cutting planes
H	— ·· — ·· —	Chain thin double	Outlines and edges of adjacent parts, outlines and edges of alternative and extreme positions of movable parts, initial outlines prior to forming, bend lines on developed blanks or patterns

Figure 4.15 Types of line and their applications

- *Dashed* lines should consist of dashes of consistent length and spacing, approximately to the proportions shown in the figure.
- *Thin chain lines* should consist of long dashes alternating with short dashes. The proportions should be generally as shown in the figure, but the lengths and spacing may be increased for very long lines.
- *Thick chain lines* should have similar lengths and spacing as for thin chain lines.

Table 4.4 Abbreviations for written statements

Term	Abbreviation	Term	Abbreviation
Across flats	A/F	Hexagon	HEX
British Standard	BS	Hexagon head	HEX HD
Centres	CRS	Material	MATL
Centre line	CL *or* ℄	Number	NO.
Chamfered	CHAM	Pitch circle diameter	PCD
Cheese head	CH HD	Radius (in a note)	RAD
Countersunk	CSK	Radius (preceding a dimension)	R
Countersunk head	CSK HD	Screwed	SCR
Counterbore	C'BORE	Specification	SPEC
Diameter (in a note)	DIA	Spherical diameter or radius	SPHERE Ø *or* R
Diameter (preceding a dimension)	Ø	Spotface	S'FACE
Drawing	DRG	Standard	STD
Figure	FIG	Undercut	U'CUT

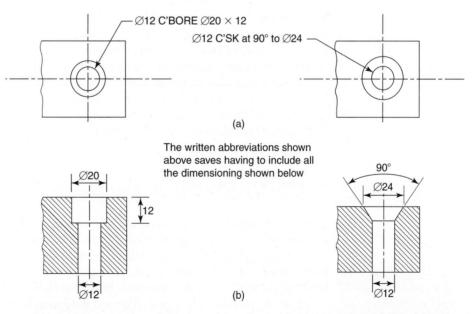

(a)

Ø12 C'BORE Ø20 × 12

Ø12 C'SK at 90° to Ø24

The written abbreviations shown above saves having to include all the dimensioning shown below

Ø20

12

Ø12

90°

Ø24

Ø12

(b)

Figure 4.16 Examples of the use of standard abbreviations (counterbored hole; countersunk hole)

■ *General.* All chain lines should start and finish with a long dash. When thin chain lines are used as centre lines, they should cross one another at solid portions of the line. Centre lines should extend only a short distance beyond the feature unless required for dimensioning or other purposes. They should not extend through the spaces between the views and should not terminate at another line of the drawing. Where angles are formed in chain lines, long dashes should meet at the corners and should be thickened as shown. Arcs should join at tangent points. Dashed lines should also meet at corners and tangent points with dashes.

4.11.2 Abbreviations for written statements

Table 4.4 lists the standard abbreviations for written statements as used on engineering drawings. Some examples of their use are shown in Fig. 4.16. Some further examples will be given when we discuss the dimensioning of drawings.

Convention symbols

Figure 4.17 shows some typical convention symbols used in engineering drawings. It is not possible, in the scope of this book, to provide the full set of conventions or to provide detailed explanations of the use. For this it is necessary to consult texts specializing in engineering drawing together with the appropriate British Standard.

4.12 Redundant views

As shown earlier, where a component is symmetrical you do not always need all of the views to provide the information required for manufacture. A ball looks the same from all directions, and to represent it by three circles arranged as a plan view, an elevation and an end view would just be a waste of time. All that is required is one circle and a note that the component is spherical. The views that can be discarded without loss of information are called *redundant views.* Figure 4.18 shows how drawing time can be saved and the drawing simplified by eliminating the redundant views when drawing symmetrical components.

4.13 Dimensioning

So far, only the shape of the component has been considered. However, in order that components can be manufactured, the drawing must also show the size of the component and the position and size of any features on the component. To avoid confusion and the chance of misinterpretation, the dimensions must be added to the drawing in the manner laid down in the appropriate British Standard.

Make sure you refer to the *current* British Standard as the Standards are constantly being updated

Figure 4.19(a) shows how projection and dimension lines are used to relate the dimension to the drawing, whilst Fig. 4.19(b) shows correct and incorrect methods of dimensioning a drawing.

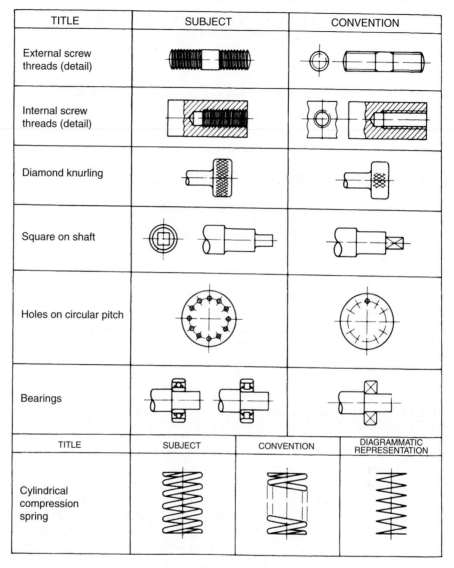

TITLE	SUBJECT	CONVENTION
External screw threads (detail)		
Internal screw threads (detail)		
Diamond knurling		
Square on shaft		
Holes on circular pitch		
Bearings		

TITLE	SUBJECT	CONVENTION	DIAGRAMMATIC REPRESENTATION
Cylindrical compression spring			

Figure 4.17 Typical conventions for some common features

4.13.1 Correct dimensioning

- Dimension lines should be thin full lines not more than half the thickness of the component outline.
- Wherever possible, dimension lines should be placed outside the outline of the drawing.
- The dimension line arrowhead must touch but not cross the projection line.
- Dimension lines should be well spaced so that the numerical value of the dimension can be clearly read and so that they do not obscure the outline of the drawing.

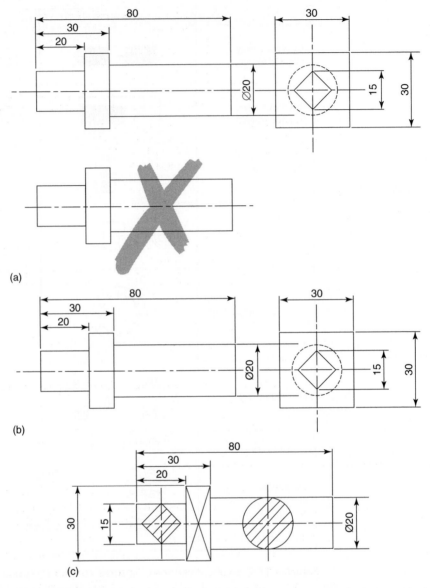

Figure 4.18 Redundant views: (a) first angle working drawing of a symmetrical component (plan view redundant); (b) symmetrical component reduced to two views; (c) working drawing reduced to a single view by using revolved sections and BS convention for the square flange

4.13.2 Incorrect dimensioning

- Centre lines and extension lines must **not** be used as dimension lines.
- Wherever possible dimension line arrowheads must not touch the outline directly but should touch the projection lines that extend from the outline.
- If the use of a dimension line within the outline is unavoidable, then try and use a leader line to take the dimension itself outside the outline.

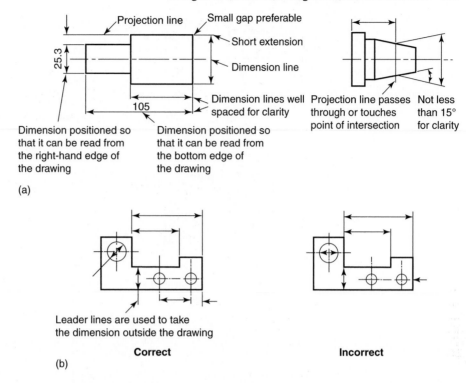

Figure 4.19 Dimensioning: (a) projection and dimension lines; (b) correct and incorrect dimensioning

4.13.3 Dimensioning diameters and radii

Figure 4.20(a) shows how circles and shaft ends (circles) should be dimensioned. It is preferable to use those techniques that take the dimension outside the circle, unless the circle is so large that the dimension will neither be cramped nor will it obscure some vital feature. Note the use of the symbol Ø to denote a diameter.

Figure 4.20(b) shows how radii should be dimensioned. Note that the radii of arcs of circles need not have their centres located if the start and finish points are known. Figure 4.20(c) shows how notes may be used to avoid the need for the full dimensioning of certain features of a drawing.

Leader lines

These indicate where notes or dimensions are intended to apply and end in either arrowheads or dots.

- *Arrowheads* are used where the leader line touches the outline of a component or feature.
- *Dots* are used where the leader line finishes within the outline of the component or feature to which it refers.

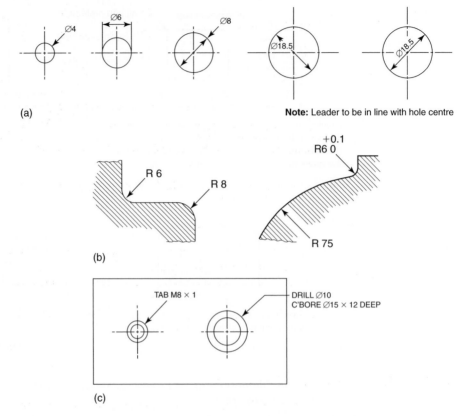

(a)

Note: Leader to be in line with hole centre

(b)

(c)

Figure 4.20 Dimensioning – diameters and radii: (a) dimensioning holes; (b) dimensioning the radii of arcs which need not have their centres located; (c) use of notes to save full dimensioning

4.14 Welding symbols

Figure 4.21 shows two types of welded joints, the butt weld and the fillet weld. Most manually welded joints are a variation on these techniques. Drawings involving welding must give all the details necessary to specify the type of weld, for instance:

- Edge preparation.
- Filler material.
- Type of joint.
- Length of run.
- Size of weld.

This information is communicated by means of British Standard drawing symbols together with any appropriate notes and dimensions. Figure 4.22 shows a selection of standard welding symbols, whilst Figs 4.23 to 4.26 inclusive shows how to apply these welding symbols to a drawing. To do this a system of arrows and reference lines must be used.

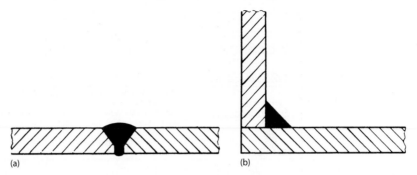

Figure 4.21 Basic welded joints: (a) butt weld; (b) fillet weld

1. When the weld symbol is *above* the reference line the weld is made on the side of the joint *opposite* the arrowhead.
2. When the weld symbol is *below* the reference line the weld is made on the *same side* of the joint as the arrowhead.
3. When the weld symbol is on *both sides* of the reference line, the welding is carried out on *both sides* of the joint.

4.14.1 Additional weld information

In addition to the use of symbols, dimensional information is required to specify the size of the weld and the length of the run. Written notes can also be added to the drawing to specify the filler material and any flux that may be needed. Some examples of this additional information are shown in Fig. 4.27.

4.15 The development of surfaces

Unlike mechanical engineers who are mostly concerned with detail drawings for solid components, sheet metal and fabrication engineers often have to employ various methods of *pattern development* in order to produce flat 'blanks' which can then be folded into the required finished shape. The three basic methods used in pattern development are:

- The parallel line method.
- The radial line method.
- The triangulation method.

The examples to be considered now will be limited to right prisms, right pyramids, right cylinders and right cones and their frustums between parallel planes. The faces of *prisms* are planes with their edges parallel. The unfolding of these faces will produce a development which takes the form of a simple rectangle.

The development of a *cylinder* is achieved by unrolling its surface, thus producing a rectangle having one side equal in length to the *circumference* of the required

SKETCH	DESCRIPTION	SYMBOL
	Square butt weld: This symbol is used to indicate a butt weld when there is no edge preparation and no fillet weld.	||
	Single-V butt weld	V
	Butt weld between flanged plates (the flanges being melted down completely)	
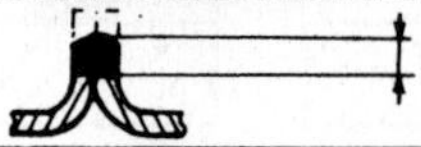	Butt welds between flanged plates not completely penetrated are symbolised as square butt welds with the weld thicknesses shown *s* is the minimum distance from the external surface of the weld to the bottom of the penetration.	*s* ||
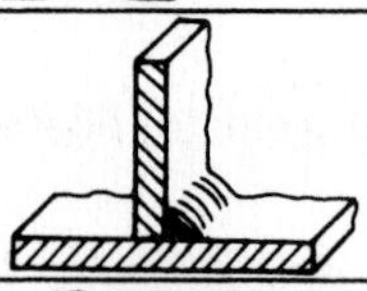	Fillet weld	◺
	Single-bevel butt weld	
	Single-V butt weld with broad root face	Y
	Single-bevel butt weld with broad root face	
	Single-U butt weld	
	Single-J butt weld	
	Backing or sealing run	
	Plug weld (circular or elongated hole, completely filled)	⊓
 (a) Resistance  (b) Arc	Spot weld (resistance or arc welding) or projection weld	○
	Seam weld	

Figure 4.22 Weld symbols

cylinder. The other side of the development is equal to the height or length of the cylinder.

The development of a *pyramid* is achieved when its surface is unfolded. This development consists, basically, of a number of triangles. The *base of each triangle*

SKETCH OF WELD   SYMBOLIC REPRESENTATION

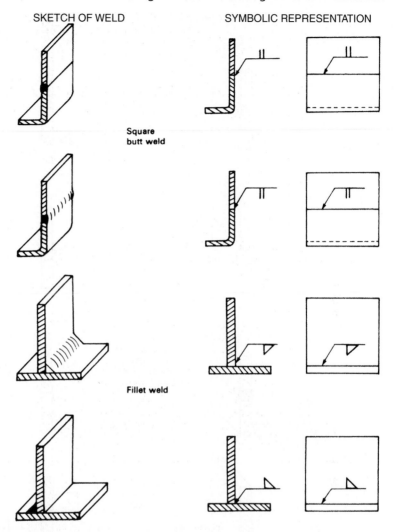

Square
butt weld

Fillet weld

Figure 4.23 Some examples showing significance of arrow and position of weld symbol in relation to the reference line

is equal in length to each side of the *base of the pyramid*. The *sides of each triangle are equal in length to the 'slant' edges of the pyramid*.

The development of a cone is achieved by unrolling its surface. The circular base of the cone unrolls around a point, which is the apex of the cone, for a distance equal to its circumference. *The radius of the arc* producing the base of the development is equal to the *'slant' height of the cone*.

In practice, a complete cone is rarely required except, perhaps for, capping stove pipes. However, in the fabrication industry conical sections are constantly required to be manufactured. These cones are part cones, often referred to as *truncated* cones. When the cone is cut off parallel with its base (when the top is removed) the remaining portion is called the *frustum* of the cone.

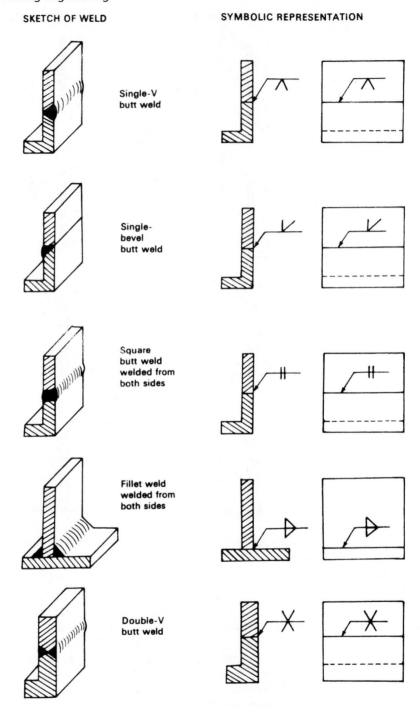

Figure 4.24 Some further examples showing significance of arrow and position of weld symbol in relation to the reference line

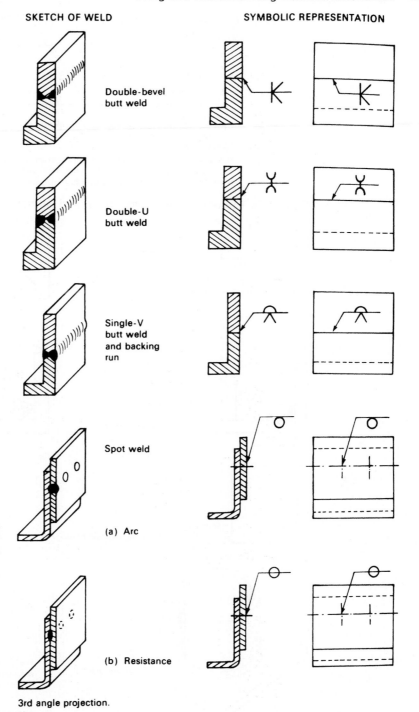

Figure 4.25 Some further examples showing significance of arrow and position of weld symbol in relation to the reference line. (a) Arc; (b) resistance

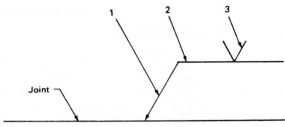

1 is the arrow line
2 is the reference line drawn parallel to the bottom
   edge of the drawing.
3 is the symbol

Note how the arrow heard is used to indicated which plate is prepared and how the disposition of symbol about reference line indicates the side from which the weld is to be made and therefore the wide part of the preparation

Figure 4.26 Use of reference lines, arrows and symbols

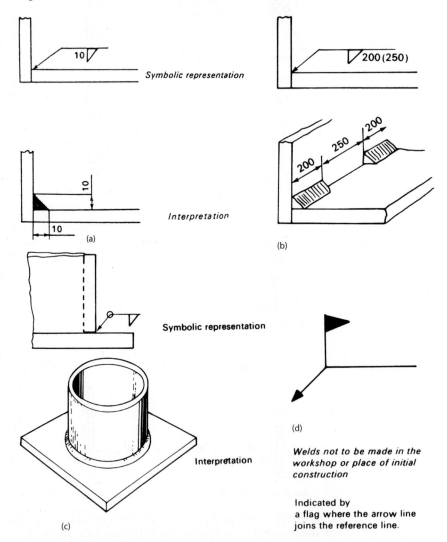

Figure 4.27 Additional weld information. (a) Size of filet; (b) intermittent weld; (c) weld all round; (d) weld on site

4.15.1 Parallel line developments

The parallel line method of development depends upon a principle of locating the shape of the pattern on a series of parallel lines. Prismatic articles or components, which have a constant cross-section throughout their length, may be developed by the parallel line method. Some simple examples of parallel line development are shown in Figs 4.28 to 4.30 inclusive. The difference between a *right-prism* and an *oblique-prism* is as follows: in a right-prism the axis of the prism is at 90° to the end faces; in an oblique prism the axis of the prism is at any angle other than 90° providing the end faces lie between parallel planes.

4.15.2 Radial line development

The radial line method of development is used for developing patterns for any article or component that *tapers to an apex*. It can also be used for articles or components that are frustums that would normally taper to an apex if the sides are produced until they intersect.

The principle of radial line development is based on the location of a series of lines which radiate down from the apex along the surface of the component to a base, or an assumed base, from which a curve may be drawn whose perimeter is equal in length to the perimeter of the base. Some simple examples of radial line development are shown in Figs 4.31 to 4.34 inclusive.

Cones and pyramids are closely related geometrical shapes. In fact, large conical shapes made from heavy metal are often formed on a press brake as if they are many-sided pyramids. Although cones and pyramids have very similar characteristics, care must be taken when developing patterns for pyramids. It is very important to recognize one specific difference between a cone and a pyramid in order to avoid mistakes in development.

A pyramid may be considered as a cone with a limited number of slant sides, whilst a cone may be considered as a pyramid with an infinite number of slant sides

Figure 4.33 shows two views of a right-pyramid which completely describe the object. The elevation shows the true slant height of each triangular face which is shown as a square in the plan view. However, the slant corners of the pyramid in the plan view are not normal (at right-angles) to the elevation. In order to establish the true length of the slant corners for the pattern development, the plan view would have to be rotated until one slant corner becomes square to the elevation. This is not possible on the drawing board, but an arc OP of radius OB may be drawn on the plan view as shown in Fig. 4.33. The distance OP can then be projected back to the elevation to give the true radius (slant corner length) which can then be used for swinging the arc for the basis of the pattern. Notice that the seam, in this example, is along the centre of one face of the pyramid. Therefore, the true length of the joint line is equal to the slant height shown in the elevation. The three full sides are marked off along the base curve in the pattern so that $A_1 B_1$ is equal in length to AB in the elevation. Similarly, $B_1 C_1$ is equal in length to BC in the elevation and $C_1 D_1$ is equal in length to the CD in the elevation. Finally, to complete the pattern the last two triangles are completed so that $A_1 S_1$ is *half the length* of AB in the plan view and $D_1 S_1$ is *half the length* of CD in the plan view. Figure 4.34 shows the pattern development for the frustum of a cone. In this case, the joint (seam) is along the slant corner $A_1 A_2$.

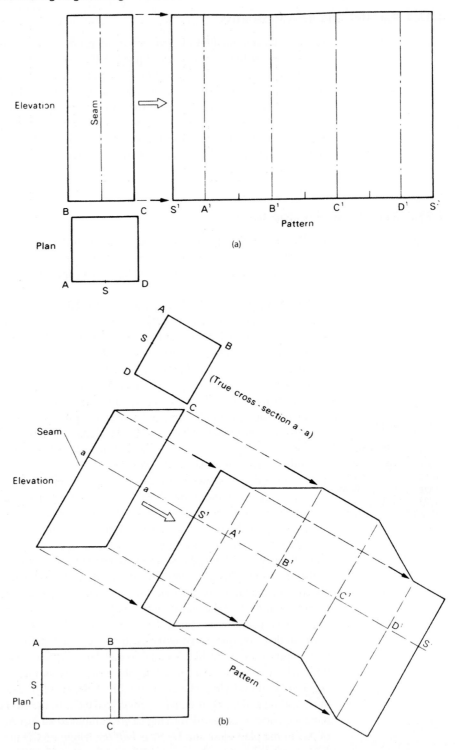

Figure 4.28 Developments of square prisms (parallel line). (a) Square right prism; (b) square oblique prism

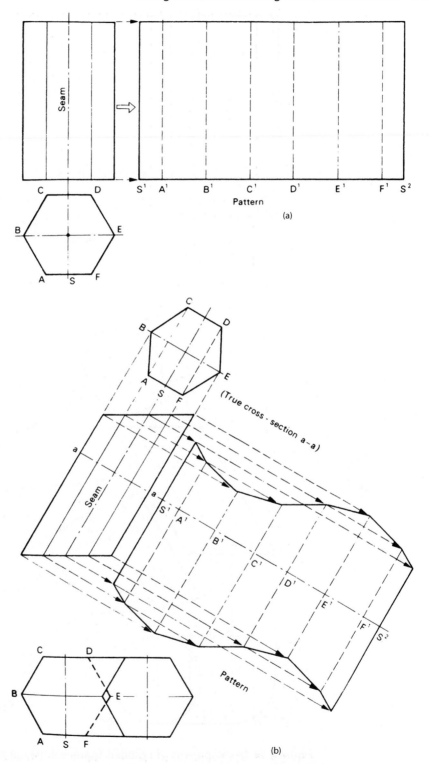

Figure 4.29 Developments of hexagonal prisms (parallel line). (a) Hexagonal right prison; (b) Hexagonal oblique prism

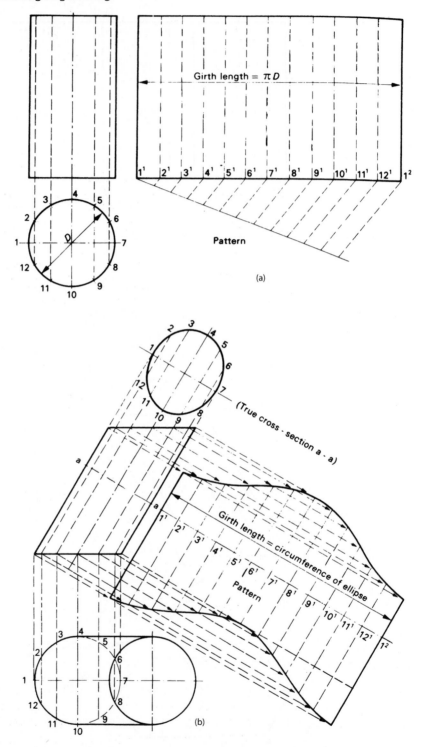

Figure 4.30 Developments of cylinders (parallel line). (a) Right cylinder; (b) oblique cylinder

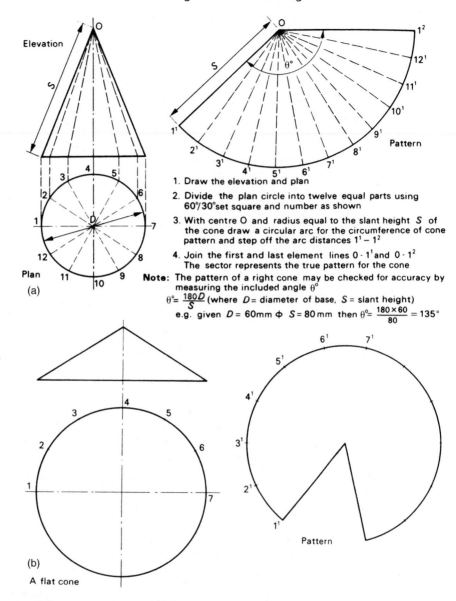

1. Draw the elevation and plan
2. Divide the plan circle into twelve equal parts using 60°/30° set square and number as shown
3. With centre O and radius equal to the slant height S of the cone draw a circular arc for the circumference of cone pattern and step off the arc distances $1^1 - 1^2$
4. Join the first and last element lines $0 \cdot 1^1$ and $0 \cdot 1^2$ The sector represents the true pattern for the cone

Note: The pattern of a right cone may be checked for accuracy by measuring the included angle $\theta°$

$$\theta° = \frac{180D}{S} \text{ (where } D = \text{diameter of base, } S = \text{slant height)}$$

e.g. given $D = 60\text{mm} \, \phi \, S = 80\text{mm}$ then $\theta° = \frac{180 \times 60}{80} = 135°$

Figure 4.31 Developments of a right cone (radial line)

4.15.3 Development by triangulation

Triangulation is by far the most important method of pattern development since a great number of fabricated components transform from one cross-section to another. A typical 'square-to-round' transformer is briefly discussed in Chapter 7. The basic principle of triangulation is to develop a pattern by dividing the surface of the component to be fabricated into a number of triangles, determine the true size and shape of each triangle, and then lay them down side by side in the correct order to produce

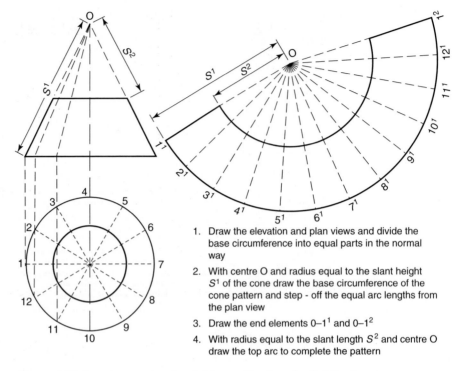

1. Draw the elevation and plan views and divide the base circumference into equal parts in the normal way

2. With centre O and radius equal to the slant height S^1 of the cone draw the base circumference of the cone pattern and step - off the equal arc lengths from the plan view

3. Draw the end elements $0-1^1$ and $0-1^2$

4. With radius equal to the slant length S^2 and centre O draw the top arc to complete the pattern

Figure 4.32 Developments of a right cone frustum (radial line)

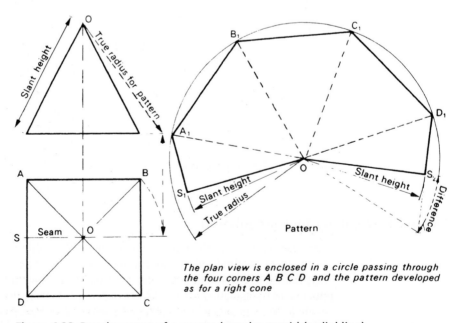

The plan view is enclosed in a circle passing through the four corners A B C D and the pattern developed as for a right cone

Figure 4.33 Development of a square-based pyramid (radial line)

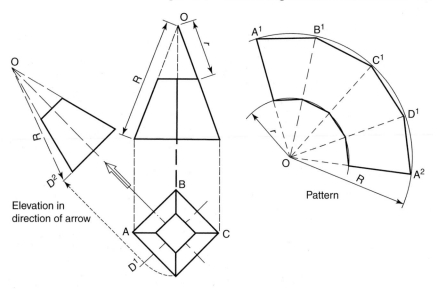

Figure 4.34 Development of a square based pyramid frustum (radial line)

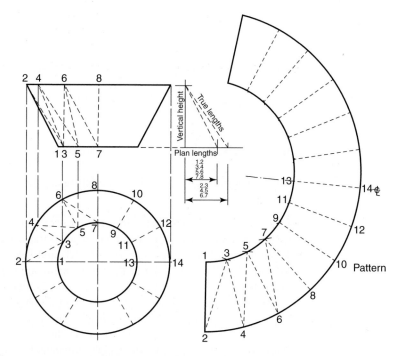

Figure 4.35 Development of a truncated cone (triangulation)

a pattern. To determine the true size of each triangle, the true length of each side must be determined and then placed in to the correct relationship to the other sides. A simple example of the development of a truncated cone (frustum) by triangulation is shown in Fig. 4.35 and a more complex example is shown in Fig. 4.36.

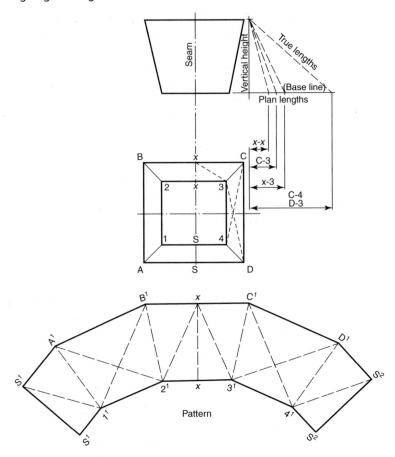

Figure 4.36 Development of a square hopper (triangulation)

Although Fig. 4.35 is largely self-explanatory, we need to consider Fig. 4.36 in greater detail.

- Draw the elevation and plan views. The corner points in the plan are lettered A, B, C, D and numbered 1, 2, 3, 4, and 5 (5 denotes the seam).
- It can be seen that the lengths AB, BC, CD and DA of the large square, and the lengths 1,2, 2,3, 3,4 and 4,1 of the small square are *true lengths* in the plan view since they lie in the same horizontal plane and therefore have no vertical height.
- For the first triangle in the pattern take the true length distance BC (in plan) and mark it on the pattern. Draw a vertical centre line *x,x*. Mark the plan length *x,x* along the base line at 90° to the vertical height and obtain its *true length* and mark it on the pattern. Obtain the true length of diagonal *x*,3 in plan and swing arcs from *x* (on BC) in the pattern. Complete the triangle in the pattern by taking true length *x*,3 (in plan) and swing arcs each side of the centre line to locate points 2 and 3.

Note: By commencing the pattern in the middle of *x – x* (i.e., opposite the seam), the whole pattern can be obtained by repeating the marking out procedure each side of the centre line. Check the pattern for symmetry – if drawn correctly the last two triangles are right-angled triangles.

- Join B,2 and C,3 on the pattern (this represents one side of the hopper); check these two sides by plotting plan length C–3 against the vertical height.
- For the next triangle mark true length arc BA. Obtain the true length of diagonal 2,A in plan by plotting it against the vertical height and swing an arc from 2 in the pattern to locate point A. Join BA in the pattern.
- For the next triangle swing true length arc 2,1 and true length arc B,1 these will intersect to locate point 1 in the pattern. Join A,1 and 2,1 to complete a second side of the hopper.
- Take 3,1 in the plan view and swing an arc from the point 1 on the pattern. Take the true length AS from the plan view and swing an arc from A in the pattern to obtain the true length 1,S by plotting its plan length against the vertical height and swing an arc from 1 on the pattern to locate the points S. The last triangle S,1,S is completed by swinging an arc from S equal to the true length of the front line. Join AS, 1,S and S,S.

4.16 Interpenetration

So far we have only considered the basic principles of pattern development for single components. It is now time to consider what happens when *interpenetration* occurs; for example, when two cylinders meet at right-angles as shown in Fig. 4.37.

4.16.1 Elbow

Figure 4.37 (in first angle projection) shows two cylinders of equal diameter meeting at right-angles to form an 'elbow'. The circle representing the cylinder is divided into twelve equal parts and lines called *generators* are projected from these points into the elevation. Since the two cylinders are of equal diameter, the same projectors will do for both cylinders. The line of intersection, in this case, is a straight line. The development of part A is best projected in line with its cylinder as shown. The divisions of the opened out cylinder may be chordal from the cylinder circle; or may be calculated from the expression $\pi \times D$ where D is the diameter of the cylinder. The labelled points of intersection 0, 1, 2, 3, 4, 5 and 6 are projected onto the cylinder development, to cut the respective generators. A fair curve drawn through the points completes the development. Since the development is symmetrical, only half of it is shown.

4.16.2 Angled co-axial interpenetration

Figure 4.38 shows a rather more complex example involving two cylinders of unequal diameters where the interpenetration occurs at any angle other than a right-angle. Commence by drawing the plan and elevation of the assembled fabrication. Then draw the geometric constructional semicircles and generators for the small cylinders B. Label the points of intersection on the plan view. Project these points onto the elevation to give the lines of intersection of the two cylinders. Project the development of the B cylinders in line with the cylinders as shown. Only half the development of the small cylinders is shown since it is symmetrical. Project the development of the larger cylinder A in line with the elevation. The chordal widths are taken from the plan for use in the development. The two apertures in the wall of the larger cylinder A are identical.

Only the interpenetration of two relatively simple examples involving cylinders are shown here to establish the basic principles but it is also possible to combine cylinders

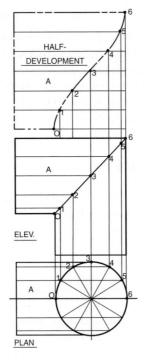

Figure 4.37 Elbow – the interpenetration of two co-axial cylinders at right-angles

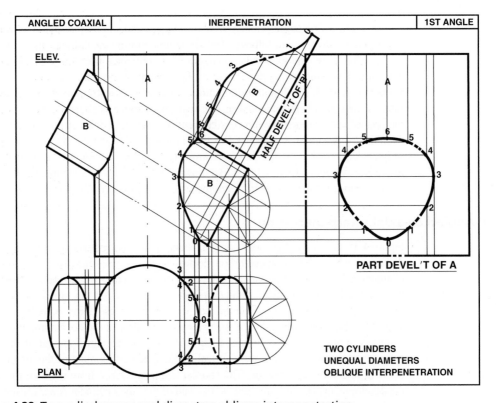

Figure 4.38 Two cylinders unequal diameters oblique interpenetration

and prisms, cylinders and cones, cylinders and pyramids, prisms and pyramids, etc. These may have axes in line (co-axial) or off-set. There axes may be at right-angles to each other or they may be oblique. The reader should consult any standard text on Geometrical Drawing for further information (a good place to start might be with the *Manual of Engineering Drawings* by Colin Simmons and Dennis McGuire, 2nd edn, Elsevier, 2003).

4.17 Pictorial views

At the start of this chapter we introduced a pictorial drawing of a clamp. In fact, it was in a style of drawing called isometric projection. It is now time to look at pictorial views in more detail starting with oblique projection.

4.17.1 Oblique projection

Figure 4.39 shows a simple component drawn in *oblique projection*. The component is positioned so that you can draw one face true to size and shape. The lines running 'into' the page are called *receding lines* and these are usually drawn at 45° to the front face as shown. To improve the proportions of the drawing and make it look more realistic, you draw the receding lines *half their true length*. For ease of drawing you should observe the following rules.

- Any curve or irregular face should be drawn true shape (front view). For example, a circle on a receding face would have to be constructed as an ellipse; whereas, if it were positioned on the front face, it could be drawn with compasses.
- Wherever possible, the longest side should be shown on the front (true view). This prevents violation of perspective and gives a more realistic appearance.
- For long circular objects such as shafts, the above two rules conflict. In this instance the circular section takes preference and should become the front view for ease of drawing, even though this results in the long axis receding.

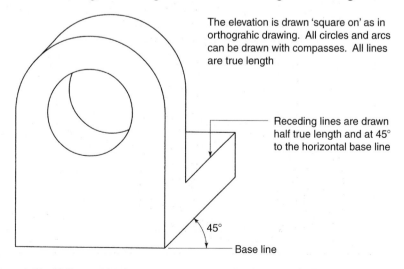

The elevation is drawn 'square on' as in orthograhic drawing. All circles and arcs can be drawn with compasses. All lines are true length

Receding lines are drawn half true length and at 45° to the horizontal base line

45°

Base line

Figure 4.39 Oblique drawing

4.17.2 Isometric projection

The bracket shown in Fig. 4.40 is drawn in *isometric projection*. The isometric axes are drawn at 30° to the base line. To be strictly accurate you should draw these receding lines to isometric scale and only the vertical lines are drawn to true scale. However, for all practical purposes, all the lines are drawn *true length* to save time.

Although isometric drawing produces a more pleasing representation than oblique drawing, it has the disadvantage that no curved profiles such as arcs, circles, radii, etc. can be drawn with compasses. All curved lines have to be constructed. You can do this by erecting a grid over the feature in orthographic projection as shown in Figure 4.41(a). You then draw a grid of equal size where it is to appear on the isometric drawing. The points where the circle cuts the grid in the orthographic drawing are transferred to the isometric grid as shown in Figure 4.41(b). You then draw a smooth curve through the points on the isometric grid and the circle appears as an ellipse. Figure 4.41(c) shows how this technique is applied to the example from the previous section. Note how the circle drawn with compasses in the oblique projection becomes an ellipse in the isometric projection.

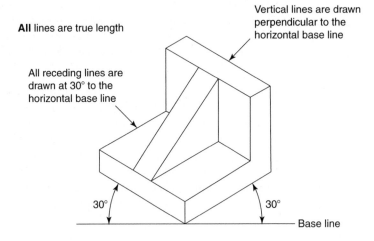

Figure 4.40 Isometric drawing

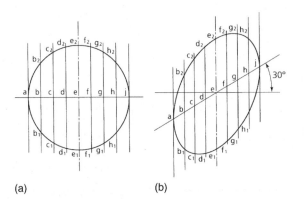

(a)   (b)

1. Construct a grid over the true circle by dividing its centre line into an equal number of parts and erecting a perpendicular at each point.

2. Construct a similar grid on the isometric centre line.

3. Step off distances b_1–b–b_2, c_1–c–c_2, etc. on the isometric grid by transferring the corresponding distances from the true circle.

4. Draw a fair curve through the points plotted.

Figure 4.41 Construction of isometric curves. (a) Orthographic circle (true size and shape); (b) isometric circle (an ellipse)

4.1 Information required to undertake work tasks
  a) You have just received the works order form for the next component you are to make. List the essential information you would expect to find on such a form.
  b) As well as the works order form, what additional and essential document do you need before you can start the task?

4.2 Interpretation of numerical information
  a) Graphs are often used for showing numerical relationships. Sketch suitable graphs to represent the following situations:
    i) The relationship between the diameter and cross-sectional area of mild steel rods of 2, 4, 6, 8, 10 and 12 mm diameter
    ii) The relationship between time and total power in watts for an office that has eight fluorescent electric lights. Each light has a power rating of 80 watts. The lights are turned on, one at a time, at ten minute intervals until they are all on
    iii) The relationship between the date and the following notable accident record at a firm:

    1994   15 accidents
    1995   24 accidents
    1996   12 accidents
    1997   7 accidents
  b) With the aid of simple examples, explain when you would use the following types of graphical representation:
    i) An ideograph (pictogram)
    ii) A pie-chart

4.3 Extraction and interpretation of engineering information
  a) The parts list of a general arrangement drawing specifies the use of a manufacturer's standard drill bush with a bore of 6 mm and an O/D of 12 mm. State where you would look for details of such bushes and list the information you would need to give to the stores so that they could purchase such a bush.
  b) What do the following initial letters stand for: BSI, ISO, EN? (Note you may find two uses of the initials EN.)
  c) Explain briefly what is meant by:
    i) A production schedule
    ii) A product specification

  d) Table 4.5 shows an abstract from some screw thread tables. What is the pitch of an M10 thread and what tapping size drill is required for tapping an internal M10 screw thread?

4.4 Evaluation of the accuracy and appropriateness of engineering information
  a) Give TWO reasons for cross-checking the accuracy of any reference books that might be lying around in your workshop.
  b) To whom should you refer for guidance as to the accuracy and relevance of reference material available in your workshop?

4.5 Recording and processing engineering information
  a) State FOUR reasons for, and the importance of, accurate record keeping in a modern factory environment.
  b) State whether it is a legal requirement to keep a log of notable accidents and, if so, who has the authority to demand access to such a log.
  c) Describe briefly how quality control is maintained in your company, or your training centre, and what records are required.

4.6 Methods of record keeping
  a) Computer files have superseded many manual filing systems. Why should back-up copies of files be kept, and how can these be kept?
  b) State the purposes for which the following methods of record keeping are used
    i) Logbooks (other than your training logbook)
    ii) Forms and schedules
    iii) Photographic (pictorial and dye-line)
    iv) Drawings and diagrams
  c) Why is it important to keep a training logbook, and why should it be kept carefully, away from dirt and oil, so that it is always clean, neat and tidy?

4.7 First and third angle projection
  a) State which of the examples shown in Fig. 4.42 are in FIRST or THIRD angle projection.
  b) Copy and complete the examples shown in Fig. 4.43. The projection symbol is placed below each example for your guidance. (Note: Sometimes complete views are missing, sometimes only lines and features.)

Table 4.5 Exercise 4.3(d)

150 metric threads (coarse series)	Minor dia. (mm)	Tensile stress area (mm2)	Tapping drill (mm)	ISO Hexagon (mm)
M0.8 × 0.2	0.608	0.31	0.68	–
M1.0 × 0.25	0.675	0.46	0.82	2.5
M1.2 × 0.25	0.875	0.73	1.0	3.0
M1.4 × 0.30	1.014	0.98	1.2	3.0
M1.6 × 0.35	1.151	1.27	1.35	3.2
M1.8 × 0.35	1.351	1.7	1.55	–
M2.0 × 0.40	1.490	2.1	1.7	4.0
M2.2 × 0.45	1.628	2.5	1.9	–
M2.5 × 0.45	1.928	3.4	2.2	5.0
M3.0 × 0.5	2.367	5.0	2.65	5.5
M3.5 × 0.6	2.743	6.8	3.1	–
M4.0 × 0.7	3.120	8.8	3.5	7.0
M4.5 × 0.75	3.558	11.5	4.0	–
M5.0 ×3 0.8	3.995	14.2	4.5	8.0
M6.0 × 1.0	4.747	20.1	5.3	10.0
M8.0 × 1.25	6.438	36.6	7.1	13.0
M10.0 × 1.50	8.128	58.0	8.8	17.0
M12.0 × 1.75	9.819	84.3	10.7	19.0
M16.0 × 2.00	13.510	157.0	14.5	24.0

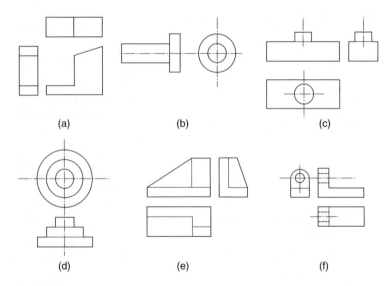

(a)   (b)   (c)

(d)   (e)   (f)

Figure 4.42 Exercise 4.7(a)

4.8 Types of line
a) Copy and complete Fig. 4.44.
b) With the aid of a sketch show what is meant by the terms:
  i) Dimension line
  ii) Leader line
  iii) Projection line

c) With reference to exercise (b):
  i) State the type of line that should be used
  ii) Indicate on your sketch where a short extension is required and where a small gap is required

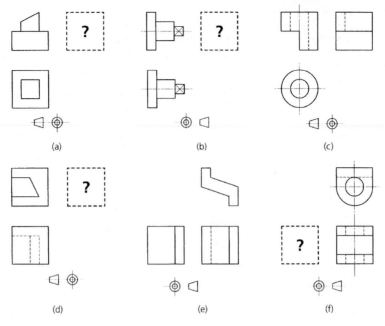

Figure 4.43 Exercise 4.7(b)

Line	Description	Application
——————	Continuous bold	
———————	Continuous fine	
∼∼∼∼∼		Limit of partial view
-------------	Fine short dashes	
——— · ———		Centre lines
	Fine chain, bold at ends and changes of direction	Cutting planes

Figure 4.44 Exercise 4.8(a)

4.9 Dimensioning
  a) With the aid of sketches show how a simple component can be dimensioned from:
   i) A pair of mutually perpendicular datum edges (or surfaces)
   ii) A datum line
   iii) A datum point
  b) With the aid of sketches show how you should dimension the following features:
   i) A circle (show FOUR methods of dimensioning and use the diameter symbol)

   ii) A radius (both convex and concave)
   iii) An angle or chamfer

4.10 Conventions
  a) Give TWO reasons why standard conventions are used on engineering drawings.
  b) Figure 4.45 relates to some commonly used welding symbols. Copy and complete the figure.

4.11 Sectioning
  a) Sketch a section through the welded bracket shown in Fig. 4.46 on the cutting plane **XX**.

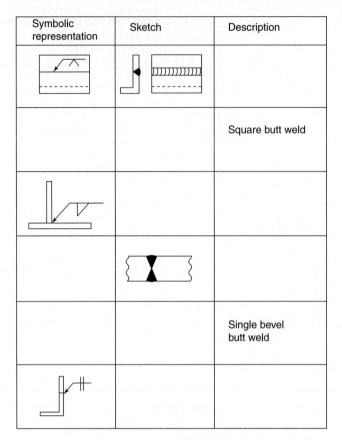

Symbolic representation	Sketch	Description
		Square butt weld
		Single bevel butt weld

Figure 4.45 Exercise 4.10(b)

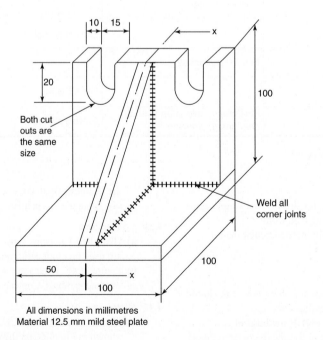

10 | 15

20

Both cut outs are the same size

100

100

Weld all corner joints

50

x

100

100

All dimensions in millimetres
Material 12.5 mm mild steel plate

Figure 4.46 Exercise 4.11(a)

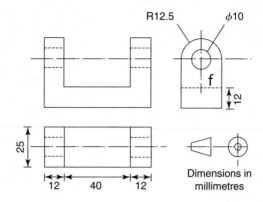

Figure 4.47 Exercise 4.12(a)

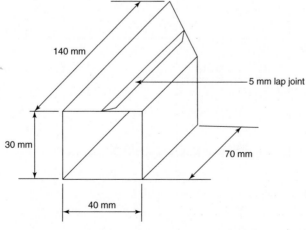

(a) Tin plate oblique frustum of a cone

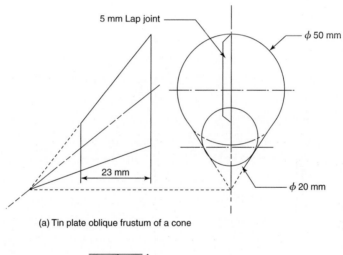

(b) Rectangular tin plate tube

Figure 4.48 Exercise 4.13(b)

4.12 Pictorial views
  a) Figure 4.47 shows a simple workpiece. Sketch it
    in:
    i) Oblique projection
    ii) Isometric projection

4.13 Developments
  a) With the aid of diagrams briefly describe what
    is meant by:
    i) Parallel line development
    ii) Radial line development
    iii) Triangulation

  b) Draw the developed blanks require to manufac-
    ture the tin-plate components shown in Fig. 4.48
    stating the method of development used. Show
    construction lines required.

4.14 Curves of interpenetration
  a) Draw the curves of interpenetration and the
    developed blanks required for the manufacture
    of the hollow components shown in Fig. 4.49.

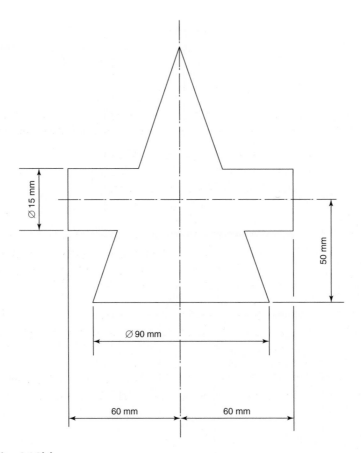

Figure 4.49 Exercise 4.14(a)

5 Measuring and marking out

When you have read this chapter, you should understand:

- What is meant by linear measurement
- How to make linear measurements
- What is meant by angular measurement
- How to make angular measurements
- The correct use of measuring equipment
- How to identify and select marking out tools for making lines
- How to identify and select marking out equipment for providing guidance
- How to identify and select marking out equipment for providing support
- How to identify and select different types of datum
- The need for templates
- The techniques for marking out templates
- The methods of manufacturing templates for sheet, plate and structural components
- What materials are used for templates
- The tools and instruments that are used for making templates
- The care of measuring and marking out equipment

5.1 Introduction

Measuring can be considered to be the most important process in engineering. Without the ability to measure accurately, we cannot:

- Mark out components
- Set up machines correctly to produce components to the required size and shape
- Check components whilst we are making them to ensure that they finally end up the correct size and shape
- Inspect finished components to make sure that they have been correctly manufactured. This is particularly important for large fabrications that have to be assembled on site

5.2 Linear measurement

Fabrication, sheet metal and welding engineers rarely have to work to the fine limits and close tolerance of size that the mechanical engineer requires: the exception being when fitted bolts are required when assembling structures. When you measure length, you measure the shortest distance in a straight line between two points, lines or faces. It doesn't matter what you call this distance (width, thickness, breadth, depth, height and diameter) it is still a measurement of length.

There are two systems for the measurement of length, the *end system* of measurement and the *line system* of measurement

The end system of measurement refers to the measurement of distance between two faces of a component, whilst the line system of measurement refers to the measurement of the distance between two lines or marks on a surface. No matter what system is used, measurement of length is the comparison of the size of a component or a feature of a component and a known standard of length. In a workshop this may be a steel rule or a micrometer calliper, for example. These, in turn, are directly related to fundamental international standards of length. Figure 5.1(a) shows a typical steel rule suitable for workshop and site use.

For measuring distances greater than 1 metre in length flexible steel tapes are used. An example of a typical steel tape is shown in Fig. 5.1(b). Woven or fabric tapes are unsuitable for engineering as they tend to stretch and always have to be used with a spring balance to apply a prescribed tension. Flexible steel tapes are available in sizes ranging from the handy pocket size (approximately 5 metres in length) up to and exceeding 30 metres.

5.2.1 Steel rules (use of)

The steel rule is frequently used in workshops and on site for measuring components of limited accuracy quickly. The quickness and ease with which it can be used, coupled with its low cost, makes it a popular and widely used measuring device. Metric rules may be obtained in various lengths from 150 mm to 1000 mm (1 metre). Imperial rules may be obtained in various lengths from 6 inch to 36 inch (1 yard). It is convenient to use a rule engraved with both systems, the inch system on the front and the metric system on the reverse side, the back.

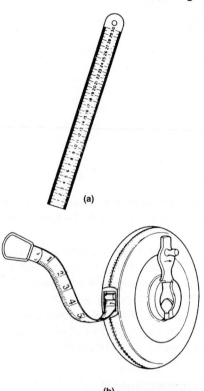

(a)

(b)

Figure 5.1 Standards of measurement – the rule and steel tape. (a) The steel rule; (b) the steel tape

Steel rules may be 'rigid' or 'flexible' depending upon their thickness and the 'temper' of the steel used in their manufacture. When choosing a steel rule the following points should be looked for. It should be:

- Made from hardened and tempered, corrosion resistant spring steel
- Engine divided. That is, the graduations should be precision engraved into the surface of the metal
- Ground on the edges so that it can be used as a straight edge when scribing lines or testing a surface for flatness
- Ground on one end so that this end can be used as the zero datum when taking measurements from a shoulder
- Satin chrome finished so as to reduce glare and make the rule easier to read, also to prevent corrosion.

No matter how accurately a rule is made, all measurements made with a rule are of limited accuracy. This is because of the difficulty of sighting the graduations in line with the feature being measured. Some ways of minimizing sighting errors are shown Fig. 5.2.

When using a rule to make direct measurements, as in Fig. 5.2(a), the accuracy of measurement depends upon the *visual alignment* of a mark or surface on the

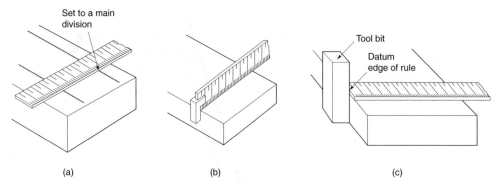

Figure 5.2 Use of a rule – measuring the distance between: (a) two scribed lines; (b) two faces using a hook rule; (c) two faces using a steel rule and a tool bit as an adjustment

work with the corresponding graduation on the rule. This may appear relatively simple but, in practice, errors can very easily occur. These errors can be minimized by using a thin rule and keeping your eyes directly above and at 90° to the mark on the work. If you look at the work and the rule from an angle, you will get a false reading. This is known as a *parallax* error. Figure 5.2(b) and (c) show two ways of aligning the datum (zero) end of the rule with the edge of the component to eliminate one source of sighting error.

5.2.2 Line and end measurement

Linear distances sometimes have to be measured between two lines, sometimes between two surfaces and sometimes between a combination of line and surface. A measurement between two lines is called *line measurement*. A measurement between two surfaces is called *end measurement*. It is difficult to convert between end systems of measurement and line systems of measurement and vice versa. For example, a rule (which is a line system measuring device) is not convenient for the direct measurement of distances between two edges; similarly, a micrometer (which is an end system measuring device) would be equally inconvenient if used to measure the distance between two lines. Therefore, a measuring device must always be chosen to suit the job in hand.

5.2.3 Calipers and their use

Calipers are used in conjunction with a rule so as to transfer the distance across or between the faces of a component in such a way as to reduce sighting errors. That is, to convert from *end measurement* to *line measurement*. Firm-joint calipers are usually used in the larger sizes and spring-joint calipers are used for fine work. Examples of internal and external calipers of both types are shown in Fig. 5.3 together with examples of their uses. The accurate use of calipers depends upon practice, experience, and a highly developed sense of feel. When using calipers, the following rules should be observed:

- Hold the caliper gently and near the joint
- Hold the caliper square (at right-angles) to the work

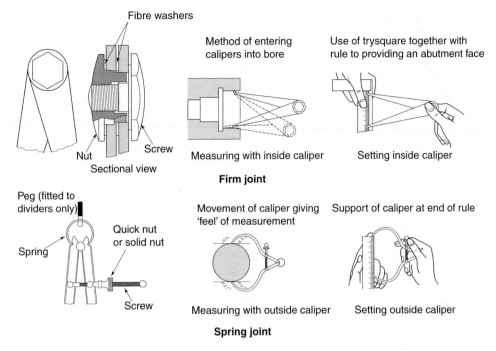

Figure 5.3 Construction and use of calipers

- No force should be used to 'spring' the caliper over the work. Contact should only just be felt
- The caliper should be handled and laid down gently to avoid disturbing the setting
- Lathe work should be *stationary* when taking measurements. This is essential for *safety* and *accuracy*.

5.2.4 The micrometer caliper (use of)

It is not usually necessary for sheet metal and fabrication engineers to have to work to fine limits of dimensional accuracy. However, it is useful to be able to use a micrometer caliper in order to check the thickness of sheet material and the diameter of drills. The constructional details of a typical micrometer caliper are shown in Fig. 5.4. This instrument depends upon the principle that the distance a nut moves along a screw is proportional to the number of revolutions made by the nut and the lead of the screw thread. Therefore, by controlling the number of complete revolutions made by the nut and the fractions of a revolution made by the nut, the distance it moves along the screw can be accurately controlled. It does not matter whether the nut rotates on the screw or the screw rotates in the nut, the principle of operation still holds good.

In a micrometer caliper, the screw thread is rotated by the *thimble* which has a scale that indicates the partial revolutions. The *barrel* of the instrument has a scale which indicates the 'whole' revolutions. In a standard metric micrometer caliper the screw has a lead of 0.5 millimetre and the thimble and barrel are graduated as in

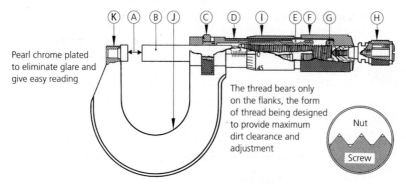

Pearl chrome plated to eliminate glare and give easy reading

The thread bears only on the flanks, the form of thread being designed to provide maximum dirt clearance and adjustment

Nut

Screw

A *Spindle and anvil faces* – Glass hard and optically flat, also available with tungsten carbide faces
B *Spindle* – Thread ground and made from alloy steel, hardened throughout, and stabilised
C *Locknut* – Effective at any position. Spindle retained in perfect alignment
D *Barrel* – Adjustable for zero setting. Accurately divided and clearly marked, pearl chrome plated
E *Main nut* – Length of thread ensures long working life
F *Screw adjusting nut* – For effective adjustment of main nut
G *Thimble adjusting nut* – Controls position of thimble
H *Ratchet* – Ensures a constant measuring pressure
I   *Thimble* – Accurately divided and every graduation clearly numbered
J *Steel frame* – Drop forged
K *Anvil end* – Cutaway frame facilitates usage in narow slots

Figure 5.4 The micrometer caliper

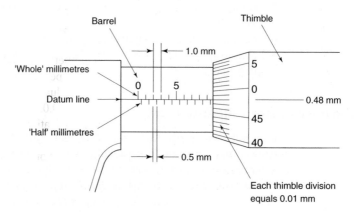

Figure 5.5 Micrometer scales (metric)

Fig. 5.5. Since the lead of the screw of a standard metric micrometer is 0.5 millimetre and the barrel divisions are 0.5 millimetre apart, one revolution of the thimble moves the thimble along the barrel a distance of one barrel division (0.5 mm). The barrel divisions are placed on alternate side of the datum line for clarity. Further, since the thimble has 50 divisions and one revolution of the thimble equals 0.5 millimetre, then a movement of *one thimble division* equals: 0.5 millimetre/50 divisions = 0.01 millimetre.

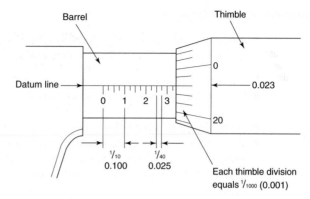

Figure 5.6 Micrometer scales

A metric micrometer caliper reading is given by:

- The largest visible 'whole' millimetre graduation visible on the barrel, *plus*
- The next 'half' millimetre graduation, if this is visible, *plus*
- The thimble division coincident with the datum line.

Therefore, the micrometer scales shown in Fig. 5.5 read as follows:

9 'whole' millimetres	= 9.00
1 'half' millimetre	= 0.50
48 hundredths of a millimetre	= 0.48
	= 9.98 mm

Figure 5.6 shows the scales for a micrometer graduated in imperial 'inch' units. The micrometer screw has 40 T.P.I. (threads per inch); therefore the lead of the screw is 1/40 inch (0.025 inch). The barrel graduations are 1/10 inch sub-divided into 4. Therefore, each subdivision is 1/40 inch (0.025 inch) and represents one revolution of the thimble. The thimble carries 25 graduations, so one thimble graduation equals a movement of 0.025 inch/25 = 0.001 inch. This is one thousandth part of an inch and is often referred to by engineers as a 'thou'. Thus, 0.015 inch could be referred to as 15 'thou'.

An inch micrometer reading is given by:

- The largest visible 1/10 inch (0.1 inch) division, *plus*
- The largest visible 1/40 inch (0.025 inch) division, *plus*
- The thimble division coincident with the datum line.

Therefore, the micrometer scales shown in Fig. 5.6 read as follows:

3 tenths of an inch	= 0.300
1 fortieth of an inch	= 0.025
23 thousandths of an inch	= 0.023
	= 0.348 inch

5.2.5 Micrometer caliper (care of)

Unless a micrometer caliper is properly looked after it will soon lose its initial accuracy. To maintain this accuracy you should observe the following precautions:

- Wipe the work and the anvils of the micrometer clean before making a measurement
- Do not use excessive force when making a measurement; two 'clicks' of the ratchet is sufficient
- Do not leave the anvil faces in contact when not in use
- When machining, stop the machine before making a measurement. Attempting to make a measurement with the machine working can ruin the instrument and also lead to a serious accident. This rule applies to all measuring instruments and all machines.

5.2.6 Vernier calipers

Vernier calipers are included here by way of introduction to the vernier principle. The vernier principle is often applied to marking out instruments, such as the trammel (beam compass) for striking large radii and the vernier caliper used for measuring angles. Although more cumbersome to use and rather more difficult to read, the vernier caliper has three main advantages over the micrometer caliper.

1. One instrument can be used for measurements ranging over the full length of its main (beam) scale. Figure 5.7(a) shows a vernier caliper.
2. It can be used for both internal and external measurements as shown in Fig. 5.7(b). Remember that for internal measurements you have to add the combined thickness of the jaws to the scale readings.

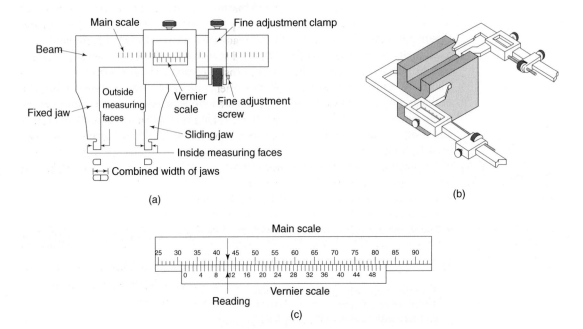

Figure 5.7 The vernier caliper: (a) construction; (b) use; (c) vernier scale (50 divisions)

3. One instrument can be used for taking measurements in both inch units and in metric dimensional systems.

The measuring accuracy of a vernier calliper tends to be of a lower order than that obtainable with a micrometer calliper because:

- It is difficult to obtain a correct 'feel' with this instrument due to its size and weight
- The scales can be difficult to read accurately even with a magnifying glass.

All vernier-type instruments have two accurately engraved scales. A main scale marked in standard increments of measurement like a rule, and a vernier scale that slides along the main scale. This vernier scale is marked with divisions whose increments are slightly smaller than those of the main scale. Some vernier callipers are engraved with both inch and millimetre scales.

In the example shown in Fig. 5.7(c) the main scale is marked off in 1.00 mm increments, whilst the vernier scale has 50 divisions marked off in 0.98 mm increments. This enables you to read the instrument to an accuracy of $1.00 - 0.98 = 0.02$ mm. The reading is obtained as follows:

- Note how far the zero of the vernier scale has moved along the main scale (32 'whole' millimetres in this example)
- Note the vernier reading where the vernier and main scale divisions coincide (11 divisions in this example. You then multiply the 11 divisions by 0.02 mm which gives you 0.22 mm
- Add these two readings together:

> 32 'whole' millimetres = 32.00 mm *plus*
> 11 vernier divisions = 00.22 mm
> therefore the reading shown in Fig. 5.7(c) = **32.22 mm**

Always check the scales before use as there are other systems available and not all vernier scales have 50 increments. This is particularly the case in some cheap instruments. Also check that the instrument reads zero when the jaws are closed. If not, then the instrument has been strained and will not give a correct reading. There is no means of correcting this error and the instrument must be scrapped.

As for all measuring instruments, vernier callipers must be cleaned before and after use. They should always be kept in the case provided. This not only protects the instrument from damage, it also supports the beam and prevents it from becoming distorted. The vernier principle can also be applied to height gauges and to depth gauges. Nowadays, both micrometer and vernier callipers are available with digital readouts for ease of use. The normal scales are retained in most cases in case of battery failure.

> Since vernier callipers are expensive, it is essential to treat them with care

5.3 Angular measurement

Angles are measured in degrees and fractions of a degree. One degree of arc is 1/360 of a complete circle. One degree of arc can be subdivided into minutes and seconds (not to be confused with minutes and seconds of time):

> 60 seconds (") of arc = 1 minute (') of arc
> 60 minutes (') of arc = 1 degree (°) of arc

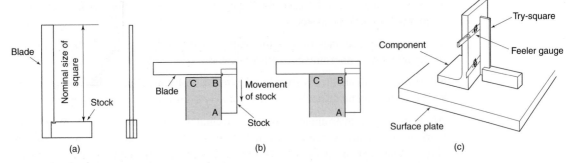

Figure 5.8 The try-square (a), its use (b) and (c)

With the introduction of calculators and computers, decimal fractions of a degree are also used. However, 1 minute of arc equals 0.0166666° recurring so there is no correlation between the two systems of sub-dividing a degree.

5.3.1 Right angles

A right angle is the angle between two surfaces that are at 90° to each other. Such surfaces may also be described as being *mutually perpendicular.* The use of engineers' try-squares and their use for scribing lines at right angles to the edge of a component will be described later in this chapter. Figure 5.8(a) shows a typical engineer's try-square.

Note that a try-square is not a measuring instrument. It does not measure the angle. It only indicates whether or not the angle being checked is a right-angle. In Fig. 5.8(b), the stock is placed against the edge AB of the work and slid gently downwards until the blade comes into contact with the edge BC. Any lack of squareness will allow light to be seen between the edge BC and the try square blade. It is not always easy or convenient to hold large work and a try-square up to the light. Figure 5.8(c) shows an alternative method using a surface plate as a datum surface. The squareness of the component face is checked with feeler gauges as shown. If the face is square to the base, the gap between it and the try-square blade will be constant.

Try-squares should be kept clean and lightly oiled after use. They should not be dropped, nor should they be kept in a draw with other bench tools that may knock up burrs on the edges of the blade and stock. They should be checked for squareness at regular intervals.

> Try-squares are precision instruments and they should be treated with care if they are to retain their initial accuracy

5.3.2 Angles other than right angles (plain bevel protractor)

Figure 5.9 shows a simple bevel protractor for measuring angles of any magnitude between 0° and 180°. Such a protractor has only limited accuracy (±0.5°).

5.3.3 Angles other than right angles (vernier protractor)

Where greater accuracy is required the vernier protractor should be used. The scales of a vernier protractor are shown in Fig. 5.10. The main scale is divided into degrees

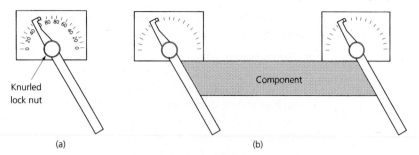

Figure 5.9 The plain bevel protractor (a), and its use in checking angles (b)

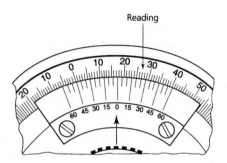

Figure 5.10 Vernier protractor scales

of arc, and the vernier scale has 12 divisions each side of zero. These vernier scale divisions are marked 0 to 60 minutes of arc, so that each division is 1/12 of 60 which equals 5 minutes of arc. The reading for a vernier protractor is given by the sum of:

- The largest 'whole' degree on the main scale as indicated by the vernier zero mark
- The reading of the vernier scale division in line with a main scale division.

Thus the reading for the scales shown in Fig. 5.10 is:

17 'whole' degrees	= 17° 00'
vernier 25 mark in line with main scale	= 00 25'
Total angle	= **17° 25'**

Vernier protractors are also available which can be read in degrees and decimal fractions of a degree.

5.4 Correct use of measuring equipment

No matter how accurately measuring equipment is made, and no matter how sensitive it is, one of the most important factors affecting the accuracy of measurement is the skill of the user. The more important procedures for the correct use of measuring equipment can be summarized as follows.

- The measurement must be made at right angles to the surface of the component.
- The use of a constant measuring pressure is essential. This is provided automatically with micrometer callipers by means of their ratchet. With other instruments such as plain callipers and vernier callipers the measuring pressure depends upon the skill and 'feel' of the user. Such skill only comes with practice and experience.
- The component must be supported so that it does not distort under the measuring pressure or under its own weight.
- Measuring instruments must be handled with care so that they are not damaged or strained. They must be cleaned and kept in their cases when not in use. Measuring instruments must be regularly checked to ensure that they have not lost their initial accuracy. If an error is detected the instrument must be taken out of service immediately so that the error can be corrected. If correction is not possible the instrument must be immediately discarded.

5.5 Marking out equipment (tools for making lines)

Marking out is, essentially, drawing on metal so as to provide guide lines for cutting out templates and blanks, and for marking where sheet metal is to be folded and holes are to be drilled or punched in the component. Usually a pencil line would not be suitable; the hard metal surface would soon make a pencil blunt and the line would become thick and inaccurate; also a pencil line is too easily wiped off a metal surface. Therefore, the line is usually scribed using a sharp pointed metal tool, such as a scriber, which cuts into the surface of the metal and leaves a fine, permanent line. The exception is when marking out tin-plate (as a scriber would cut through the protective film of tin and allow the steel base metal to corrode). Here, bend lines are always drawn with a soft pencil so as to protect the tin coating. Only cutting lines are drawn with a scriber.

5.5.1 Scriber

This is the basic marking out tool. It consists of a handle with a sharp point. The pointed end is made from hardened steel so that it will stay sharp in use. Engineers' scribers usually have one straight end and one hooked end, as shown in Fig. 5.11. It is essential that the scribing point is kept sharp. Scribing points should not be sharpened on a grinding machine – the heat generated by this process tends to soften the point of the scriber so that it soon becomes blunt. Instead, the scribing point should be kept needle sharp by the use of an oil stone.

5.5.2 Centre, dot and nipple punches

Typical centre, dot and nipple punches are shown in Fig. 5.12. They are used for making indentation in the surface of the metal. Figure 5.12(a) shows a dot punch. This has a relatively fine point of about 60° or less and is used for locating the legs

Figure 5.11 Scriber

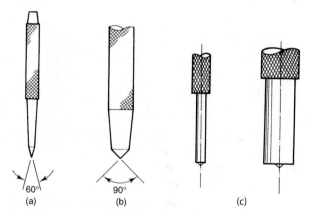

Figure 5.12 Punches: (a) dot; (b) centre punch; (c) nipple punch

of such instruments as dividers and trammels. It is also used to preserve scribed lines and to provide witness marks. Figure 5.12(b) shows a centre punch. This is heavier than a dot punch and has a less acute point (usually 90° or greater). It is used to make a heavy indentation suitable for locating the point of a twist drill. Figure 5.12(c) shows a typical nipple punch. Unlike centre and dot punches, the parallel diameter of a nipple punch has to be a slide fit in the template so as to locate the indentation accurately in the work without the need for marking out.

The correct way to using dot and nipple punches is shown in Fig. 5.13. First, let's consider the use of the dot punch. Usually the position for making a dot mark is at the junction of a pair of scribed line at right-angles to each other.

■ You hold the punch so that it is inclined away from you. This enables you to see when the point of the punch is at the junction of the scribed lines as shown in Fig. 5.13(a).

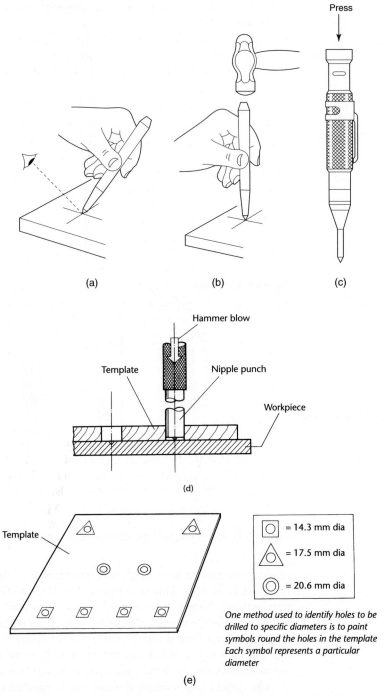

Figure 5.13 Dot, centre and nipple punches: (a) and (b) correct way to use dot and centre punches; (c) automatic dot punch; (d) correct way to use a nipple punch; (e) template identification symbols

- You then carefully bring the punch up to the vertical taking care not to move the position of the point.
- You then strike the punch lightly and squarely with a hammer as shown in Fig. 5.13(b).
- Check the position of the dot with the aid of a magnifying glass. Draw the dot over if it is slightly out of position.

For rough work you can use a centre punch in the same way but you need to hit it harder with a heavier hammer if you are to make a big enough indentation to guide the point of a drill. Because of the difficulty in seeing the point of a centre punch, it is preferable to make a dot punch mark and, when you are satisfied that it is correctly positioned, you can enlarge the dot mark with a centre punch. The centre punch is correctly positioned when you feel its point 'click' into the mark left by the dot punch.

Figure 5.13(c) shows an automatic dot punch. This has the advantage that it can be used single-handed and it is less likely to skid across the surface of the work. The punch is operated by downward pressure that releases a spring loaded hammer in its body. No separate hammer is required.

Figure 5.13(d) shows how a nipple punch is used in conjunction with a template. The plain, parallel portion of the punch must match the hole in the template. The mark left by the nipple punch is relatively small and has to be enlarged with a heavier centre punch. The sizes of the holes for various diameters are identified by symbols scribed or drawn around the centre punch marks as shown in Fig. 5.13(e).

5.5.3 Dividers and trammels

These instruments are used for marking out circles and arcs of circles. A typical pair of dividers and the names of its component parts are shown in Fig. 5.14(a). Dividers are used to scribe circular lines as shown in Fig. 5.14(b). They are set to the required radius as shown in Fig. 5.14(c). They are also used for stepping off equal distances (such as hole centres along a line or round a pitch circle) as shown in Fig. 5.14(d). The leg about which the dividers pivot is usually located in a fine centre dot mark. To locate the point of this leg accurately it is essential to use a sharp dot punch as shown in Fig. 5.14(e).

Trammels are used for scribing large diameter circles and arcs that are beyond the range of ordinary divider. They are also called beam compasses when the scribing points are located on a wooden beam as shown in Fig. 5.14(f). Trammels have a metal beam usually in the form of a solid rod or a tube. This often carries a scale and one of the scribing points is fitted with a vernier scale and a fine adjustment screw for accurate setting as shown in Fig. 5.4(g).

5.5.4 Hermaphrodite callipers

These are usually called *odd-leg* callipers or *jenny* callipers. They consist of one calliper leg and one divider leg and are used for scribing lines parallel to an edge as shown in Fig. 5.15(a). They are set to the required size as shown in Fig. 5.15(b).

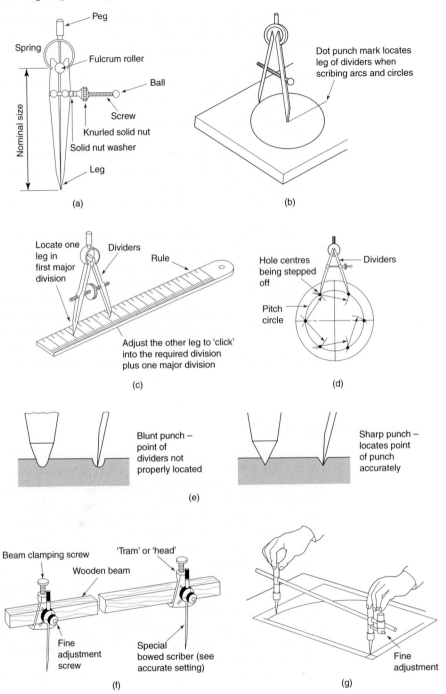

Figure 5.14 Dividers and trammels: (a) parts of a divider; (b) scribing a circle; (c) setting a required radius; (d) stepping off hole centres; (e) location of divider point; (f) trammel or beam compass; (g) adjustment of trammel

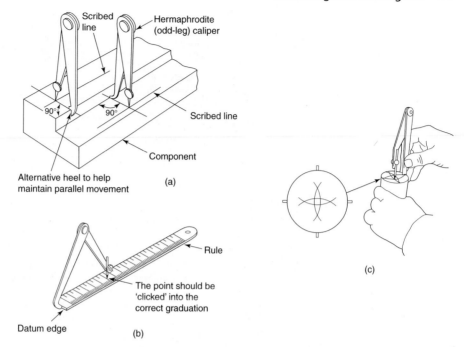

Figure 5.15 Hermaphrodite (odd-leg) calipers: (a) scribing lines parallel to an edge; (b) setting odd-leg calipers (c) finding the centre of a bar

5.5.5 Scribing block

A scribing block or surface gauge is used for marking out lines parallel to a datum surface or a datum edge. The parts of a typical scribing block are shown in Fig. 5.16(a) and some typical applications are shown in Fig. 5.16(b). Normally the scribing point is set to mark a line at a given height above the base of the instrument. This line will be marked parallel to the surface along which the base of the instrument is moved. When a line parallel to a datum edge is required, the edge pins are lowered. These pins are then kept in contact with the datum edge as the scribing block is moved along the work.

5.6 Marking out equipment (tools for providing guidance)

You cannot draw a straight line with a scriber without the help of some form of straight edge to guide the scriber. Let's now consider the tools that provide guidance for the scribing point.

5.6.1 Rule and straight edge

Where a straight line is required between two points, a rule can be used or, for longer distances, a straight edge. The correct way to use a scriber is shown in Fig. 5.17(a).

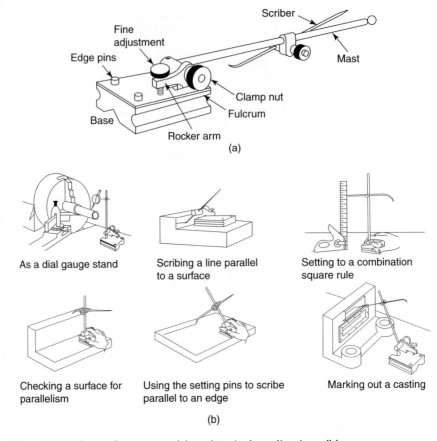

Figure 5.16 The surface gauge (a) and typical applications (b)

The scriber is always inclined away from any guidance edge. Its point should always trail the direction of movement to prevent it 'digging-in' to the metal surface so that it produces a poor line and damages the scribing point.

5.6.2 Box square

This is also known as a *key seat rule*. It is used for marking and measuring lines scribed parallel to the axis of a cylindrical component such as a tube. A typical box square and its method of use is shown in Fig. 5.17(b).

5.6.3 Try-square

When you need to scribe a line at 90° to a datum edge a try-square is used as shown in Fig. 5.18. A line scribed at 90° to an edge or another line is said to be at *right-angles* to that edge or line or it is said to be *perpendicular* to that edge or line. They both mean the same thing.

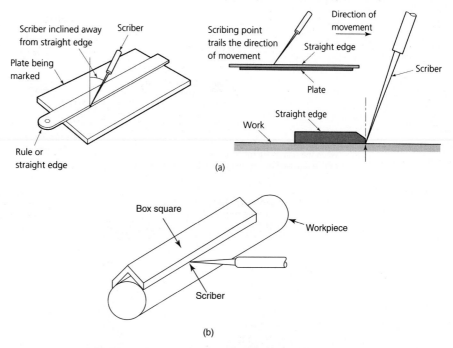

Figure 5.17 Scribing straight lines: (a) scribing a straight line using a rule as a straight edge; (b) scribing a straight lime using a box square

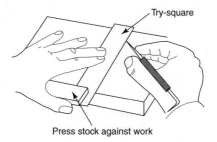

Figure 5.18 Scribing a straight line perpendicular to an edge

5.6.4 Combination set

This is shown in Fig. 5.19(a). It consists of a strong, relatively thick and rigid rule together with three 'heads' that are used individually but in conjunction with the rule.

- The square head can be clamped to the rule at any point along its length. It can either be used as a try square (90°) or as a mitre square (45°) as shown in Fig. 5.19(b).
- The centre head or centre finder can also be clamped to the rule at any point along its length. The edge of the blade that passes through the centre of the centre finder also passes through the centre of cylindrical work-piece. The centre of the cylindrical work-piece is found by scribing two lines at right-angles to each other as shown in Fig. 5.19(c). The lines intersect at the centre of the work-piece.

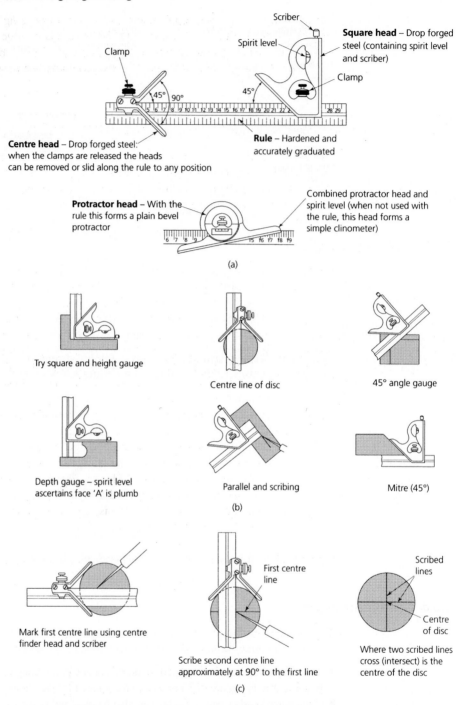

Figure 5.19 The combination set: (a) construction; (b) uses; (c) finding the centre of a circular component

- A protractor head is also supplied and this is used for marking out lines that are at any angle other than at 90° or 45° to the datum surface or edge.
- The square head and the protractor head are supplied with spirit (bubble) levels for setting purposes. However, they are only of limited accuracy.

5.7 Marking out equipment (tools for providing support)

When marking out a component, it is essential that the blank is properly supported. As well as keeping the work-piece rigid and in the correct position, the supporting surface may also provide a datum from which to work. A datum is a line, surface or edge from which measurements are taken.

5.7.1 Surface plate and tables

Surface plates are cast from a stable cast iron alloy and are heavily ribbed to make them rigid. An example is shown in Fig. 5.20(a). They are used on the bench to provide a flat surface for marking out small work pieces.

> Surface plates are very heavy and should only be moved with care, preferably by two or more persons

Surface tables (marking out tables), such as the one shown in Fig. 5.20(b), are used for providing a support and datum surface when marking out larger work-pieces. A marking out table is of heavy and rigid construction. The working surface may by of cast-iron machined or ground flat. Plate glass and granite are also used because of their smoothness and stability. They do not give such a nice 'feel' as cast iron when moving the instruments upon them. This is because cast iron is self-lubricating.

The working surface must be kept clean and in good condition. Nothing must be allowed to scratch or damage the table and heavy objects must be slid gently onto the table from the side. Clean the table before and after use and make sure all sharp corners and rough edges are removed from the work-piece before it is placed on the table. Keep the table covered when it is not in use. Oil the working surface of the table if it is not to be used for some time.

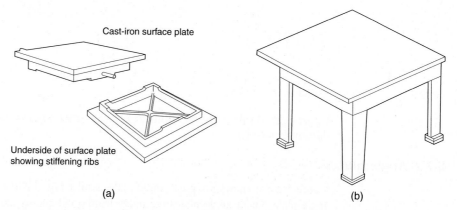

Cast-iron surface plate

Underside of surface plate showing stiffening ribs

(a)

(b)

Figure 5.20 Surface plate (a) and marking-out table (b)

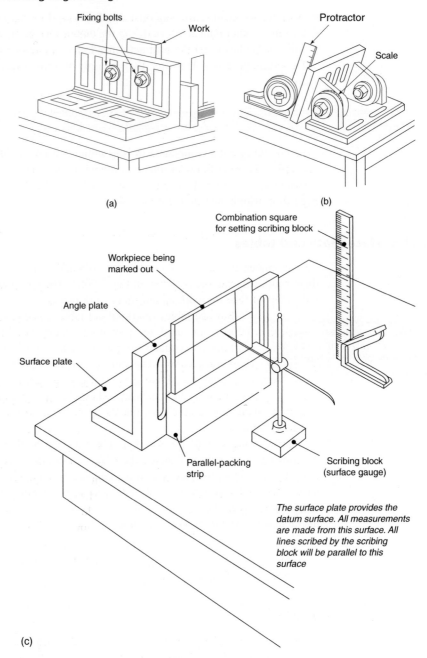

Fixing bolts

Work

Protractor

Scale

(a)

(b)

Combination square
for setting scribing block

Workpiece being
marked out

Angle plate

Surface plate

Parallel-packing
strip

Scribing block
(surface gauge)

*The surface plate provides the
datum surface. All measurements
are made from this surface. All
lines scribed by the scribing
block will be parallel to this
surface*

(c)

Figure 5.21 Angle plate (a), adjustable angle plate (b), marking-out from a
datum surface (c)

5.7.2 Angle plates

Figure 5.21(a) shows a typical angle plate, whilst Fig. 5.21(b) shows an adjustable
angle plate. These angle plates are made from good quality cast iron and the working faces are machined at right-angles to each other. The ends are also machined

so that the angle plate can be stood on end when it is necessary to turn the work clamped to it through 90°. Figure 5.21(c) shows a typical angle plate being used to support work perpendicular to the datum surface of a marking out table.

5.8 Techniques for marking out

Finally, this chapter will consider a range of techniques used in the marking out in the sheet metal and fabrication industry, including the preparation and use of templates and the setting out of pipe work.

5.8.1 Types of datum

The term datum has already been used several times in this chapter. It has also been described as a point, line or edge from which measurements are taken. Let's now examine the different types of datum in more detail.

- *Point datum.* This is a single point from which dimensions can be taken when measuring and marking out. For example, the centre point of a pitch circle.
- *Line datum.* This is a single line from which, or along which, dimensions are taken when measuring and marking out. It is frequently the centre line of a symmetrical component.
- *Edge datum.* This is also known as a *service edge.* It is a physical surface from which dimensions can be taken. This is the most widely used datum for marking out. Usually two edges are prepared at right-angles to each other. They are also referred to as *mutually perpendicular* datum edges. These two edges ensure that the distances marked out from them are also at right-angles to each other.
- *Surface datum.* For example, this can be the working surface of a surface plate or a marking out table. It provides a common datum to support the work and the measuring and marking out equipment in the same plane. If, for example, you set your work with its datum edge on the surface datum of the marking out table, and you set your surface gauge or scribing block to 25 mm, then the line you scribe on your work will be 25 mm from its datum edge. This is because the datum surface of the foot of the surface gauge and the datum surface of your work are both being supported in the same plane by the surface plate or marking out table, as shown previously in Fig. 5.21.

5.8.2 Surface preparation

- Before commencing to mark out a metal surface, the surface must be cleaned and all oil, grease, dirt and loose material removed.
- A dark pencil line shows up clearly on white paper because of the colour contrast. Since scribed lines cut into the metal surface there is very little colour contrast and they do not always show up clearly.
- To make the line more visible, the metal surface is usually coated in a contrasting colour. Large plates are usually whitewashed locally where the mark is to be made, but smaller steel and non-ferrous components are usually coated with a quick drying layout 'ink'.

- Avoid using the old-fashioned technique of copper plating the surface of a steel component with a solution of copper sulphate containing a trace of sulphuric acid. Although it leaves a very permanent coating, it can only be used on steels and it is corrosive if it gets on marking out instruments. The coating can only be removed by using emery cloth or by grinding.
- Layout ink is available in a variety of colours and can be readily applied to a smooth surface using an aerosol can. The ink should be applied thinly and evenly. Two thin coats are better than one thick coat. Wait for the ink to dry before marking out. The ink can be removed with a suitable solvent when the component is finished.
- **Safety**. Direct the spray only at the work piece, never at your workmates. Obey the maker's instructions at all times. Use only if there is adequate ventilation. Avoid breathing in the solvent and the propellant gas.

5.8.3 Marking out a large template or work-piece (general)

Components marked out on sheet metal and on plate are often large by normal engineering standards. This makes conventional engineering techniques using straight edges, protractors rules and try-squares difficult and sometimes impossible in practice. It is often easier to construct the shape required geometrically using dividers and/or trammels.

Large sheet metal fabrications and plate-work jobs have to be marked out 'in the flat'. It is often convenient to cut a sheet or plate to the correct shape and dimensions before marking the position of rivet or bolt holes if these are around the edge only, otherwise the entire job must be lad out. If the required sheet or plate is not too large, a *datum line* may be scribed adjacent to one edge with the aide of a straight edge and scriber. For large steel plates with a hot-rolled finish a *chalk-line* may be used as shown in Fig. 5.22. The white line shows up clearly against the dark metal background of the oxidized steel.

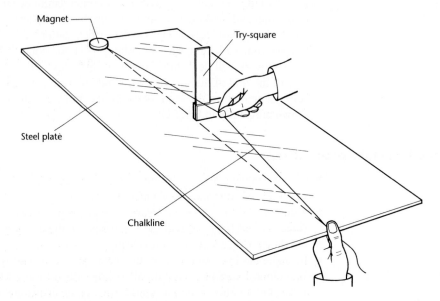

Figure 5.22 The use of a chalk line for marking-out

After the line has been thoroughly chalked, it is secured at one end whilst being stretched out firmly and held against the work in line with witness *marks* previously made with a centre punch. The line is then 'flicked' against the metal plate, as shown, when it will leave a white chalk line in the correct position.

5.8.4 Use of trammels and steel tapes

These are used for striking lines at right-angles to each other and for measuring distances accurately as shown in Fig. 5.23(a). The required line is to be marked at right-angles to a datum line at point D. Points A and B are established on the datum line at any convenient distance either side of D providing they are at the same distance. They may be measured with a rule or steel tape depending on the size of the work or they may be stepped off with dividers, a beam compass or trammels. Arcs are then struck from A and B, as shown, using trammels so that they intersect at C. The line DC will then be at right angles (perpendicular) to the datum line.

The theorem of Pythagoras may also be used for establishing two lines at right-angles as shown in Fig. 5.23(b). The theorem states that in any right-angled triangle the square on the hypotenuse is equal to the sum of the squares on the other two sides. In practice, the lengths of the sides used are usually 3, 4 and 5 units long as they are easy to remember and are convenient to use. In the example shown, the distance AB is set along the datum line using a steel rule or tape so that AB is 4 units in length. With centre A, and with the trammels set to 3 units of length, strike an arc. Then with centre B, and with the trammels set to 5 units, a second arc is struck so that the two arcs intersect at C. The line AC will be perpendicular to the line AB. The triangle should be as large as possible to ensure maximum accuracy. For very large work two steel rules are frequently used to establish the distances AC and BC.

5.8.5 The flat square

The engineer's try-square is made with a thick stock and a thin blade so that a line may be scribed perpendicular to an edge. A sheet-metal worker will use a flat square of uniform thickness so that it can be laid on a sheet or plate with out rocking. Two examples are shown in Fig. 5.24. The one shown in Fig. 5.24(a) is fabricated from well seasoned hardwood for lightness in the larger sizes, whilst the hardened and tempered steel one shown Fig. 5.24(b) is normally used for small and medium sized work.

5.8.6 Setting out a steel plate

Figure 5.25 shows how to mark out a steel plate which is to be 1.58 m × 1.58 m with square corners using a steel tape and trammels. *A steel tape is used for all measurements.*

1. Select an existing straight side of the plate which can be used as a base line (datum). This is the edge AB in Fig. 5.25. Set the trammels to the required width of the finished plate ($R = 1.58$ m) and with any two points 'a' and 'b' as centres on the base line AB strike two arcs as shown. Using the same centres and with the trammels set to approximately half this dimension (radius r) two more arcs are struck as shown.

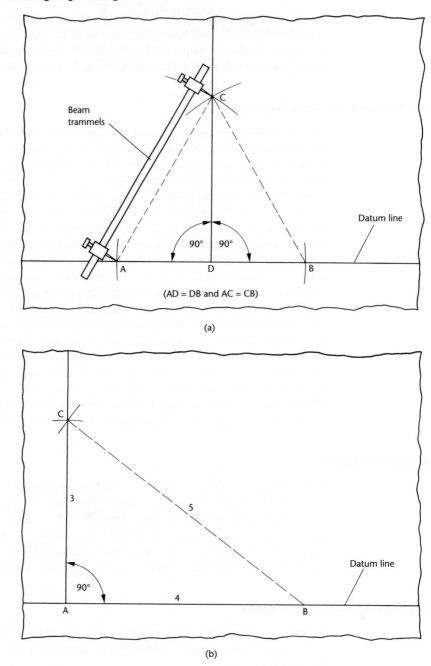

Beam trammels

Datum line

90° 90°

A   D   B

(AD = DB and AC = CB)

(a)

C

3

5

4

90°

A   B

Datum line

(b)

Figure 5.23 Applications of beam trammels and steel tape (marking-out). (a) Use of trammels to construct a right angle; (b) use of trammels and steel tape to construct a right angle

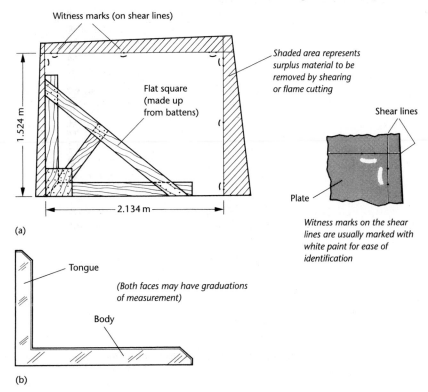

(a)

(b)

Figure 5.24 The flat square. The fabricated flat square; (b) the steel flat square (one piece)

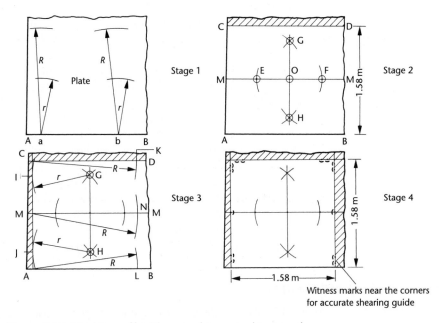

Figure 5.25 Marking-off with a steel tape and trammels

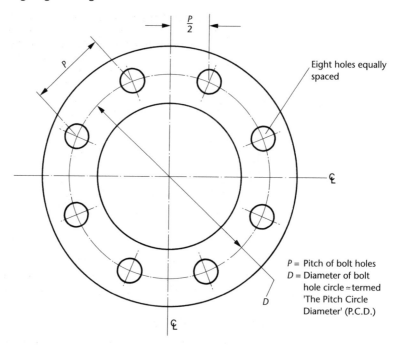

Figure 5.26 Marking-out bolt holes for flanges

2. Parallel lines CD and MM are marked with a chalk line so that they are tangential to the arcs struck in stage 1 above. A light centre punch mark is made at O positioned so that it is approximately half way along the line MM. From the point O on the line MM construct a perpendicular GH and mark with a chalk line. Lightly centre-punch the points G and H. The points G, H and O are used to check whether or not the edges of the plate are straight and parallel to this line of points.

3. If the edges of the plate are not straight and parallel as shown in Fig. 5.25 then the trammels are set to radius *r*, where *r* is *half* the required distance of 1.58 m. With centres G and H arcs are struck to provide a suitable shearing margin at points I and J. The end shearing line is made with a chalk line held tangentially to these arcs. The plate edge measurements for the length of the plate are made from this line (through I and J). The trammels are set to $R = 1.58$ m and arcs are struck from the line passing through I and J. The final end chalk line is made tangential to K, N and L as shown in Fig. 5.25, stage 3.

4. The shear lines are protected by witness marks made with a centre punch in case the chalk mark becomes indistinct. The witness marks are identified by dashes of white paint as shown in Fig. 5.25, stage 4. The finished outline should be checked for squareness by measuring the diagonal distances.

5.8.7 Marking out flanges

Figure 5.26 shows a flange with eight holes equally spaced. The holes lie on a circle which is known as a *pitch circle*. Note that bolt holes never lie on the vertical centre

Table 5.1 Constants for bolt hole location (flanges*)*

Number of bolt holes	Constant (to be multiplied) by Bolt Circle Diameter or PCD
4	0.4071
8	0.3827
12	0.2588
16	0.1951
20	0.1564
24	0.1305
28	0.1120
32	0.0980
36	0.0872

line. This is to prevent any seepage at the bottom of the flange from corroding the lowest bolt. Such corrosion may weaken the bolt and also cause difficulty in removing the nut and bolt when maintenance is required. The distance between adjacent holes is referred to as the *pitch*. Table 5.1 shows the constants used to calculate the pitch of adjacent holes for various numbers of bolt holes.

For example, if the eight holes in Fig. 5.26 are to be drilled on a pitch circle of 406 mm then the pitch of adjacent holes may be calculated as follows.

The pitch or chord distance of adjacent holes = the PCD × the constant for
8 holes
= 406 mm × 0.3827
= **155.76 mm**

This pitch distance can be stepped off around the PCD using dividers.

Sometimes the number of holes required is not given in the table of constants. In this case, to obtain the constant, multiply the sine of half the subtended angle between any pair of adjacent holes by the PCD, where the subtended angle = 360/(number of holes required). For example, determine the constant for 15 holes.

The subtended angle is 360/15 = 24°
Thus the half angle is 24°/2   = 12°
Therefore the constant   = sine 12° = **0.2079** for 15 holes

Having determined the constant, calculate the pitch as in the previous example.

5.8.8 Further marking out aids

The bevel

This is a useful marking out tool that is frequently used in fabrication work for the marking out of angles or mitres on steel sections. This simple tool, as shown in Fig. 5.27(a), consists of a blade and a base which are set in the required position and to the required angle by a locking screw. The bevel is used for checking, transferring and marking out angles. It is not a measuring device.

The pipe square

This is used to check the correct alignment of pipe flanges where these are required to be perpendicular to the axis of the pipe as shown Fig. 5.27(b). Care must be taken to ensure that the pipe square does not come into contact with the arc welding electrode or earthing clamp during assembly operations.

The scratch gauge

Two examples are shown in Fig. 5.27(c). A simple scratch gauge can be made from a piece of scrap steel of suitable thickness. If more than one line is to be scribed,

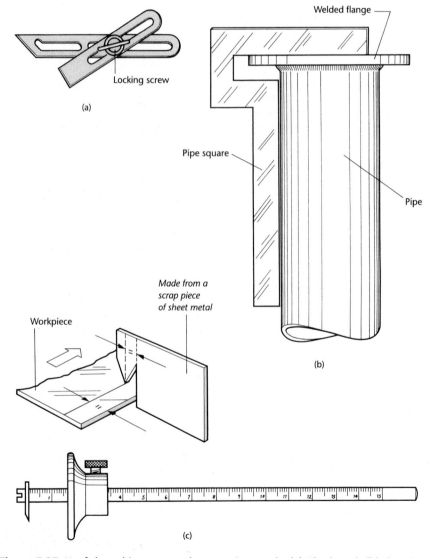

Figure 5.27 Useful marking-out and measuring tools. (a) The bevel; (b) the pipe square; (c) scratch gauges

gauge plate should be used so that the scratch gauge can be hardened and tempered to retain its scribing point. The second example is adjustable and can be set to the required length using the scale on the beam. The scribing point is replaceable and adjustable and is made from hardened and tempered high-carbon steel.

The plumb line

Although this is a marking out aid, a plumb line is indispensable in steel fabrication work for lining up sub-assemblies, as shown in Fig. 5.28. The bottom vessel is securely bolted into position and the position of the top vessel is adjusted so that the top and bottom flanges are equidistant from the plumb line. The flanges should also be checked against the plum line for parallelism.

The tensioned wire

For large fabricated components, a tensioned wire may be used to check straightness and alignment. Piano wire (high carbon spring wire) or stainless steel wire approximately 0.5 mm diameter should be used. When not in use it should be kept on a large diameter reel to prevent it from kinking. Figure 5.29 shows a tensioned wire in use. The wire is supported on pulleys or round steel bar but never over a sharp corner. It is tensioned by hanging weights on both ends of the wire. Sometimes the wire is clamped at one end and only one weight is used.

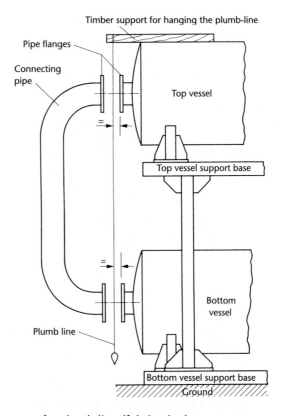

Figure 5.28 The use of a plumb-line (fabrication)

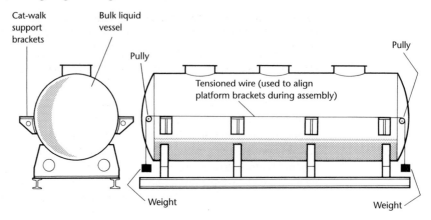

Figure 5.29 Use of a tensioned wire

5.9 The need for templates

Templates or patterns are used in sheet metal and plate-work for the following reasons:

- To avoid repetitive measuring and marking out when making a batch of parts to ensure they are all identical
- To guarantee economical use of material when marking out a sheet of material for a number of components. Various layouts can be tried before cutting takes place
- To act as a guide when flame cutting by hand or for controlling automated profile flame cutting machines
- For checking bend angles and bend radii and contours during forming and rolling operations
- As a precise method of marking out the positions of holes on sheet metal fabrications, plate-work and structural steel-work such as roof trusses.

5.10 The manufacture of templates

Most sheet metal workers make their own templates for small one-off jobs from time to time. However, templates for large fabrications such as bridge, ship, aircraft and structural steel work are made by highly skilled specialist template makers. Such template makers work in an area of the factory known as the template shop pattern shop or loft. A sheet metal pattern shop should not be confused with a casting pattern shop in a foundry although many of the tools used are the same.

The *setting out floor* is normally of wooden floorboards laid across joists. Unlike conventional floors where the floorboards are laid at right angles to the joists, the floorboards of the pattern shop (loft) are laid diagonally. Since the templates for large components are laid out full-size on the pattern shop floor, joints between the diagonal boards cannot be mistaken for layout lines; neither can layout lines coincide with the joints. The pattern shop floor is generally painted matt black so that

Table 5.2 Tools used by template makers

Tools or items	Remarks
Carpenters' saws, planes, hand-brace and bits	For making wooden templates
Joiner's marking gauge	Use for scribing scrieve lines on batten templates for steelwork
Steel tap to measure about 15 m	Steel tapes are available for measuring up to 50 m. Used for marking out large plates and long batten templates
Various size compasses or dividers	These are used for marking small-diameter circles, and for dividing lengths on templates for pitch of hole centres
A pair of trammel heads	These may be used with any length of beam for maring out large radii
A protractor	For measuring and marking angles
Back gauges	The adjustable type are more suitable, used for marking the positions of tail holes at standard 'back mark' dimensions from the heel of the section
Engineer's squares and flat squares	For checking the squareness of two planes or marking a line square to another
A steel straight edge	For marking straight lines up to 2.5 m in length
Hammers, centre and nipple punches	For marking hole centres, and making witness marks
Chalk line and soft chalk. French chalk	For marking long straight lines. French chalk is generally supplied in sticks abut 10 mm square and 100 mm long
Coloured and indelible pencils/crayons	Used for marking instructions and information on templates

chalk-line marks show up clearly. Laying out templates on the pattern shop floor is called *lofting*. Patterns for small sheet metal fabrications are usually laid out on large marking out tables. Typical tools used in the manufacture of templates are listed in Table 5.2.

5.10.1 Materials used for templates

Examples of the materials used in pattern making and some typical applications are listed in Table 5.3. To minimize the weight of templates and for ease in cutting them, they are frequently made from wood or aluminium. However, these are soft materials and if they are to be used repeatedly for positioning drilled holes they will need to be reinforced with hardened steel bushes.

Suitable battens of straight, well-seasoned timber in convenient widths and lengths and approximately 10 mm to 12 mm in thickness are cut to represent the steel members outlined on the template floor. These battens are then laid on the appropriate lines on the loft floor together with paper, hardboard or medium density

Table 5.3 Materials for templates

Material	Applications
Template paper	Outlines for small bent shapes, such as brackets, small pipe bends and bevelled cleats, may be set out on template paper. Used for developing patterns for sheet metal work
Hardboard	Templates for gusset plates to be produced in small quantities
Timber	Used in considerable quantities for steel-work templates. Easy to drill and cut to shape. Whitewood timber strips (battens) up to 153 mm wide and 12.7 mm thickness are used to represent steel members. Plywood used for making templates for use with oxy-fuel gas profiling machines
Sheet metal	Used for making patterns for repetition sheet metal components. Templates for checking purposes. Steel, 3.2 mm thick is used for profiling templates on oxy-fuel gas profiling machines fitted with a magnetic spindle head
Steel plate	Light steel plate fitted with drilling bushes is used as templates for batch drilling of large gusset plates

fibre board (MDF) patterns representing gusset plates and cleat angle connections. All are temporarily nailed to the floor in their exact positions to represent the particular steel structure.

The centres of holes required for making bolted or riveted connections are marked on the assembled templates which are then removed from the floor to be drilled and have the necessary fabrication instructions marked on them. After the holes have been drilled, the whole assembly is replaced on the loft floor for a final check and then carefully stored away until required in the workshop or on site. In the interests of economy, wooden templates and patterns are frequently cut down and re-used when no longer required. Holes can be plugged and drilled in the new positions and instructions can be planed off the surface of the wood.

5.10.2 Information given on templates

Written information for the users of the templates is usually marked on the templates and on the finished parts. For wooden, hardboard or MDF templates the information can be marked on with an indelible pencil or a felt tipped marker pen. Various holes having the same diameter are often distinguished by marking them with rings, triangles or squares around the holes. These symbols are often colour coded. Steel templates are often whitewashed or painted matt-white where information is be marked on them. Typical information may be the:

- Job or contract number
- Size and thickness of plate
- Steel section and length
- Quantity required
- Bending or folding instructions
- Drilling requirements

- Cutting instructions
- Assembly reference marks
- 'Hand' of the part.

5.11 The use of templates

Simple components of a structure do not require to be set out on the template floor (lofted) but can be marked out directly from the drawings at the bench in the fabrication shop. However, even for simple jobs, templates need to be made where a number of identical components are required to ensure uniformity. Let's now consider some examples of the use of templates in greater detail.

5.11.1 Templates for checking

Templates for checking are made out of metal or wood, depending upon their size and life expectancy. Figure 5.30(a) shows some templates suitable for checking angular feature, whilst Fig. 5.30(b) shows a template suitable for checking a corner radius. Fig. 5.30(c) shows a fabricated wooden template being use to check the radius of a rolled steel plate. Finally, Fig. 5.30(d) shows a disc template suitable for checking the diameter and roundness of a sheet metal pipe (ductwork).

5.11.2 Templates for setting out sheet-metal fabrications

Light gauge sheet metal and template-making paper are the materials most frequently used for making templates (patterns) for sheet metal fabrications. For many sheet metal developments it is only necessary to use part patterns which are aligned with datum lines. This not only economizes on the amount of template material used but also economizes on the time taken to make the templates. Figure 5.31 shows examples of the use of the templates used to manufacture a chimney cowl. The finished cowl is shown in Fig. 5.31(a). It is to be manufactured from 1.2 mm sheet steel and hot-dipped galvanized after manufacture and assembly. For safety, the edges of the open ends are wired using a 3.2 mm diameter wire to prevent injuries during installation. The connection flanges are 12 mm wide for spot welding and the side seams are should be 6 mm grooved. Basically, the cowl is a combination of 'tee' pieces comprising of cylinders of equal diameter and is an exercise in the development of *curves of interpenetration*.

Only a part template is required and this is shown as shaded in Fig. 5.31(b) and (c). Since the templates are symmetrical, this part template can be used for both left and right handed curves and for the curves (contours) of all the intersection joint lines for the parts A, B and C whose developed sizes are marked-out in the flat with the appropriate datum lines. The stages in the development of the individual components of the finished assembly are shown in Fig. 5.31(b)(c).

Figure 5.32(a) shows an isometric view of a square to round transformer. Such transformers are used to connect square ducting to circular ducting of equal cross-sectional area. In this example the diameter of the circular duct is 860 mm, the length of one side of the square duct is 762 mm and the distance between the two

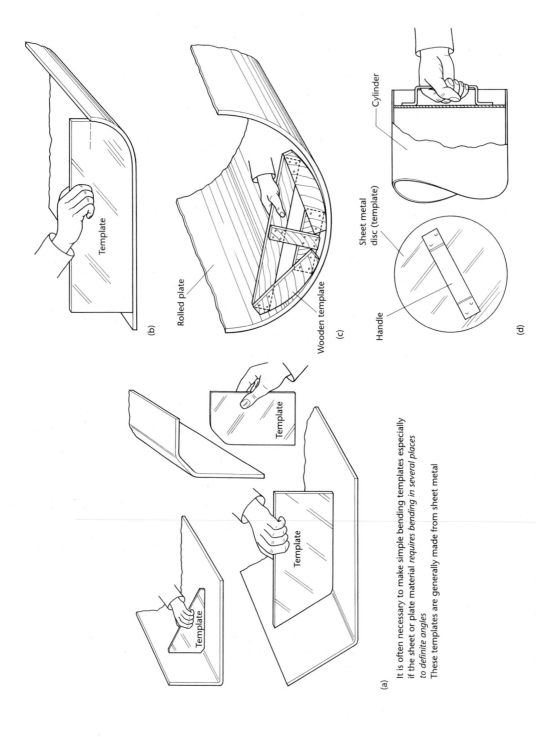

It is often necessary to make simple bending templates especially if the sheet or plate material *requires bending in several places to definite angles*

These templates are generally made from sheet metal

Figure 5.30 The use of templates as a means of checking. (a) Checking angles with a template; (b) checking the contour of a radiused corner; (c) checking the contour of a rolled plate; (d) template used for checking contour of cylindrical work such as ductwork

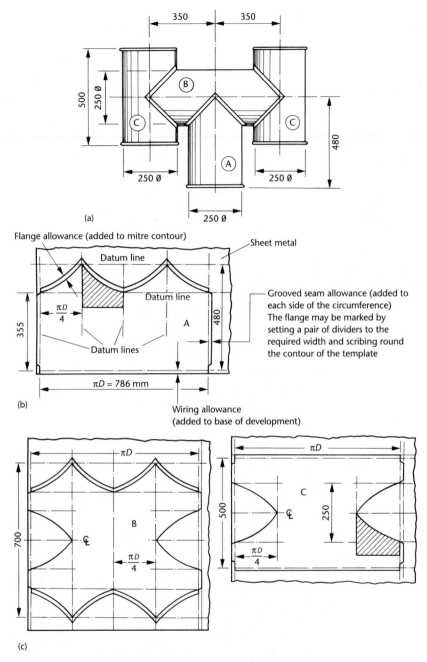

Figure 5.31 An example of the use of templates. (a) Chimney smoke cowl; (b) the developed layout for the part marked 'A' (the template is shown shaded); (c) layout for parts 'B' and 'C'

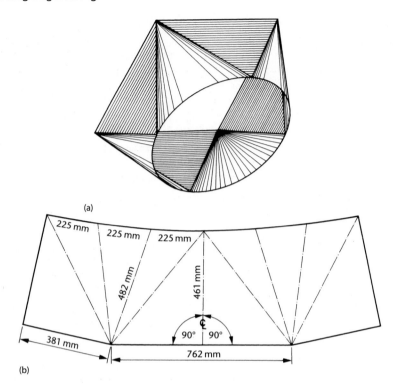

Figure 5.32 Square to round transformer. (a) The complete transformer; (b) pattern or half-template without joint allowances

ducts is 458 mm. The transformer is to be made from galvanized steel sheet 1.2 mm thick. Figure 5.32(b) represents a scale development pattern on which are marked the full-size ordinate dimensions. Such drawings are provided by the drawing office for use by the craftsperson for marking out purposes. Since the transformer is symmetrical only a half template is required. Any necessary allowances for seams and joints must be added to the layout (two required).

5.11.3 Box templates

These are made from wood or MDF and are essentially two flange templates joined together at right-angles as shown in Fig. 5.33. They are used for marking out the positions of the bolt or rivet holes in batches of identical purlin cleats used in structural steelwork. Box templates made from wood or MFD are too soft to be used as drill jigs. They are merely used to guide a suitable nipple punch for marking out the hole centres ready for drilling as shown.

5.11.4 Steel templates (ordinary and bushed)

These are used for the manufacture of steel plate components such as gusset plates in structural steelwork. One plate is accurately marked out with the required hole

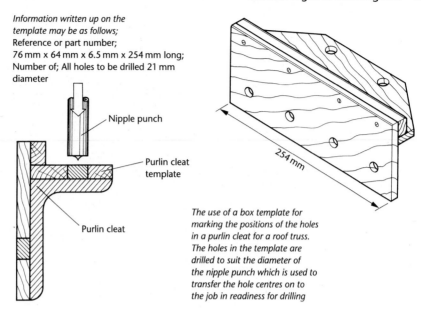

Information written up on the
template may be as follows;
Reference or part number;
76 mm x 64 mm x 6.5 mm x 254 mm long;
Number of; All holes to be drilled 21 mm
diameter

Nipple punch

Purlin cleat
template

Purlin cleat

254 mm

The use of a box template for
marking the positions of the holes
in a purlin cleat for a roof truss.
The holes in the template are
drilled to suit the diameter of
the nipple punch which is used to
transfer the hole centres on to
the job in readiness for drilling

Figure 5.33 Box templates for purlin cleats

positions centre punched ready for drilling. This plate is mounted on top of the stack of plates to be drilled and they are all clamped together ready for drilling as shown in Fig. 5.34(a). For subsequent stacks of plates the top plate becomes a drill jig. After the final stack has been drilled the top plate can be used as part of the finished structure, thus saving on material.

If the template (top plate) is to be retained and used frequently, the holes should be reinforced with hardened steel bushes as shown in Fig. 5.34(b) in order to prevent wear.

5.11.5 Marking-off holes in angle sections

Angle sections are usually cut to length and mitred, where applicable, before the hole positions are marked. Figure 5.35 shows how the hole positions are marked on angle sections.

■ First, a *batten template* is made as shown in Fig. 5.35(a).
■ Second, the template is laid on the larger flange with the heel line of the template on the heel line of the angle as shown in Fig. 5.35(b). The holes marked 'A' on the template are marked through on the surface of the flange with a nipple punch.
■ Third, the angle section is then turned over in the gantry and the template (bottom face up) is laid on the surface of the smaller flange with the heel line of the template in line with the heel of the angle as shown in Fig. 5.35(c), and the *tail holes* marked 'B' on the template are marked through.
■ Alternatively, when the tail holes are not drilled in the template, their positions can be marked off using a set square, chalk and a *back-gauge* as shown in Fig. 5.35(d) and (e)and their centres marked with a centre punch. The back-gauge

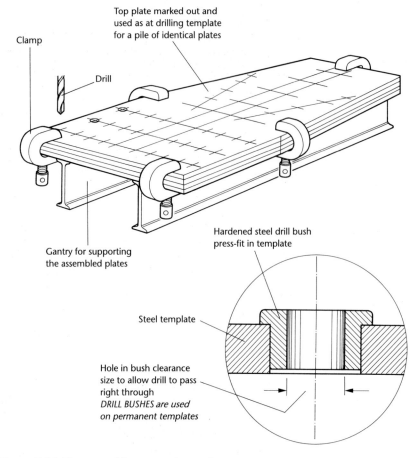

Figure 5.34 The use of large steel templates

shown is of the adjustable type which may be set to a standard back-mark dimension from the heel of the angle. These standard dimensions are usually supplied by the drawing office.

5.11.6 Marking-off holes in channel sections

Channel sections, cut to the required lengths, are placed on a simple gantry with the web horizontal. The wooden template is placed in position, with the information uppermost, and clamped to the channel section. Figure 5.36(a) shows the method of marking off holes in the channel section. The hole positions in the web are marked through the template using a nipple punch as shown in Fig. 5.36(b). Whilst the template is clamped in position, the positions of the tail holes are marked using a set square and chalk, i.e. 'square-off' on the faces of both flanges. When the template has been removed a back-gauge and chalk are used to mark the position of the tail holes from the heel of the flanges as shown on Fig. 5.36(c) and their centres are then marked with a centre punch.

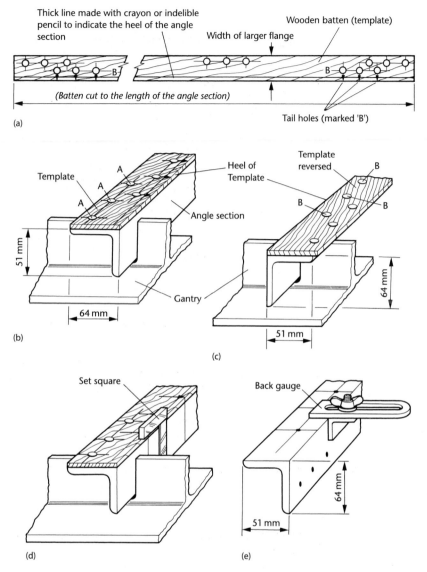

Figure 5.35 Marking-off hole positions in angle sections. (a) Typical template for angle sections; (b) marking-out 'A' holes; (c) marking-out 'B' holes; (d) marking-out 'tail holes' with a try square; (e) use of a 'back gauge'

5.11.7 Marking-off holes in T-sections

One bottom template is generally used to mark off the hole positions on both the flange and the web or stalk, as shown in Fig. 5.37(a). Before positioning the template, a centre line representing half the thickness of the web or stalk are marked on both ends of the T-section. The template (with the instructions uppermost) is laid on the surface pf the flange with the centre line of the template aligned with the centre lines marked on each end of the T-section as shown in Fig. 5.37(b). The template is

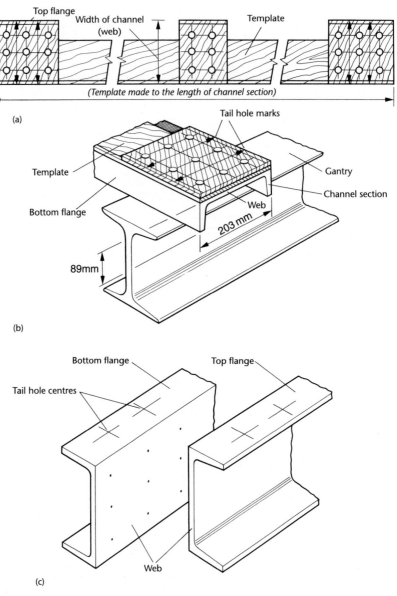

Figure 5.36 Marking-off hole positions in channel sections. (a) Typical template for channel sections; (b) marking-out the web; (c) marking-out the flange

clamped to the T-section so that it can't move whilst the holes are marked through with a nipple punch.

The T-section is then turned on its side on the gantry with the stalk horizontal, and a back-marked line is marked on the face of the stalk as shown in Fig. 5.37(c). The template is laid on the face of the stalk with the centre-line of the tail holes in line with the back-marked line and clamped in position. The tail holes are marked through to the face of the stalk with a nipple punch.

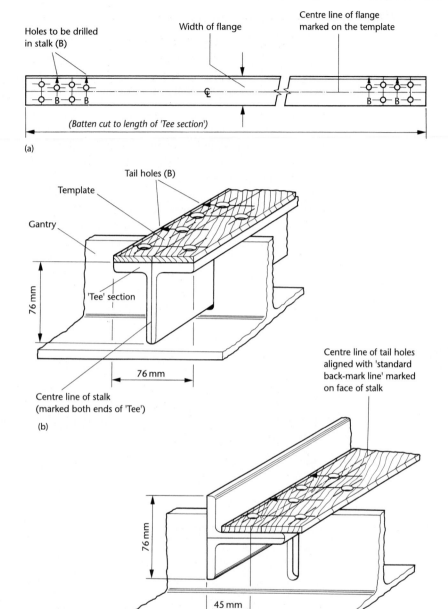

Figure 5.37 Marking-off hole positions in 'Tee' sections. (a) Typical template for Tee sections; (b) marking-out the flange; (c) marking-out the web (stalk)

5.11.8 Marking-off holes in beams and columns

The procedure for marking off hole positions in beams and columns is shown in Fig. 5.38. Beams lie horizontally and have a *top flange* and a *bottom flange*, whilst column stands vertically and has an *outside flange* and an *inside flange*. Figure 5.38(a) shows a typical wooden template for marking out columns and universal beams.

■ First, centre lines are marked on the web at both ends of the section for the purpose of aligning with the centre line marked on the template for location purposes.
■ Second, the template is laid on the surface of the top flange with the respective centre lines aligned as shown in Fig. 5.38(b) and clamped securely in position whilst hole positions other than those marked 'A' are transferred with a nipple punch.
■ Third, the above procedure is repeated for the bottom flange.
■ Finally, the beam is laid on its side with the web horizontal and the hole positions marked 'A' are transferred onto the web as shown in Fig. 5.38(c) with a nipple punch.

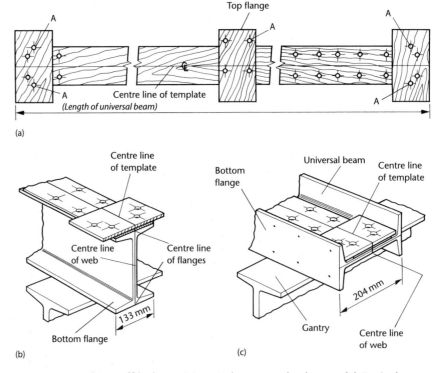

Information marked on template: Plain holes and A holes to be drilled on the top flange. Plain holes and B holes to be drilled on the bottom flange C holes to be drilled in web. All flange holes to be drilled 14mm diameter. Web holes to be drilled 17.5mm diameter

Figure 5.38 Marking-off hole positions in beams and columns. (a) Typical template for columns and universal beams; (b) marking-out the flange; (c) marking-out the web

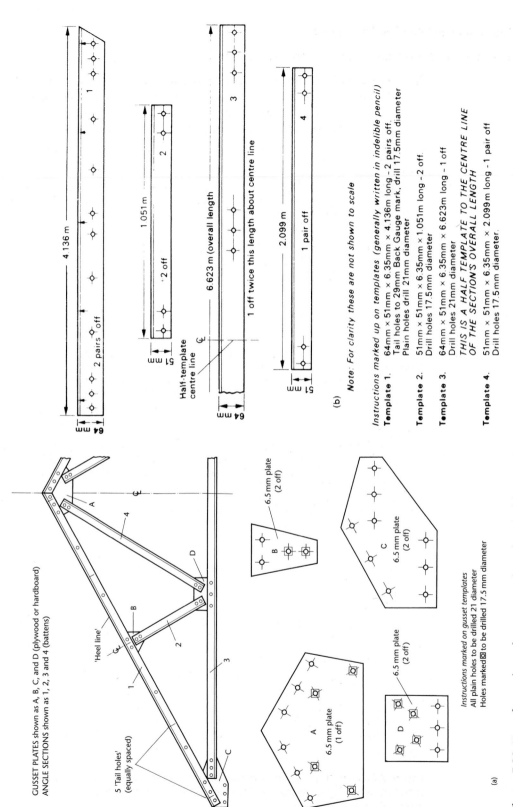

GUSSET PLATES shown as A, B, C, and D (plywood or hardboard)
ANGLE SECTIONS shown as 1, 2, 3 and 4 (battens)

'Heel line'

5 'Tail holes'
(equally spaced)

6.5 mm plate
(2 off)
B

6.5 mm plate
(2 off)
C

6.5 mm plate
(1 off)
A

6.5 mm plate
(2 off)
D

Instructions marked on gusset templates
All plain holes to be drilled 21 diameter
Holes marked ⊠ to be drilled 17.5 mm diameter

(a)

4.136 m

2 pairs off

64 mm

1.051 m

'2 off'

51 mm

6.623 m (overall length)

1 off twice this length about centre line

Half-template
centre line

64 mm

2.099 m

1 pair off

51 mm

(b)

Note: *For clarity these are not shown to scale*

Instructions marked up on templates (generally written in indelible pencil)

Template 1. 64mm × 51mm × 6.35mm long – 2 pairs off.
Tail holes to 29mm Back Gauge mark, drill 17.5mm diameter
Plain holes drill 21mm diameter

Template 2. 51mm × 51mm × 6.35mm ×1.051m long – 2 off.
Drill holes 17.5 mm diameter

Template 3. 64mm × 51mm × 6.35m × 6.623m long – 1 off
Drill holes 21mm diameter
*THIS IS A HALF TEMPLATE TO THE CENTRE LINE
OF THE SECTION'S OVERALL LENGTH*

Template 4. 51mm × 51mm × 6.35mm × 2.099m long – 1 pair off
Drill holes 17.5 mm diameter

Figure 5.39 Use of templates for structural steelwork: (a) Typical layout of templates for simple roof truss; (b) Batten templates for the angle sections

5.11.9 Laying out templates for a roof truss

From the information supplied on the drawings, lines representing the members of the roof truss are marked out on the floor of the lofting shop. To ensure proper alignment of the holes through the templates, the battens (templates representing the angle sections) are drilled and placed on the plate templates in the correct position on the lines on the floor. The holes in the plate templates are marked from the hole positions in the battens and then drilled. The 'tail' or 'back' holes are marked in position for the purlin cleats and shoe connections.

After the templates have been checked for accuracy by replacing all of the parts on the layout lines, they are marked up ready for the erection crew on site. The edge of each batten template to be set against the heel of the angle section is marked with a line next to that edge. Figure 5.39 shows a layout for a simple roof truss together with information on how the various templates employed are marked up. Box templates are used for the purlin cleats as described in section 5.12.3.

5.12 Condition and care of equipment

Marking out equipment should be kept in good condition if inaccuracies are to be avoided.

■ As has been mentioned previously, the points of scribers and dividers should be kept needle sharp by regular dressing with a fine oil slip. This is shown in Fig. 5.40(a). Do not sharpen by grinding, as the heat generated will soften the scribing point.

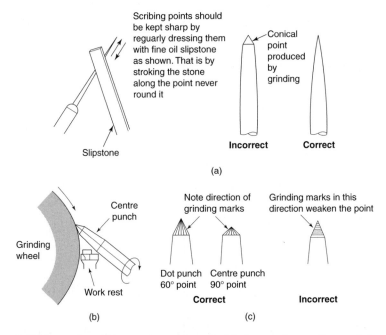

Figure 5.40 Care of marking-out equipment: (a) sharpening scriber points; (b) sharpening a centre punch. And dot punches; (c) correct and incorrect dot and centre punch point configurations

- When sharpening the point of a dot punch or a centre punch, the punch is presented to the abrasive wheel as shown in Fig. 5.40(b). This ensures that the grinding marks run down the point as shown in Fig. 5.40(c) and not round it.
- Steel rules should be kept clean. The datum end should be protected and *never* used as a makeshift screwdriver or for shovelling swarf out of the T-slot of a machine tool bed. The edges of a rule must also be kept in good condition if it is to enable straight lines to be scribed or it is to be used as a straight edge.
- Try squares must also be treated carefully and cleaned and boxed when not in use. They should never be dropped, mixed with other tools or used for any purpose other than that for which they are designed.
- Angle plates must be kept clean and free from bruises. Bruises not only prevent proper contact between the angle plate and the work it is supporting, they also cause damage to the surface of the marking out table on which they are supported.
- Surface plates and marking out tables must also be treated with care as they provide the datum from which other dimensions are taken.

5.1 Measuring equipment

   a) List the most important features of a steel rule. State briefly how it should be cared for to maintain its accuracy.

   b) Briefly describe the main causes of reading error when using a steel rule and how these errors can be minimized.

   c) Give an example where it would be appropriate to use a steel rule and where it would be appropriate to use a steel tape.

   d) Describe the difference between a *line-system* and an *end-system* of measurement.

   e) With the aid of sketches show how an engineer's try-square is used to check the squareness of a rectangular sheet metal blank.

   f) With the aid of sketches show how a plain bevel protractor is used to measure angles.

5.2 Measuring instruments

   a) Sketch a micrometer caliper and:

      i) Name its more important features

      ii) Show how the scales are arranged for metric readings

      iii) Show how the scales are arranged for 'inch' readings

   b) Write down the micrometer readings shown in Fig. 5.41(a).

   c) Sketch a vernier caliper that can be used for internal and external measurements, and also depth measurements. Name its more important features.

   d) Write down the vernier readings shown in Fig. 5.41(b).

5.3 Marking out and marking out equipment

   a) List the reasons for marking out components ready for manufacture.

   b) List the advantages and limitations of manual marking out in terms of accuracy and possible damage to the surfaces of the workpiece. Explain why it is sometimes better to use a pencil than a scriber.

   c) With the aid of sketches explain the difference between dot, centre and nipple punches and give examples where each would be used.

   d) i) Sketch a pair of dividers, name the more important features and explain with the aid of simple sketches typical uses of dividers.

      ii) Sketch an odd-leg caliper, name the more important features and explain with the aid of sketches typical uses of this instrument.

      iii) Sketch a typical trammel (beam compass), name the more important features and explain with the aid of sketches typical uses of this instrument.

5.4 Use of tools providing guidance and support

   a) With the aid of sketch examples of the following items of equipment and explain when and how they are used:

      i) Straight-edge (rule)

      ii) Box square

      iii) Try square.

   b) With the aid of a sketch show how a thing plate template can marked out on a surface plate using a scribing block and an angle plate to support the plate perpendicular to the surface plate which provides a datum surface.

   c) List and describe the various types datum used as a basis for marking out, giving examples where each would be used.

   d) With the aid of sketches explain the difference between a *try-square* and a sheet metal worker's *flat-square.*

5.5 Techniques for marking out

   a) When marking out a large steel plate describe, with the aid of sketches, how:

      i) Trammels may be used to construct a right-angle

      ii) A steel tape may be used to construct a right-angle

   b) Describe with the aid of sketches how a rectangle 1.5 m × 2.0 m can be marked out on a steel plate that has only one straight edge and no corners that are right-angles, using a steel tape and trammels.

   c) With reference to Table 5.1 calculate the chordal distance between the centres of adjacent holes for 12 holes on a pitch circle of 525 mm diameter.

   d) Using the technique described in section 5.8.7, calculate the chordal distance between the centres of adjacent holes for 15 holes on a pitch circle of 575 mm diameter.

   e) With the aid of sketches illustrate the following marking out devices and show how they are used.

      i) Pipe square

      ii) Scratch gauge

      iii) Plumb line

      iv) Tensioned wire.

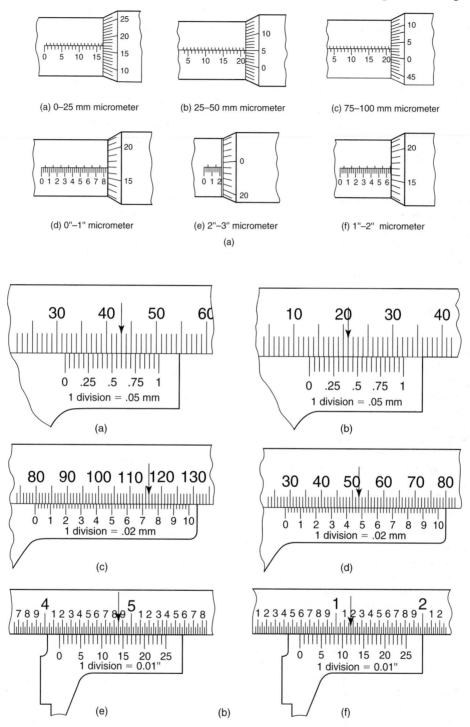

(a) 0–25 mm micrometer

(b) 25–50 mm micrometer

(c) 75–100 mm micrometer

(d) 0"–1" micrometer

(e) 2"–3" micrometer

(f) 1"–2" micrometer

(a)

0   .25   .5   .75   1
1 division = .05 mm

(a)

0   .25   .5   .75   1
1 division = .05 mm

(b)

0 1 2 3 4 5 6 7 8 9 10
1 division = .02 mm

(c)

0 1 2 3 4 5 6 7 8 9 10
1 division = .02 mm

(d)

0   5   10   15   20   25
1 division = 0.01"

(e)

(b)

0   5   10   15   20   25
1 division = 0.01"

(f)

Figure 5.41

5.6 Templates (patterns)

a) List FIVE advantages for using templates in sheet metalwork.

b) List the essential information that should appear on a template.

c) With the aid of sketches briefly describe THREE different types of template and show how they are used.

d) Describe how the profile of the template can be transferred to workpiece material.

e) With the aid of sketches briefly describe how a template can be used for checking.

5.7 Minimizing inaccuracies when marking out

a) Explain what is meant by the term *parallax errors* when marking out using a steel rule. How can such errors be minimized?

b) Describe two ways in which a scribed line can be made to show up more clearly.

c) Describe TWO ways (other than those in (a) and (b)) by which marking out inaccuracies can be minimized.

5.8 Care of marking out tools and equipment

a) With the aid of sketches describe how the scribing points of the following marking out tools should be sharpened:

  i) Divider points

  ii) Dot-punch point.

b) Describe how marking out tools and measuring instruments should be cared for and stored in order to maintain their accuracy and to maintain them in good condition.

6 Material removal

When you have read this chapter, you should understand:

- The principles of metal-cutting and the importance of cutting tool angles
- The types, uses and care of hand tools
- The power hacksaw and its use
- The types and uses of drilling machines
- The types and uses of twist drills and associated cutting tools used in conjunction with drilling machines
- The types and uses of routing machines
- The principles of cutting sheet metal by shearing
- The types and uses of hand shears and shearing machines
- The types and safe use of portable power tools
- Abrasive wheel-cutting machine
- The correct selection, use and care of grinding wheels
- Principles of blanking (stamping) and piercing sheet metal in hand (fly) and power presses
- Flame-cutting (manual and machine)

6.1 Cutting tool principles

The tools used for cutting sheet metal, plate and structural sections will depend upon the thickness of the metal being cut and the quantities (batch size) required. We will commence by discussing the basic principles of metal-cutting tools.

6.1.1 The metal-cutting wedge

Possibly the first controlled cutting operation you performed was the sharpening of a pencil with a penknife. In the absence of formal instruction you soon found out (by trial and error) that the knife had to be presented to the wood at a suitable angle if success was to be achieved, as shown in Fig. 6.1.

If the blade is laid flat on the wood it just slides along without cutting. If you tilt the blade at a slight angle, it will bite into the wood and start to cut. If you tilt the blade at too steep an angle, it will bite into the wood too deeply and it will not cut properly. You will also have found that the best angle will vary between a knife that is sharp and a knife that is blunt. A sharp knife will penetrate the wood more easily at a shallower angle and you will have more control. A blunt knife would need to be presented at a steeper angle before it will cut, causing it to 'dig in' out of control. But look at that knife blade – it is the shape of a wedge. In fact, all cutting tools are wedge-shaped (more or less).

6.1.2 The angles of a wedge-shaped cutting tool and their terminology

So let's now look at the angles of a typical metal-cutting tool and how they relate one to another.

Clearance angle

We have already seen that for our knife to cut, we need to incline it at an angle to the surface being cut, and that we have to control this angle carefully for effective cutting. This angle is called the *clearance angle* and we give it the Greek letter 'beta' (β). All cutting tools have this angle. It has to be kept as small as possible to prevent the tool from 'digging in'. At the same time it has to be large enough to allow the tool to penetrate the workpiece material. The clearance angle will vary slightly depending upon the cutting operation and the material being cut. It is usually about 5° to 7°.

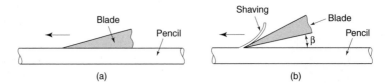

(a)   (b)

Figure 6.1 The clearance angle (β): (a) no clearance ($\beta = 0$) – the blade skids along the pencil without cutting; (b) clearance ($\beta > 0$) – the blade bites into the pencil and cuts

Wedge angle

If, in place of our pencil, we used our knife to cut a point on a piece of soft metal such as copper our knife would soon become blunt. If you examined this blunt edge under a magnifying glass, you would see that the cutting edge has crumbled away. To cut metal successfully, the cutting edge must be ground to a less acute angle to give it greater strength when cutting metal, as shown in Fig. 6.2.

The angle to which the tool is ground is called the *wedge angle* or the *tool angle* and it is given the Greek letter 'gamma' (γ). The greater the wedge angle, the stronger will be the tool. Also the greater the wedge angle the quicker the heat generated by the cutting process will be conducted away from the cutting edge. This will prevent the tool overheating and softening, and help to prolong the tool life. Unfortunately, the greater the wedge angle, the greater will be the force required to make the tool penetrate the workpiece material. The choice of the wedge angle becomes a compromise between all these factors.

Rake angle

To complete the angles associated with cutting tools, reference must be made to the *rake angle*. This is given the Greek letter 'alpha' (α). This rake angle is very important, for it alone controls the geometry of the chip formation for any given material and, therefore, it controls the mechanics of the cutting action of the tool. The relationship of the rake angle to the angles previously discussed is shown in Fig. 6.3.

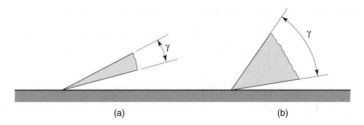

(a)   (b)

Figure 6.2 Wedge (tool) angle (γ): (a) blade sharpened for cutting wood; (b) blade sharpened for cutting metal

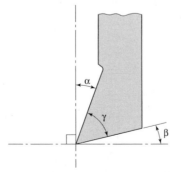

Material being cut	α
Cast iron	0°
Free-cutting brass	0°
Ductile brass	14°
Tin bronze	8°
Aluminium alloy	30°
Mild steel	25°
Medium-carbon steel	20°
High-carbon steel	12°
Tufnol plastic	0°

Figure 6.3 Cutting tool angles: α = rake angle; β = clearance angle; γ = wedge or tool angle

Increasing the angle increases the cutting efficiency of the tool and makes for easier cutting. Since increasing the rake angle reduces the wedge angle, increased cutting efficiency is gained at the expense of tool strength. Again a compromise has to be reached in achieving a balance between cutting efficiency, tool strength and tool life. Let's now consider how the basic principles of the metal-cutting wedge can be applied to a range of tools used by fabrication engineers and sheet-metal workers.

6.2 The application of basic cutting tool angles to hand tools

When using hand tools at the bench, it is necessary to hold the workpiece material securely so that the cutting forces can be resisted and the cutting process accurately controlled. To do this the workpiece is usually secured in a vice.

6.2.1 The bench vice

Figure 6.4(a) shows a typical parallel jaw vice. It is often fitted with a quick-release device that frees the screw from the nut so that the vice can be opened and closed quickly when changing between components of different widths in order to save time. To keep the vice in good condition the following rules should be observed:

- Clean and oil the screw and nut regularly.
- Clean and oil the slideways regularly.
- Ensure that the vice is substantial enough for the work in hand.
- Heavy hammering and bending should be confined to the anvil and not performed on the vice.
- When cutting with a cold chisel the thrust of the chisel should be towards the fixed jaw.
- Never hammer on the top surface of the slide.

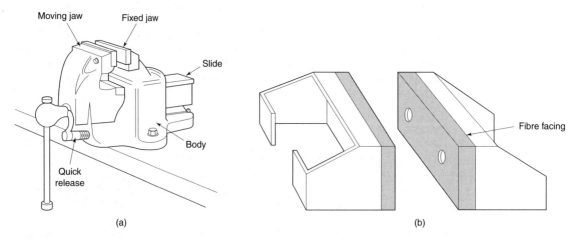

Figure 6.4 Fitter's vice (a), vice shoes (b)

6.2.2 Vice shoes

The jaws of a vice are serrated to prevent the work from slipping. However, these serrations can mark and spoil a finished surface. If the vice is only to be used for light work some sheet-metal workers have the jaws surface ground smooth. However, if the vice is to be used for both rough and fine work then *vice shoes* should be used. These can either be cast from soft metal such as lead or faced with sheet fibre as shown in Fig. 6.4(b). These vice shoes are clipped over the serrated jaws as and when required for fine work and removed for rough work.

6.2.3 Using a vice

The vice should be securely bolted to the work bench and should be positioned so that the fixed jaw is just clear of the edge of the bench. This allows long work to hang down clear of the bench. Work should be positioned in the vice so that the major cutting forces acting upon the work are directed towards the fixed jaw. The work should always be held in the vice with a minimum of overhang as shown in Fig. 6.5. There is always a possibility that work protruding too far out of the vice will bend under the force of the cut and also that the work will vibrate and produce an irritating squealing sound.

6.2.4 Cold chisels

The basic wedge angle described above applies to all metal-cutting tools. Figure 6.6(a) shows how the cutting edge of a chisel forms a metal-cutting wedge with rake

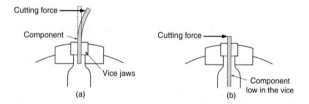

Figure 6.5 Positioning work in the vice: (a) incorrect – if the cutting force is applied too far from the vice jaws, it will have sufficient 'leverage' to bend the component; even when the force is too small to bend the component it will make it vibrate and give off an irritating squealing noise; (b) correct – when the component is held with the least possible overhang, the cutting force does not have sufficient 'leverage' to bend the component or to make it vibrate

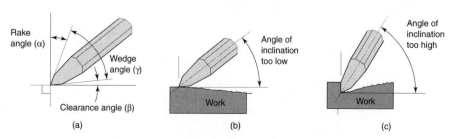

Figure 6.6 The cold chisel

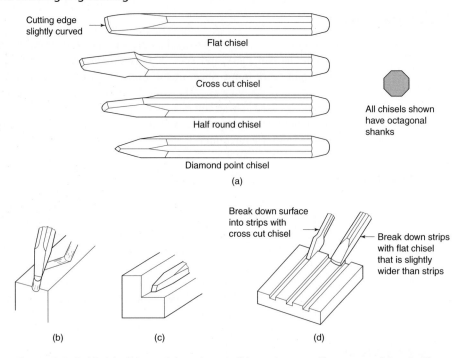

Figure 6.7 Cold chisel types (a) and uses: (b) cutting an oil groove with a half-round chisel; (c) squaring out a corner with a diamond point chisel; (d) chipping a flat surface

and clearance angles, and how the angle at which the chisel is presented to the work (angle of inclination) affects the cutting action of the chisel.

Figure 6.6(b) shows that when the chisel is presented to the work so that the angle of inclination is too small the rake angle becomes larger and the clearance angle disappears. This prevents the cutting edge of the chisel from biting into the workpiece and the cut becomes progressively shallower until the chisel ceases to cut.

Figure 6.6(c) shows that when the chisel is presented to the work so that the angle of inclination is too large, the rake angle is reduced and the clearance angle becomes larger. This results in the cutting edge of the chisel 'digging in' so that the cut becomes progressively deeper.

6.2.5 Types and use of cold chisels

Figure 6.7 shows a selection of cold chisels and some typical 'chipping' applications. Chisels are used for rapidly breaking down a surface by hand but the finish is poor and the accuracy is low. The safe use of chisels was discussed in Section 1.10.4.

When used for cutting sheet metal, the flat chisel must be held at a slight angle to the line of cut as shown in Fig. 6.8(a). The reasons for inclining the chisel are:

- To provide a shearing angle.
- To make the chisel move along the line of cut smoothly and continuously.

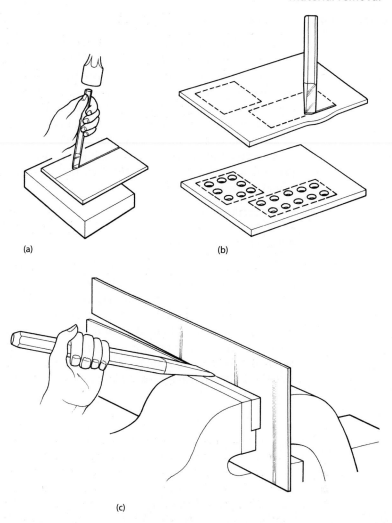

Figure 6.8 The use of a chisel for cutting thin plate. (a) Cutting sheet metal with a flat chisel; (b) cutting slots or apertures; (c) Cutting sheet metal supported in a vice

If the chisel is held vertically, a separate cut is made each time a hammer blow is delivered and the 'line' becomes a series of irregular cuts. A block of soft cast iron is generally used to support the sheet metal whilst it is being cut. The chisel is also inclined when cutting slots or aperture of various shapes and sizes. In this case the removal of the material is simplified by punching or drilling a series of holes (*chain drilling*) as near together as possible, before the chisel is used. The advantages of pre-drilling or pre-punching the sheet metal are shown in Fig. 6.8(b).

Figure 6.8(c) shows how sheet metal may be cut between a chisel and the vice jaws to create a shearing action. Care must be taken to ensure that the line along which the cut is to be made is as near to the top of the vice jaws as possible to prevent the sheet metal being bent and the cut edge badly burred over.

6.2.6 Hammers

In the previous section, we saw that hammers were used to drive the chisel through the material being cut. There are various types and sizes of hammer use in metal-working and the parts of the hammer are shown in Fig. 6.9(a). The most commonly used type of hammer used in metal-working is the ball-pein hammer as shown in Fig. 6.9(b). If a hammer is too big, it will be clumsy to use and proper control cannot be exercised. If a hammer is too small it has to be wielded with too much effort, so that again proper control cannot be exercised. In both these instances the use of the incorrect size of hammer will result in an unsatisfactory job, possible damage to the work and possible injury to the user. Before using a hammer you must check it to make sure of the following:

- The handle is not split.
- The head is not loose.
- The head is not cracked or chipped.

When using a hammer never 'strangle' it by holding it too near the head. It should be held as shown in Fig. 6.9(e). A hammer is usually used to strike other tools such as chisels, drifts, and centre punches as shown in Fig. 6.9(f). When a hammer is used to strike a component, such as a key or a location dowel, directly as shown in Fig. 6.9(f), care must be taken so that the component is not bruised. In order to avoid damage to the component either a soft metal drift (brass or aluminium) must be inserted between the component and the hammer head or, alternatively, a soft-faced hammer should be used.

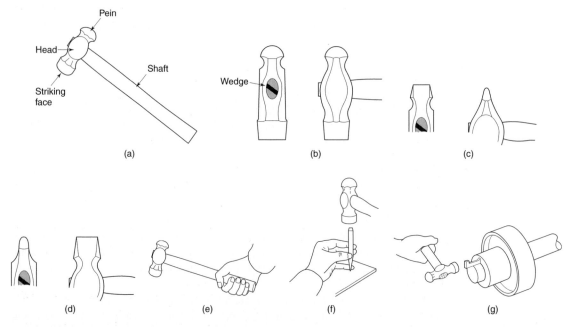

Figure 6.9 Hammer construction (a), ball pein type (b), cross pein type (c), straight pein type (d), correct grip (e), used with another tool (f), used directly (g)

6.2.7 Files

Files are used for a wide range of operations from 'roughing down' to final finishing on a wide variety of components and surfaces. For this reason there is a wide variety files available in different shapes and sizes. The main features of a file are shown in Fig. 6.10. The *tang* is provided for the purpose of fitting the file into a wooden handle.

The point, body and shoulder of the file are hardened and tempered, whilst the tang is usually left soft to prevent it from inadvertently being snapped off.

A file is specified by its *length, grade of cut*, and *shape*. The grade or cut of a file depends upon its length: the shorter the file, the smaller will be the pitch of its teeth. Table 6.1 shows the pitch range and application of normally available files.

> For safety reasons files should never be used without a handle of the correct size being fitted

There is a wide variety of files available for metal-working purposes; the type and shape selected is generally governed by a particular application. Figure 6.11 shows a range of files and typical uses. Like all cutting tools, files must have teeth with correctly formed cutting angles. A *single-cut* file as shown in Fig. 6.12(a) has a series of parallel teeth formed at an angle of about 70° to the axis of the file and is usually used for soft materials. The most widely used files are *double-cut* as shown in Fig. 6.12(b). These have a second series of parallel teeth formed in the opposite direction and crossing the first set of teeth at an angle of about 45°. Double-cut files reduce the effort required to remove a given volume of metal from the workpiece.

6.2.8 Filing techniques

In order to generate a plane surface by filing, the file must be moved parallel to the plane of the required surface. This carefully controlled movement depends solely on

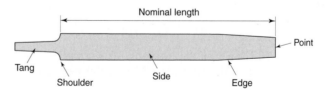

Figure 6.10 Engineer's file

Table 6.1 File grades

Grade	Pitch (mm)	Use
Rough	1.0–1.3	Soft metals and plastics
Bastard	1.6–0.65	General roughing out
Second cut	1.4–0.60	Roughing out tough materials. Finishing soft materials
Smooth	0.8–0.45	General finishing and draw filing
Dead smooth	0.5–0.25	Not often used except on tough die steels where high accuracy and finish is required

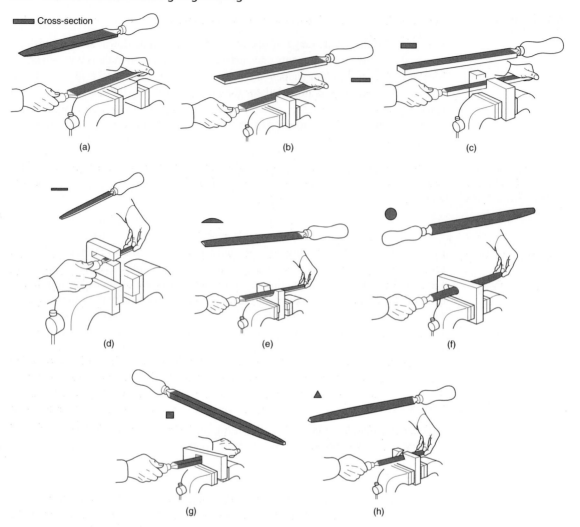

Figure 6.11 Types of file and their applications: (a) flat file; (b) hand file; (c) pillar file; (d) ward file; (e) half-round file; (f) round file; (g) square file; (h) three-square file

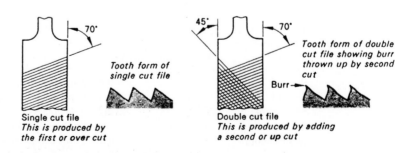

Figure 6.12 Types of file cut

the muscular co-ordination of the craftsperson. This involves the following essential basic principles.

Height of the vice

For correct control, the top of the vice jaws should be level with the forearm of the craftsperson when the forearm is held parallel to the ground as shown in Fig. 6.13(a).

Correct stance

The file can only be properly controlled when the body of the user is correctly balanced. Figure 6.13(b) shows how the user's feet should be positioned relative to the vice and workpiece in order to achieve this.

Filing (downward) pressure should only be exerted by the craftsperson when the file is pushed on the forward stroke. This filing pressure is the released whilst the file is returned ready for the next forward stroke

Application of pressure

An important point to remember when filing metal components is that files are designed to cut in a forward direction. Therefore wherever possible, try to use the whole length of the file to ensure it wears evenly.

Correct grip

The fingertips play an important part in keeping the file straight whilst filing a flat surface Care must be taken not to 'rock' the file during the forward stroke otherwise a convex surface, rather than a plane surface, will be generated. Figure 6.14 shows how the file should be held for some typical operations. To ensure flatness remember that during each cutting stroke of the file, the weight must be gradually transferred from the front hand to the back hand.

Draw filing

As draw filing is employed as a finishing operation, the teeth of the file must be free from being loaded with particles of metal otherwise scratches will occur on the workpiece surface. Regular use of a *file card* (a stiff wire brush) will keep the teeth of the file clean and free from particles of metal. For final finishing it is advisable to

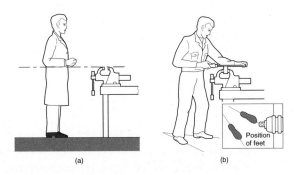

(a)   (b)

Figure 6.13 Use of a file: (a) top of the vice should be in line with the forearm held parallel to the ground; (b) position of feet and balance

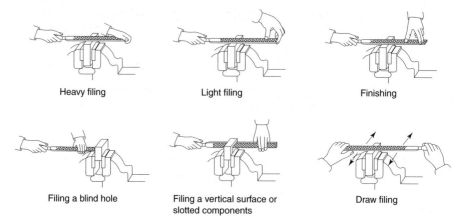

Heavy filing

Light filing

Finishing

Filing a blind hole

Filing a vertical surface or
slotted components

Draw filing

Figure 6.14 Correct grip for different file applications

rub chalk on the file in order to fill the spaces between the teeth. This helps to prevent scratching and also gives a finer finish.

6.2.9 Care of files

Files should be treated with care. Files that are badly treated are hard to use and leave a poor finish and poor accuracy.

- Keep all files in a suitable rack. Do not jumble them up in a draw or keep them with other tools as this will damage the teeth.
- Keep your files clean with a *file card* before putting them away so that they are always ready for use. The file card removes particles of metal trapped between the teeth which would otherwise reduce the rate of metal removal and score the surface of the workpiece.
- Never use new files on steel. This will chip the teeth and render the file useless. Always 'break in' a new file on a softer metal such as brass or bronze.
- Never file quickly; this only wears out the file and the user. Slow, even strokes using the full length of the file are best.
- As stated earlier, files cut only on the forward stroke. The downward pressure should be relieved on the return stroke to reduce wear on the teeth. Do not lift the file from the work on the return stroke. Keeping the file in contact with the work helps to remove any particles of metal that lie between the teeth and also maintains your balance and rhythm that are essential to the production of a flat surface.

6.2.10 Hacksaws

Figure 6.15(a) shows a typical metal-worker's hacksaw with an adjustable frame that will accept a range of blade sizes. For best results the blade should be carefully selected for the work in hand. It must be correctly fitted and correctly used.

Figure 6.15(b) shows the main features and dimensions of a hacksaw blade. Like all metal-cutting tools, the teeth of a hacksaw form a metal-cutting wedge with rake and

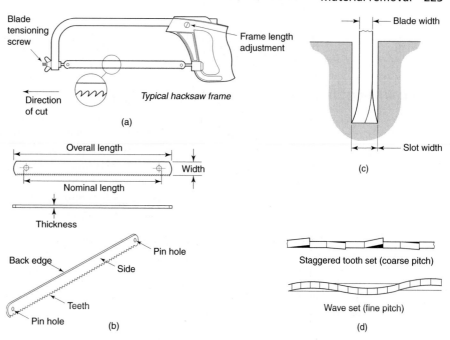

Figure 6.15 The hacksaw and its blades: (a) engineer's hacksaw showing typical hacksaw frame; (b) hacksaw blade; (c) the effect of set; (d) types of set

clearance angles. To prevent the blade jamming in the slot that it is cutting, side clearance must be provided by giving the teeth of the blade a 'set' as shown in Fig. 6.15(c).

There are two ways in which this set may be applied. For coarse pitch blades for general workshop use, the teeth are bent alternatively to the left and right with each intermediate tooth being left straight to clear the slot of swarf. Some blades leave every third tooth straight. For fine tooth blades used for cutting sheet metal and thin walled tube, the edge of the blade is given a 'wave' set. Both types of set are shown in Fig. 6.15(d).

6.2.11 Use of the hacksaw

- The coarser the pitch of the teeth the greater will be the rate of metal removal and the quicker the metal will be cut. However, there must always be a minimum of three teeth in contact with the metal as shown in Fig. 6.16(a).
- Thick material should be broken down into shorter surfaces as shown in Fig. 6.16(b).
- 'Rigid' or 'all-hard' high-speed steel blades give the best results but tend to break easily in unskilled hands. 'Flexible' or 'soft-back' blades are best for persons when they are undergoing training.
- The teeth of the blade should face the direction of cut and the blade should be correctly tensioned. After the slack has been taken up, the wingnut should be given at least one more full turn.
- The correct way to hold and use hacksaw is shown in Fig. 6.16(c).
- The rate of sawing should not exceed 50 to 60 strokes per minute.

■ With use, the blade gradually loses its 'set' and the slot being cut will become narrower. For this reason never use a new blade in a slot started with an old blade. The new blade will jam and break. Always start a new cut with a new blade.

6.2.12 Sawing sheet metal

The depth to which a hacksaw can cut is limited to the depth of the frame. Long narrow cuts are often required in sheet metal and for this purpose the blade can be turned through 90° as shown in Fig. 6.17(a). It is not so easy to exert a downward force on the blade with the saw in this position but this is not so important when cutting sheet material of limited thickness.

An ordinary hacksaw blade is useless for cutting profiles and for this application a *tension file* should be used. This is a long, thin, round file that is kept in tension

Material	Pitch (mm) solid metal	Pitch (mm) tube and sheet
Ferrous metal	1.4–1.6	0.8
Non-ferrous metal	1.8–2.1	1.0–1.2

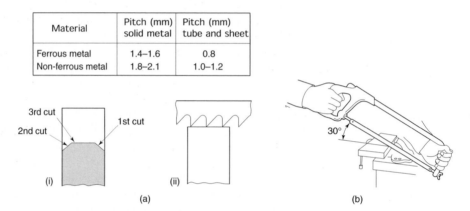

Figure 6.16 The hacksaw blade: (a) blade selection – (i) a wide component should be broken down in a series of short cuts; (ii) the pitch of the blade should be chosen so that at least three teeth are in contact with the workpiece all the time; (b) use of a hacksaw

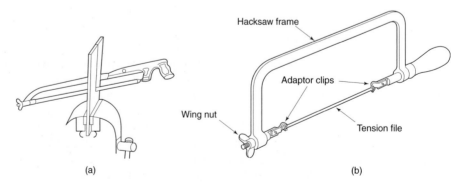

Figure 6.17 Cutting sheet metal: (a) the blade turned through 90° to cut sheet metal; (b) tension file – when the wingnut is tightened, the frame distorts and is put into a state of stress; in trying to spring back to its original shape it exerts a tensile (pulling) force on the blade or file, which is now in a state of tension

by the saw frame as shown in Fig. 6.17(b). The tension file is held in the frame by means of adaptor clips.

6.2.13 Power hacksaw

A typical power hacksaw is shown in Fig. 6.18. Unlike the fitter's hacksaw, where cutting takes place on the forward stroke and the teeth of the blade face forward, in most power hacksaws cutting takes place on the return stroke and the teeth of the blade face backwards. By pulling the blade through the work, the blade is kept in a state of tension. This prevents the blade buckling and snapping.

The blade is raised on the forward (non-cutting) stroke to prevent it rubbing. If the blade were allowed to rub, it would quickly become blunt. A dashpot is provided to prevent the blade being dropped onto the work. There is no power feed mechanism and the rate of in-feed is dependent on the force of gravity acting on the bow and blade assembly, the resistance of the work material to cutting and the area of the cut under the blade.

A coolant should always be used to keep the blade cool and to flush away the chips. These might otherwise clog in the cut and the teeth of the blade, causing the blade to jam and break. The workpiece material should be securely fastened in the machine vice provided before cutting commences.

Since the saw is only cutting on the return stroke it is only doing useful work for 50% of its operating time. For larger machines used for cutting heavy structural members, horizontal bandsaws are used. The blade runs continuously in the same direction so that cutting is taking place for 100% of the operating time.

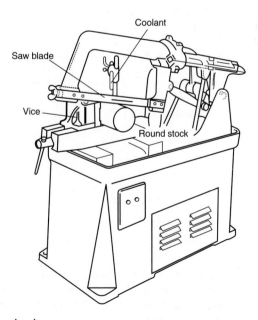

Figure 6.18 Power hacksaw

6.3 Drilling, drilling machines and routing

The only conventional machine tools widely used in fabrication engineering are the *drilling machine* and the *router*. Drilling machines are used for cutting holes and routers are used for profiling blanks.

6.3.1 The bench (sensitive) drilling machine

The simplest type of drilling machine is the bench drilling machine as shown in Fig. 6.19(a). It is capable of accepting drills up to 12.5 mm (0.5 inch) diameter. Generally these machines have the chuck mounted directly onto the spindle nose. Variation in spindle speed is achieved by altering the belt position on the stepped pulleys.

For normal drilling the spindle axis must be perpendicular to the work-table. However, if the hole is to be drilled at an angle to the workpiece, the table can be tilted as shown in Fig. 6.19(b).

The feed is operated by hand through a rack-and-pinion mechanism. This type of feed mechanism enables the operator to 'feel' the progress of the drill through the material being cut so that the operator can adjust the feed rate to suit the cutting conditions. It is from this close control that the operator has over the feed of the drill, that this type of drilling machine gets its name of a *sensitive drilling machine*. Some sensitive drilling machines have an elongated column so that they can be floor-standing instead of bench-mounted. Otherwise they are essentially the same machine.

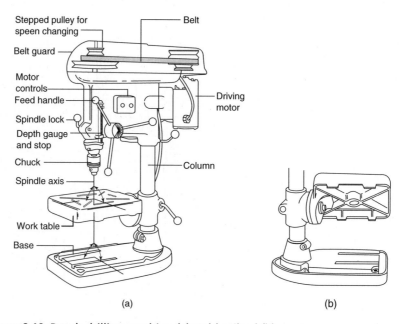

(a)   (b)

Figure 6.19 Bench drilling machine (a); table tilted (b)

6.3.2 The pillar drilling machine

Figure 6.20(a) shows a typical pillar drilling machine. It can be seen that it is an enlarged and more powerful version of the machine just described. It is floor-mounted and much more ruggedly constructed. The spindle is driven by a more powerful motor, and speed changing is accomplished through a gearbox instead of belt changing. Sensitive rack-and-pinion feed is provided for setting up and starting the drill. Power feed is provided for the actual drilling operation. The feed rate can also be changed through an auxiliary gearbox. The spindle is always bored with a morse taper to accept taper shank tooling as well as a drill chuck.

Figure 6.20(b) shows that the circular work-table can be rotated as well as swung about the column of the machine. This allows work clamped to any part of the table to be brought under the drill by a combination of swing and rotation. This allows all the holes in the workpiece to be drilled without it having to be unclamped and repositioned on the work-table. The work-table can also be raised and lowered up and down the column to allow for work of different thicknesses and drills or tooling of different lengths. Holes up to 50 mm in diameter can be drilled from the solid on this type of machine.

6.3.3 The column-type drilling machine

This is similar to the pillar-type drilling machine except that the machine elements are mounted on a cast, box-type column as shown in Fig. 6.21(a). This has the advantage that the work-table can be raised and lowered without loss of axial alignment when using tooling of different lengths. To enable the work to be positioned easily under the drill on column-type machines, they are often fitted with a *compound table*. When the table lock is released the table is free to move on anti-friction slides both to and from the column and from sided to side. This is shown in Fig. 6.21(c).

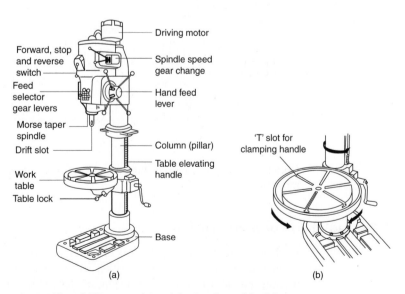

(a)   (b)

Figure 6.20 Pillar drilling machine (a); circular table (b)

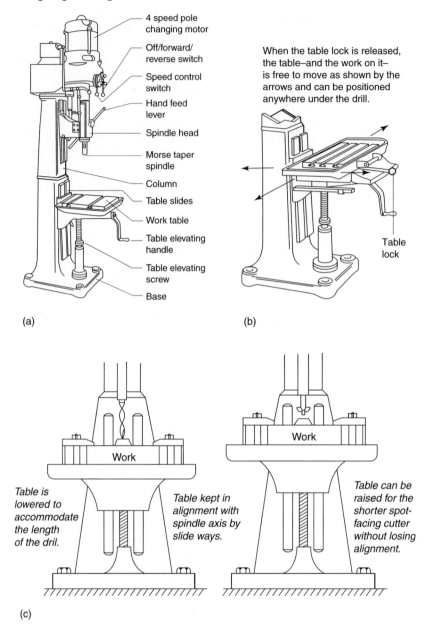

Figure 6.21 Column type drilling machine (a); compound table (b); table movement (c)

6.3.4 The radial arm drilling machine

For heavy and very large work it is often more convenient to move the drilling head about over the work than to reposition and reset the work. The radial arm drilling machine provides such a facility and an example is shown in Fig. 6.22(a). Radial arm drilling machines are the most powerful drilling machines available, often drilling holes

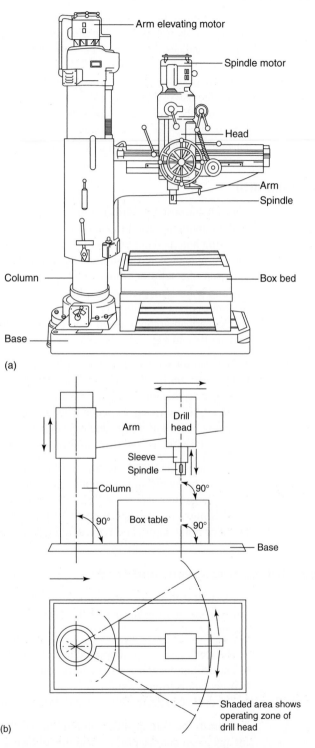

Figure 6.22 Radial arm drilling machine. (a) Radial drilling machine; (b) direction of movement

up to 75 mm in diameter from the solid. Powerful drive motors are geared directly into the head of the machine and a wide range of power feed rates are available as well as sensitive and geared manual feed facilities. The arm is raised and lowered by a separate motor mounted on the column. The arm can be swung from side to side around the column and locked in position. The drilling head can be run back and forth along the arm by a large hand wheel operating a rack-and-pinion mechanism. Once positioned correctly over the work the drilling head can also be locked in position. The spindle motor is reversible so that power tapping attachments can be used. Figure 6.22(b) shows the range of movements of such a machine.

6.3.5 The twist drill

The sole purpose of a drill is to remove the maximum volume of metal from a hole in the minimum period of time. A drill does not produce a precision hole. When a precision hole is required, it is first drilled slightly under size and opened up to size using a reamer. The two-flute twist drill is the most common type of drill used. It cuts from the solid or, under satisfactory conditions, can be employed to 'open up' existing holes. A typical taper shank twist drill is shown in Fig. 6.23 together with the names of its most important features.

As with any other cutting tool, the twist drill must be provided with the correct tool angles and Fig. 6.24 shows how these compare with a single point lathe tool.

- The *clearance angle* of a twist drill can be adjusted during sharpening the point.
- The *rake angle* of a twist drill is formed by the helix angle of the flutes and is set during manufacture. The rake angle can be reduced during re-grinding but it cannot be increased. It needs to be reduced to prevent the drill 'grabbing' when cutting some soft metals such as copper, brass and bronze. Drills with some alternative rake angles and their applications are shown in Fig. 6.25.

In addition to the basic cutting angles of a twist drill, its performance can also be improved by modifying the point angle from the standard 118° for certain materials. Where large numbers of drills of the same size are being purchased for long production runs, the web and the land can also be varied for different materials. Some modified point angles and typical applications are shown in Fig. 6.26.

6.3.6 Twist drill cutting speeds and feeds

For a twist drill to cut satisfactorily it must be used at the correct peripheral speed (cutting speed in m/min) and the correct rate of feed. For optimum results it is essential that:

- The work is securely supported and clamped.
- The drilling machine is in good condition.
- A coolant is used if required.
- The drill is correctly selected and ground to suit the material being cut.

To convert the cutting speed (m/min) to the spindle speed of the machine (rev/min) for any given diameter of the drill the following formula can be used:

$$N = 1000S/\pi d$$

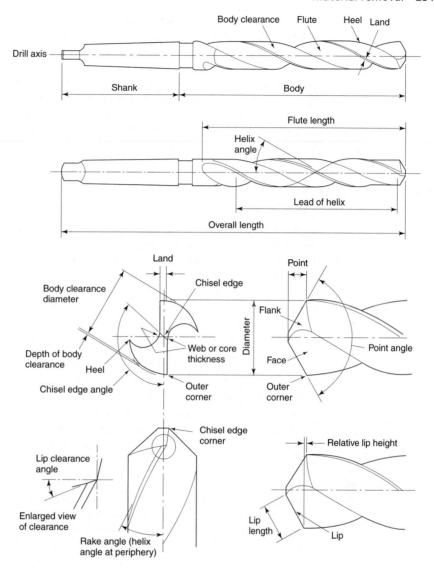

Figure 6.23 Twist drill elements

where N = spindle speed (rev/min), S = cutting speed (m/min), d = drill diameter (mm) and π = 3.14.

Table 6.2 gives a range of cutting speeds suitable for jobbing work and Table 6.3 gives some typical rates of feed. The rates of feed and cutting speeds for twist drills are lower than for most other machining operations. This is because:

- A drill is relatively weak compared with other cutting tools.
- In deep holes it is difficult for the drill to reject the chips resulting in clogging and drill breakages.
- It is difficult to keep the point and cutting edge cool when they are encased in the hole being cut.

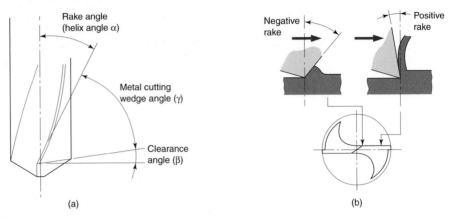

Figure 6.24 Twist drill cutting angles: (a) cutting angles applied to a twist drill; (b) variations of rake angle along lip of a drill – note that the rake angle at the periphery is equal to the helix angle of the flute

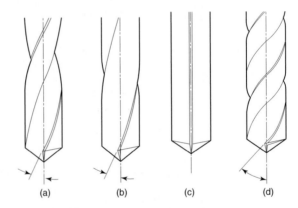

Figure 6.25 Helix angles: (a) normal helix angle for drilling low and medium tensile materials; (b) reduced or 'slow' helix angle for high tensile materials; (c) straight flute for drilling; free-cutting brass – to prevent the drill trying to draw in; (d) increased or 'quick' helix for drilling light alloy materials

6.3.7 Miscellaneous operations

As well as drilling holes the following operations may also be performed on a drilling machine:

- Trepanning.
- Countersinking.
- Counterboring.
- Spotfacing.

Trepanning

Not only is it dangerous to try to drill large holes in sheet metal with a twist drill, but the resulting hole will be unsatisfactory. Sheet metal and thin plate have insufficient

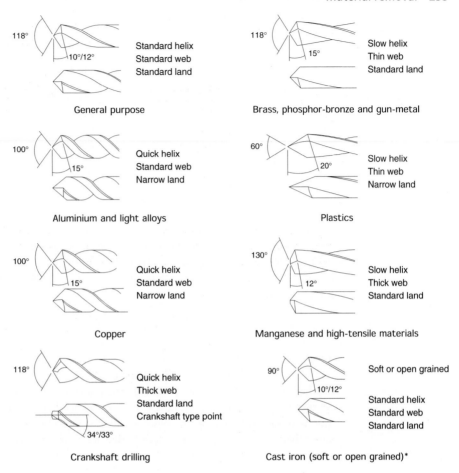

General purpose
- Standard helix
- Standard web
- Standard land
118° 10°/12°

Brass, phosphor-bronze and gun-metal
- Slow helix
- Thin web
- Standard land
118° 15°

Aluminium and light alloys
- Quick helix
- Standard web
- Narrow land
100° 15°

Plastics
- Slow helix
- Thin web
- Narrow land
60° 20°

Copper
- Quick helix
- Standard web
- Narrow land
100° 15°

Manganese and high-tensile materials
- Slow helix
- Thick web
- Standard land
130° 12°

Crankshaft drilling
- Quick helix
- Thick web
- Standard land
- Crankshaft type point
118° 34°/33°

Cast iron (soft or open grained)*
- Soft or open grained
- Standard helix
- Standard web
- Standard land
90° 10°/12°

* For medium or close grain use a standard drill
  For harder grades of alloy cast iron it may be necessary to use a manganese drill

Figure 6.26 Point angles

Table 6.2 Cutting speeds for high-speed steel (HSS) twist drills

Material being drilled	Cutting speed (m/min)
Aluminium	70–100
Brass	35–50
Bronze (phosphor)	20–35
Cast iron (grey)	25–40
Copper	35–45
Steel (mild)	30–40
Steel (medium carbon)	20–30
Steel (alloy – high tensile)	5–8
Thermosetting plastic*	20–30

*Low speed due to abrasive properties of the material.

Table 6.3 Feeds for HSS twist drills

Drill diameter (mm)	Rate of feed (mm/rev)
1.0–2.5	0.040–0.060
2.6–4.5	0.050–0.100
4.6–6.0	0.075–0.150
6.1–9.0	0.100–0.200
9.1–12.0	0.150–0.250
12.1–15.0	0.200–0.300
15.1–18.0	0.230–0.330
18.1–21.0	0.260–0.360
21.1–25.0	0.280–0.380

thickness to guide the drill point and resist the cutting forces. This will result in the drill 'grabbing' so that the hole will have torn and jagged edges and it will be out of round. Further, the metal in which the hole is being drilled will also become buckled and twisted round the hole.

One way of overcoming this problem is to use a *trepanning cutter*. Instead of cutting all the metal in the hole into swarf, the trepanning cutter merely removes a thin annulus of metal. This leaves a clean hole in the stock and the waste metal is a circular blank slightly smaller in diameter than the required hole. This is shown in Fig. 6.27. A small diameter pilot hole is drilled in the stock at the required position in which the centre pin can be located. The diameter of the hole to be trepanned is controlled by the amount the cutter bar protrudes from the cutter head and shank. The main disadvantage of this type of cutter is that the cutting action is one-sided and unbalanced. Its main advantage is that holes of any size can be cut with the same tool.

Hole-sawing

Where a number of holes of a standard size are to be cut in sheet metal or thin plate a hole saw, such as the example shown in Fig. 6.28, can be used. The balanced cutting action of a hole saw leaves a hole that is more truly round than a trepanning cutter and the finish will be better. Again, a small diameter pilot hole is required to guide the hole saw.

Countersinking

Figure 6.29(a) shows a typical countersink bit. Since the bit is conical in form it is self-centring and does not require a pilot to ensure axial alignment. Although countersinking is used primarily for providing a recess for countersunk head screws as shown in Fig. 6.29(b), less deep countersinking can be used to remove the sharp edge (burr) from a previously drilled hole.

Counterboring

Counterboring produces a cylindrical recess concentric with a previous drilled hole to house the head of a *cheese head* or a *cap head* screw. Figure 6.30(a) shows a

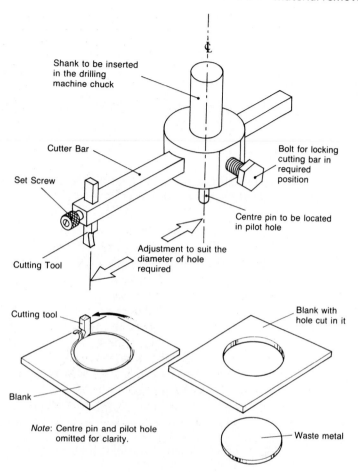

Shank to be inserted
in the drilling
machine chuck

Cutter Bar

Set Screw

Cutting Tool

Bolt for locking
cutting bar in
required
position

Centre pin to be located
in pilot hole

Adjustment to suit the
diameter of hole
required

Cutting tool

Blank with
hole cut in it

Blank

Waste metal

Note: Centre pin and pilot hole
omitted for clarity.

Figure 6.27 Principle of trepanning

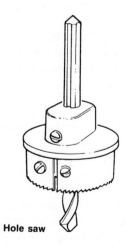

Hole saw

Figure 6.28 Hole saw

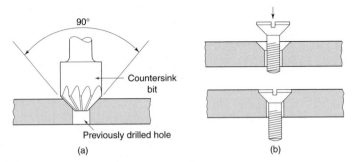

Figure 6.29 Countersinking: (a) cutting a countersink; (b) the countersink allows head of the screw to be recessed

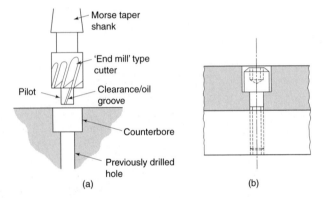

Figure 6.30 Counterboring: (a) cutting a counterbore; (b) cap head screw recessed into a counterbore

typical counterbore cutter and the recess it produces, whilst Fig. 6.30(b) shows a cap screw within the recess. The type of cutter used is called a *piloted counterbore* and is similar in appearance to a short, stubby end mill with a pilot. The pilot ensures that the counterbored hole is concentric with the previously drilled hole. (*Concentric* means that both holes have a *common axis*.)

Spot-facing

The purpose of spot-facing is to produce a flat surface as shown in Fig. 6.31(a). This provides a seat for a bolt head or a nut in an otherwise uneven surface. Bolt heads and nuts must always sit on a surface that is smooth and square to the axis of the bolt hole so that the shank of the bolt does not become bent. The spot-facing cutter is similar to a counterbore cutter except that the cutter diameter is much larger relative to the diameter of the pilot that fits in the previously drilled hole. This is because the spot-facing cutter has to provide a seating large enough to clear the corners of the hexagon bolt or nut for which it is providing a seating as shown in Fig. 6.31(b).

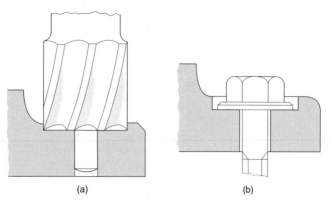

Figure 6.31 Spot facing: (a) cutting a spot facing on a casting; (b) spot facing provides a seating for the bolt head

6.3.8 Routing

The choice of technique for cutting out sheet-metal blanks is governed by such factors as size and shape, quantity required, material properties and equipment available. Routing, possibly, provides the most economical method of cutting sheet-metal blanks required in small batches, and is widely used in the aircraft industry for aluminium and its alloys. The equipment used has been adapted from the woodworking industry. This method of material removal is extremely rapid and can be used for blanks up to approximately 1.2 metres wide by any length. The cutting tool used is similar to an end mill and it revolves at very high speeds (up to 10000 rev/min), cutting a number of blanks simultaneously from a stack of sheets. Figure 6.32 illustrates the principles of material removal by three types of routing machines, two of which are the fixed cutter type.

Figure 6.32(a) shows the simplest arrangement where the cutter projects vertically through the centre of a horizontal table. At the lower end of the cutter there is a stationary collar of the same diameter as the cutting edges. The stack of sheets is secured to the wooden template by screws passing through some or all of the holes required in the finished blanks that will have been pre-drilled. The work is pressed against the cutter until the collar (which is under the cutter) encounters the template. Sideways pressure is then applied to keep the template against the guide collar whilst the exterior edges of the blanks are machined.

Figure 6.32(b) shows a similar method of material removal except that the cutter projects downwards in a fixed position. The bottom end of the cutter has a pilot which locates in a collar that does not rotate. The collar guides the rotating cutter around the template as in Fig. 6.32(a).

Figure 6.32(c) shows a type of machine that reverses the principles employed in Fig. 6.32(a) and Fig. 6.32(b). This time the work is stationary and the cutter is traversed around the template. The cutter is held in the spindle of the cutter head which, in turn, is supported on an arm that can slide easily back and forth through a pillar. The pillar is free to rotate so that the cutter can easily follow the template as shown in Fig. 6.33. The cutter head can be manipulated around the work profile with very little effort on the part of the operator. This time the template is on top of the

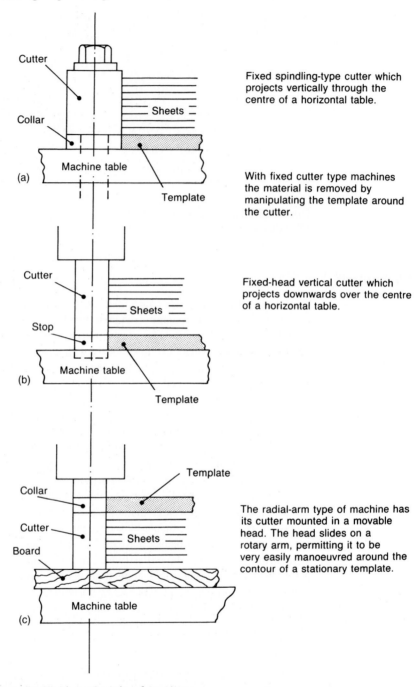

Fixed spindling-type cutter which projects vertically through the centre of a horizontal table.

With fixed cutter type machines the material is removed by manipulating the template around the cutter.

Fixed-head vertical cutter which projects downwards over the centre of a horizontal table.

The radial-arm type of machine has its cutter mounted in a movable head. The head slides on a rotary arm, permitting it to be very easily manoeuvred around the contour of a stationary template.

Figure 6.32 The principle of routing

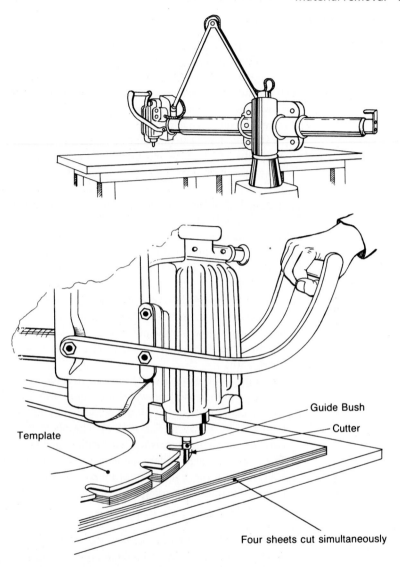

Template

Guide Bush

Cutter

Four sheets cut simultaneously

Figure 6.33 Radius arm type of routing machine

stack of sheets and the stationary guide collar is at the top (shank end) of the cutter. The template and stack of blanks are securely fixed to the work-table of the machine.

6.4 Shearing sheet metal

Sheet metal and thin plate is usually cut to shape by the use of shears. Hand shears (snips) are used for tin plate and thin sheets. Motorized power shears are used for thicker sheets and thin plate. The principle of shearing is the same whichever type of shearing equipment is used.

6.4.1 Principles of shearing

In a shearing machine used for straight-line cutting, be it hand or power-driven, one blade is fixed (normally the lower blade) and an upper, moving blade (inclined to the fixed blade) is brought down to meet the fixed blade as shown in Fig. 6.34(a). If the blades were arranged to be parallel to each other, the area under shear would be the cross-section of the material to be cut, i.e. length × thickness, as shown in Fig. 6.34(b). This would require a massive shearing force and a very substantial and costly machine, so the shearing force is reduced by inclining the top blade at an angle of approximately 5° so that the length of cut and, therefore, the area under shear at any given moment in time is greatly reduced and so is the shearing force required. The inclination of the top blade is called the shearing angle. This is shown in Fig. 6.34(c).

Figure 6.35 shows how the shearing action is used to cut metal. The shear blades have a *rake angle* of approximately 87°. There is no clearance angle in this instance but a clearance gap of approximately 5–10% of the metal thickness between the blades depending upon the properties of the metal being cut. The importance of this clearance will be discussed later in this section.

1. The top cutting blade is brought down until the stock material to be cut is trapped between the top and bottom blades as shown in Fig. 6.35(a). As the top cutting blade moves downwards and brought to bear on the stock metal with continuing pressure, the top and bottom surfaces of the metal are *deformed* as shown in Fig. 6.35(b).
2. As the pressure increases, the internal structure (crystal structure) of the metal is subjected to increasing *plastic deformation* prior to shearing.
3. After a certain amount of plastic deformation the cutting blades start to penetrate the metal as shown in Fig. 6.35(c). The uncut metal between the blades *work-hardens* due to the increasing deformation, and becomes brittle.
4. The work-hardened metal commences to fracture from the cutting edges of the blades. When the fracture lines meet, the stock metal is sheared and separates despite the fact that complete separation has not occurred. This is shown in Fig. 6.35(d).

The importance of blade clearance has already been referred to. Blade clearance should be set to suit the thickness and properties of the material being cut to ensure optimum shearing results. Figure 6.36 shows the results of incorrect and correct setting of the shear blade clearance. Table 6.4 relates metal thickness and properties to blade clearance.

6.4.2 Guillotine shear (types)

Figure 6.37 shows a treadle-operated guillotine shearing machine capable of cutting tin plate and thin sheet steel up to 1.6 mm thick by 1260 mm in length. This type of shearing machine is only suitable for straight line cutting. They are usually used for cutting large rectangular sheets into smaller rectangular blanks. For thicker sheets and thin plate work power-driven machines (mechanical and hydraulic) are available for cutting sheets up to 2600 mm in length. Heavy-duty power-operated shears have a power-operated pressure pad ('hold-down') which descends just before the cutting blade in order to hold the sheet down tightly whilst it is being cut.

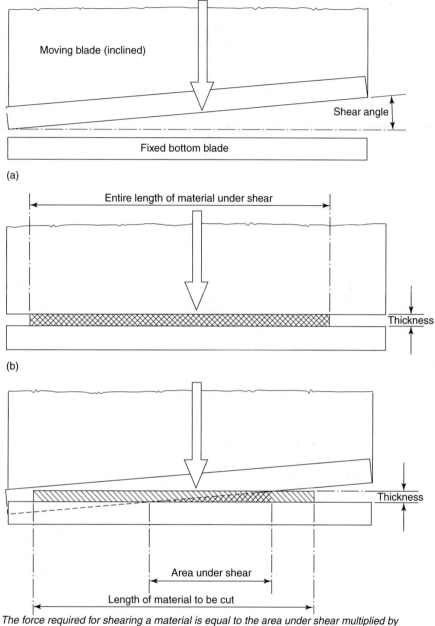

Moving blade (inclined)

Shear angle

Fixed bottom blade

(a)

Entire length of material under shear

Thickness

(b)

Thickness

Area under shear

Length of material to be cut

The force required for shearing a material is equal to the area under shear multiplied by the 'shear strength' of the material

(c)

Figure 6.34 The effect of shear angle (shearing machine): (a) shear blade movement; (b) paralled cutting blades; (c) top cutting blade inclined

6.4.3 Guillotine shear (setting)

Treadle and power guillotine shears are fitted with front, side and back gauge bars (stops). The fixed side gauge, which is at 90° to the fixed cutting blade, is often referred to as a squaring guide. The front gauge bar is adjustable across the bed or

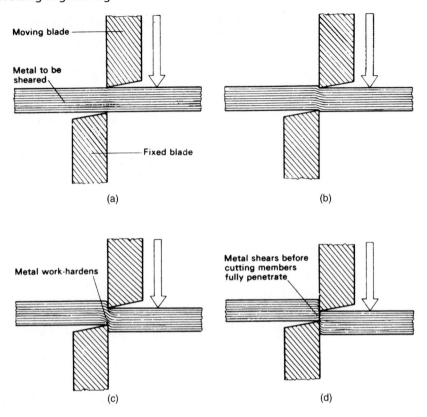

Figure 6.35 The action of shearing metal

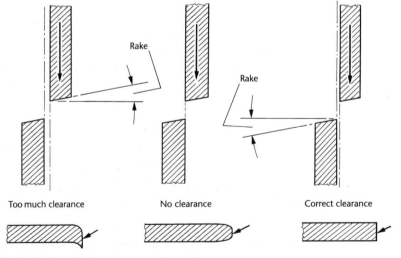

Too much clearance   No clearance   Correct clearance

1. Excessive clearance, causes a burr to form on the underside of the sheet
2. With no clearance, overstrain is caused, the edge of the sheet becomes flattened on the underside
3. With the correct clearance optimum shearing results are obtained

Figure 6.36 Results of incorrect and correct setting of shear blades

Table 6.4 Blade clearances for optimum cutting

Metal thickness		Blade clearance*			
(in)	(mm)	Low tensile strength (e.g. Brass)		High tensile strength (e.g. Steel)	
		(in)	(mm)	(in)	(mm)
0.015	0.381	0.000 3	0.007 5	0.000 5	0.013
0.032	0.813	0.001 5	0.038	0.001 8	0.046
0.065	1.651	0.002 0	0.051	0.002 5	0.064
0.100	2.540	0.002 2	0.056	0.003 0	0.076
0.125	3.175	0.003 0	0.076	0.004 0	0.10
0.250	6.350	0.005 5	0.14	0.007 0	0.18

*Tested by Feeler gauges

Figure 6.37 Guillotine shears

table of the machine and further along the extension bars (slotted arms fitted to the front of the machine. Figure 6.38(a) shows a plan view of the side and front gauges, whilst Fig. 6.38(b) and Fig. 6.38(c) show how the front gauge is set.

Most cutting operations are performed using the back gauge. There are various types of these but the simplest and most usual type consists of an angle iron guide bar that is adjustable. Two types of adjustable back gauge are shown in Fig. 6.39.

For angular cutting (i.e. gusset plates) a bevel gauge is used. This is secured to the tee-slots in the machine table as shown in Fig. 6.40. The bevel gauge can be turned over for left-hand or right-hand cutting.

6.4.4 Bench shears and hand shears (snips)

Unlike the cutting action of the guillotine-type shear describe above, bench shears and hand shears (snips) have blades that rotate about a pivot (like a pair of scissors) and it is this pivoting action that creates the shear angle. With all shearing machines, sufficient force must be applied to the moving blade to overcome the *shear strength* of the material being cut. Bench shears and hand shears (snips) rely on the *mechanical advantage* gained from a system of levers to magnify the effort exerted by the operator to overcome the shear strength of the material. Figure 6.41(a) shows a typical bench shearing machine whilst Fig. 6.41(b) shows how this mechanical advantage is achieved by carefully positioning the pivot points (fulcrums) and proportioning the system of levers of a hand-operated bench shearing machine.

Figure 6.42(a) shows some typical hand shears whilst Fig. 6.42(b) shows the cutting action of hand shears.

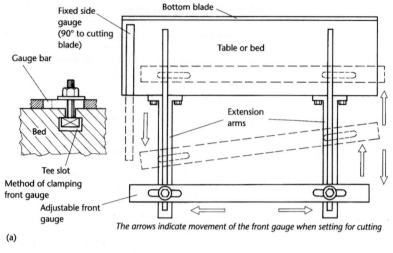

Fixed side gauge (90° to cutting blade)

Bottom blade

Gauge bar

Table or bed

Bed

Extension arms

Tee slot

Method of clamping front gauge

Adjustable front gauge

The arrows indicate movement of the front gauge when setting for cutting

(a)

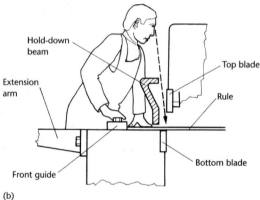

Hold-down beam

Top blade

Extension arm

Rule

Front guide

Bottom blade

(b)

1. Make sure machine is switched off
2. Place rule between blades
3. Position front gauge against end of rule, and sight as shown opposite

Set each end of bar, and lightly tighten bolts. Check each end then fully tighten in position

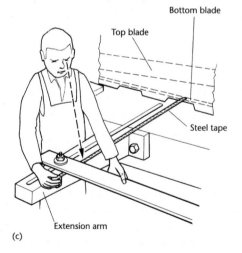

Bottom blade

Top blade

Steel tape

1. Slide the end of the tape between the blades and hook it against the bottom blade
2. Roughly position front gauge and slightly tighten clamping bolts
3. Adjust bar to correct dimensions, check each end for parallelism and fully tighten bolts

Extension arm

(c)

Figure 6.38 Front and side gauges (guillotines). (a) Plan view of guillotine table showing front gauges; (b) setting the front gauge – using a rule; (c) setting the front gauge – using a steel tape

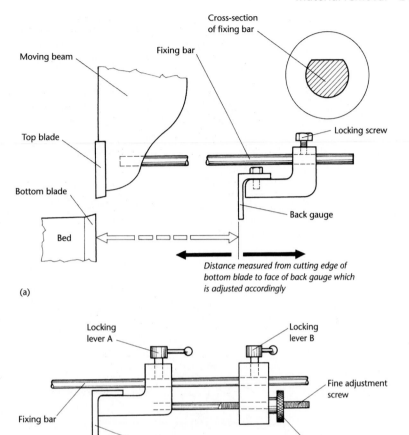

Figure 6.39 Details for setting back gauges (guillotines). (a) Simple back gauge (angle back bar); (b) simple fine adjustment for back gauge

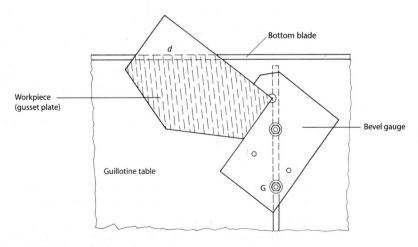

Figure 6.40 Use of a bevel gauge

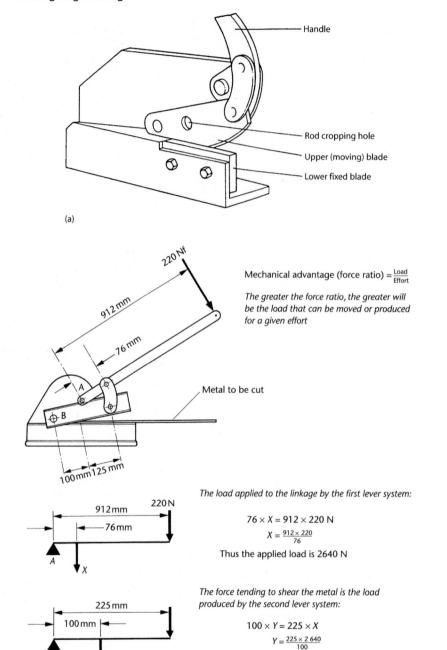

(a)

Handle

Rod cropping hole

Upper (moving) blade

Lower fixed blade

Mechanical advantage (force ratio) $= \frac{Load}{Effort}$

The greater the force ratio, the greater will be the load that can be moved or produced for a given effort

Metal to be cut

The load applied to the linkage by the first lever system:

$$76 \times X = 912 \times 220 \text{ N}$$
$$X = \frac{912 \times 220}{76}$$

Thus the applied load is 2640 N

The force tending to shear the metal is the load produced by the second lever system:

$$100 \times Y = 225 \times X$$
$$Y = \frac{225 \times 2\,640}{100}$$

Thus the force tending to shear the metal is 5940 N

(b)

Figure 6.41 Hand operated bench shears. (a) Hand lever bench shears; (b) the lever system of hand-operated bench shears

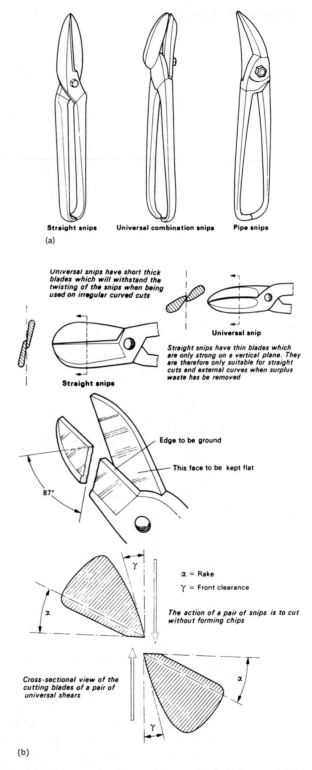

Straight snips   Universal combination snips   Pipe snips

(a)

Universal snips have short thick
blades which will withstand the
twisting of the snips when being
used on irregular curved cuts

Universal snip

Straight snips have thin blades which
are only strong on a vertical plane. They
are therefore only suitable for straight
cuts and external curves when surplus
waste has be removed

Straight snips

Edge to be ground

This face to be kept flat

87°

α = Rake

γ = Front clearance

The action of a pair of snips is to cut
without forming chips

Cross-sectional view of the
cutting blades of a pair of
universal shears

(b)

Figure 6.42 Hand shears snips. (a) Basic types of hand shears; (b) details of the
cutting action of hand shears

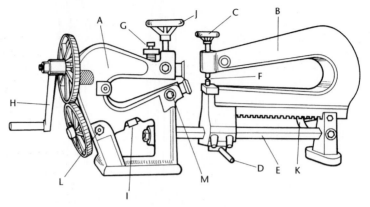

Circle cutting machine (hand operated)

A — Cutting head
B — Bow
C — Pallet adjustment wheel
D — Clamping handle (bow)
E — Bar
F — Clamping pallets

G — Stop screw
H — Operating handle
I — Lower cutter adjusting screw
J — Handwheel
K — Handwheel (circle diameter control)
L, M — Lower cutter bearing adjustment nuts

A good practical method of adjusting the cutters is to aim at cutting a true circle in paper. If a machine will do this in a satisfactory manner, then it will shear sheet metal without burring the edge of the disc

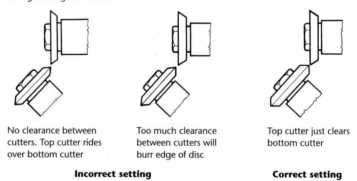

No clearance between cutters. Top cutter rides over bottom cutter

Too much clearance between cutters will burr edge of disc

Top cutter just clears bottom cutter

Incorrect setting

Correct setting

Figure 6.43 Circle cutting machine

6.4.5 Rotary shears

Manually operated bench-mounted rotary shears employ a pair of *rotary cutters* in place of the conventional flat blades employed in the bench shears and hand shears previously described. The advantage of rotary shears is that there is no restriction on the length of cut, and a further advantage is that both straight line and circular cuts can be taken. Figure 6.43 shows a rotary-shear circle-cutting machine. The cutters rotate, producing a continuous cutting action with very little distortion of the material. Heavy-duty, floor-mounted, power-driven rotary shears are also available for thin plate work. These are two basic types of rotary shearing machines for straight line cutting as shown in Fig. 6.44.

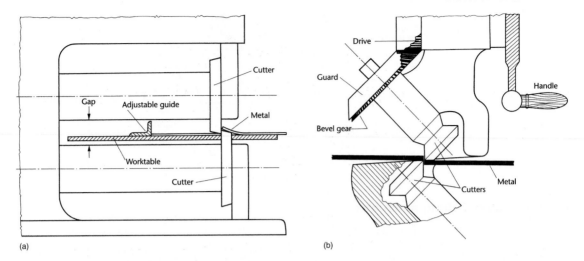

Figure 6.44 Rotary shears. (a) Bench rotary machine for straight-line cutting (horizontal spindles); (b) hand-operated rotary throatless shear (inclined spindles)

6.4.6 Essential requirements of shearing machines

- All shearing machines must have two cutting blades between which shearing of the stock material takes place.
- The operational clearance between the cutting blades is very important to achieve optimum shearing conditions. This will depend upon the thickness and properties of the material being cut.
- The sharpness of the cutting edges: the cutting members must always be kept sharp to produce a more intense local strain and a cleaner cut with minimum burr.
- Most cutting members have a face shear angle of about 2° (rake angle) sloping away from the cutting edge. This, together with blade sharpness, assists the cutting action by increasing the local strain.
- The rigidity and robustness of the shearing machine should be sufficient to withstand the cutting forces and so avoid deflection of the cutting blades. The cutting blades must maintain their correct position relative to each other throughout the cutting cycle.
- Easy and satisfactory means of blade adjustment should be provided. This enables the cutting clearance to be adjusted and the positions of the cutting members to be maintained to suit the cutting conditions.

6.5 Portable power tools

These may be electrically or pneumatically powered. All the examples described below are electrically powered and can be listed as:

- Portable drilling machines.
- Portable shear-type 'nibbling' machines.
- Portable punch-type 'nibbling' machines.
- Portable grinding machines.

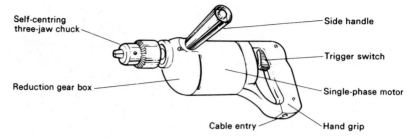

Figure 6.45 Portable electric drill

6.5.1 Portable drilling machines

Figure 6.45 shows a typical portable drilling machine. For safety, portable, electrically powered machines should be used on a 110 volt supply via a step-down transformer from the normal (UK) single-phase supply (see also Section 1.4, Electrical hazards). Note that single-wound 'auto-transformers' are unsuitable for this purpose since they do not provide the isolation from the supply that is essential for safety. In addition, the mains supply to the transformer should be protected by a circuit breaker with a residual current detector (RCD). The slightest leak of electricity to earth through any path will result in the breaker tripping and the appliance being isolated. However, plastic-cased (double insulated) and battery-powered drills need not be earthed.

> Metal case drills must always be earthed

6.5.2 Portable shear-type machines

Portable shear-type machines can be used for the rapid, straight-line cutting or curved-line cutting operations. They can be used on material up to 4.5 mm thick. Shear-type nibblers are essentially short-stroke power shears fitted with a rapidly reciprocating cutting blade working against a fixed blade so that each stroke makes a cut of approximately 3 mm in length. The speed of cutting is between 1200 and 1400 strokes per minute, and the linear cutting speed for material up to 1.62 mm in thickness is approximately 10 metres per minute, reducing to 4.5 metres per minute for metal of maximum thickness.

The shear-type nibbler is fitted with a pair of very narrow, flat blades. Generally these blades have a very pronounced rake angle to permit piercing of the material for internal cutting. Since the blades are so narrow, the sheet material can be easily manoeuvred during cutting. The top blade is fitted to the moving member and the bottom blade is fixed on a spiral extension or 'U' frame. This extension is shaped like the body of a *throatless shear* in order to part the material as it is cut. Therefore, there is no limit to the length of the cut. Lighter machines have a minimum cutting radius of 16 mm, and the heavier machines have a minimum cutting radius of 50 mm.

There is usually provision for vertical adjustment of the blades to allow for re-sharpening by grinding, and provision for lateral adjustment behind the bottom blade for setting the cutting clearance. A typical shear-type 'nibbler' is shown in Fig. 6.46.

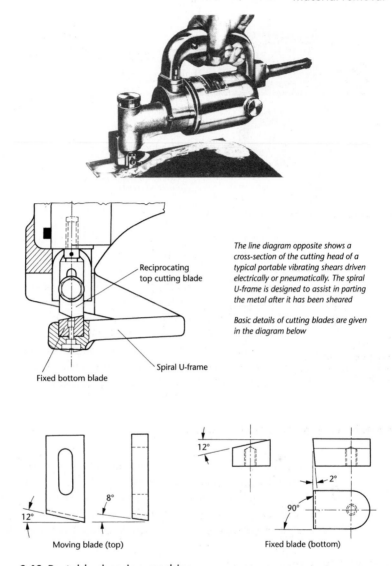

The line diagram opposite shows a cross-section of the cutting head of a typical portable vibrating shears driven electrically or pneumatically. The spiral U-frame is designed to assist in parting the metal after it has been sheared

Basic details of cutting blades are given in the diagram below

Reciprocating top cutting blade

Spiral U-frame

Fixed bottom blade

12°   8°

Moving blade (top)

12°   2°   90°

Fixed blade (bottom)

Figure 6.46 Portable shearing machine

6.5.3 Portable punch-type 'nibbling' machines

An example of a portable punch-type 'nibbling' machine and its principle of operation is shown in Fig. 6.47. A punch and die is employed in this type of machine instead of a pair of shearing blades in order to cut out the required shape. The advantage of this type of machine is that it can perform certain operations that cannot be accomplished on other shearing machines. For example, it may be used to cut out apertures. A further advantage is that the cutting process causes less distortion of the work. Like the shear-type nibbler, the top cutting tool (a punch) reciprocates with fast, short strokes over a fixed die so as to produce a series of overlapping holes. The principles of piercing holes are dealt with more fully in Section 6.5.1. Unfortunately,

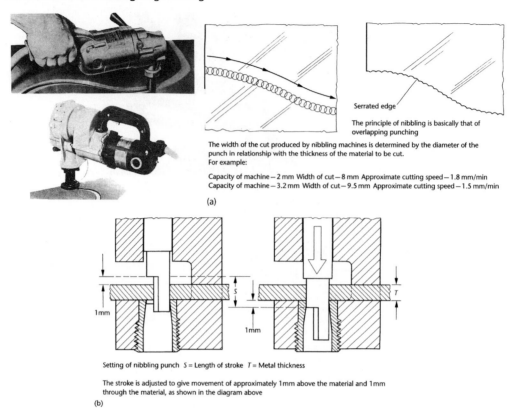

The width of the cut produced by nibbling machines is determined by the diameter of the punch in relationship with the thickness of the material to be cut. For example:

Capacity of machine – 2 mm Width of cut – 8 mm Approximate cutting speed – 1.8 mm/min
Capacity of machine – 3.2 mm Width of cut – 9.5 mm Approximate cutting speed – 1.5 mm/min

(a)

Serrated edge
The principle of nibbling is basically that of overlapping punching

Setting of nibbling punch   S = Length of stroke   T = Metal thickness

The stroke is adjusted to give movement of approximately 1 mm above the material and 1 mm through the material, as shown in the diagram above

(b)

Figure 6.47 Portable nibbling machine

this leaves a slightly serrated edge that requires finishing by filing or grinding. Light-weight machines are used for metal up to 3.2 mm thick whilst heavy-duty machines can be used for metal up to 6.35 mm thick. Standard punches of 4.5, 6.5 and 9.5 mm diameter are employed, depending on the thickness of the metal being cut. The maximum linear cutting speed is approximately 1.8 metres per minute.

6.5.4 Portable grinding machines

Portable grinding machines are used for cleaning up welded joints. Figure 6.48(a) shows a straight-type portable grinding machine and Fig. 6.48(b) shows an angle-type portable grinding machine. Angle-type grinding machines can be fitted with depressed centre, reinforced grinding wheels. Such wheels have a honeycomb of small holes to ventilate the wheel and work to prevent overheating. The angle-type grinding machine can also be fitted with cutting-off grinding wheels. For dressing curved surfaces the angle-type grinding machine can be fitted with flexible abrasive discs as shown in Fig. 6.48(c). Portable grinding machines and all abrasive wheels above 55 mm in diameter must be marked with the maximum safe operating speed by the manufacturer as shown in Fig. 6.48(d). When using a grinding machine:

- Always wear safety goggles.
- Always wear a dust mask.

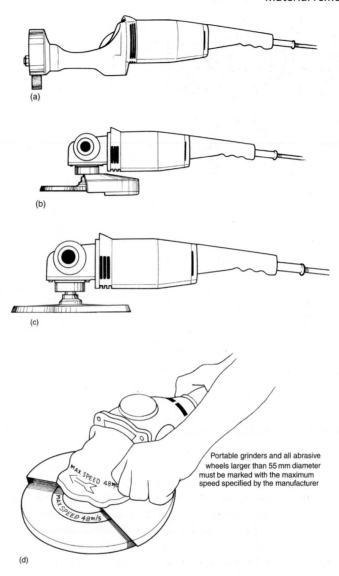

Figure 6.48 Portable electric grinding machines. (a) Straight grinder; (b) angle grinder; (c) angle sander; (d) abrasive wheel with markings showing

- Do not overheat the wheel and work or apply excessive force against the wheel.
- Always move the wheel back and forth to even out the wear.

6.5.5 The double-ended off-hand grinding machine

Figure 6.49(a) shows a typical floor-mounted double-ended off-hand grinding machine widely used in workshops for sharpening cutting tools, such as chisels, and general grinding operations where the work is small enough to be held in the hand.

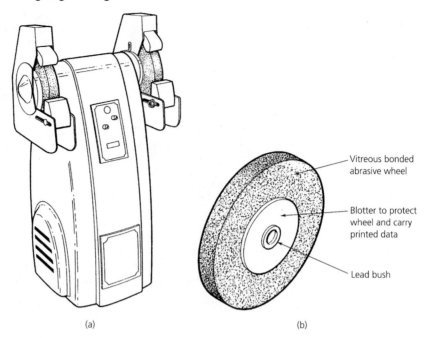

Vitreous bonded abrasive wheel

Blotter to protect wheel and carry printed data

Lead bush

(a)   (b)

Figure 6.49 Double-ended, off-hand grinding machine

It uses plain cylindrical grinding wheels of the type shown in Fig. 6.49(b). Usually a coarse grit wheel is fitted on one end of the spindle for roughing down and a fine grit wheel is fitted on the opposite end of the spindle for finishing operations.

Because of its apparent simplicity, this type of grinding machine comes in for more than its fair share of abuse. For safe and efficient metal-cutting the grinding wheel must be correctly selected and mounted and correctly used. Remember that under the Abrasive Wheel Regulations only certificated personnel and trainees under the direct supervision of a certificated person may change a grinding wheel. The names of the parts of the wheel-mounting assembly are shown in Fig. 6.50(a) and the specification of a grinding wheel printed on the 'blotter' is shown in Fig. 6.50(b). For safe operation the grinding wheel must be fitted with a guard of substantial construction not only to prevent the operator from coming into contact with the rapidly rotating wheel but also to protect the operator in the event of the wheel failing (bursting), that is, *burst containment*.

6.5.6 Grinding wheel use

Grinding wheels do not rub the metal away; they cut and produce swarf like any other multi-tooth cutter such as a milling cutter. However, because of the large number of fine abrasive grains from which the wheel is manufactured, this swarf is very small in size and the surface produced is very smooth. This 'dross' consists of blunt abrasive particles stripped from the grinding wheel together with metallic swarf cut from the workpiece. To cut correctly, grinding wheels must be trued and dressed.

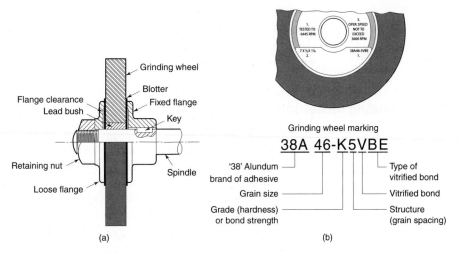

Figure 6.50 Mounting a grinding wheel: (a) the wheel mounting; (b) checking the new wheel (reproduced courtesy of Norton Grinding Wheel Co.)

Figure 6.51 Huntington wheel dresser

Truing

Grinding wheels should be trued to ensure that they run concentrically with the spindle and maintain their balance. This is necessary to prevent vibration damage to the machine bearings, to allow accurate dimensional control, and to allow the work rest to be kept as close to the wheel as possible to prevent the work being trapped between the work rest and the wheel, which would cause the wheel to burst.

Dressing

Dressing is required to restore the cutting efficiency of a grinding wheel that has become *glazed* or *loaded*. Under correct cutting conditions the abrasive grains of the grinding wheel should break away from the bond when they become blunt, thus exposing new, sharp grains. If the bond is too strong the blunt grains are not removed and the grinding wheel tales on a shiny appearance. *It is said to be glazed*.

Figure 6.51 shows a Huntington-type wheel dresser. The star wheels of the dresser dig into the grinding wheel and break out the blunt grains and any foreign matter that may be clogging the grinding wheel. Since the star wheels rotate with the grinding wheel, little abrasive action takes place and wear of the star wheels is minimized.

Loading occurs when soft material such as a non-ferrous metal is ground with an unsuitable wheel; the spaces (voids) between the abrasive particles become clogged with metal particles. Under such conditions the particles of metal can often be seen embedded in the grinding wheel. Loading is detrimental to the cutting efficiency of the grinding wheel since it destroys the clearance between the grains, causing them to rub rather than cut.

6.5.7 Grinding wheel selection

To prevent loading and glazing as far as possible it is necessary to select a grinding wheel of the correct specification for any given job. Basically, a grinding wheel consists of *abrasive grains* that do the cutting and a *bond* that holds the grains together. Let's now see how grinding wheels are specified.

The specification of a grinding wheel indicates its construction and its suitability for a particular operation. For example, let's consider a wheel carrying the marking:

$$38A60\text{-}J5V$$

This is interpreted as follows:

- 38A is the *abrasive type* (see Table 6.5).
- 60 is the *grit size* (see Table 6.6).
- J is the *grade* (see Table 6.7).
- 5 is the *structure* (see Table 6.8).
- V is the *bond material* (see Table 6.9).

Therefore, a wheel carrying the marking 38A60-J5V has an aluminium-oxide-type abrasive. The abrasive grit has a medium to fine grain size and the wheel is soft. The structure has a medium spacing and the grains are held together by a vitrified bond.

Let's now consider the coding of grinding wheels in greater detail.

Abrasive

This must be chosen to suit the material being cut. As a general classification:

- 'Brown' aluminium oxide is used for grinding tough materials.
- 'White' aluminium oxide is used for grinding hardened carbon steels and high-speed steels.
- Silicon carbide (green grit) is used for very hard and brittle materials such as tungsten carbide.

Abrasive coding may follow the British Standard system or follow manufacturers own systems. Table 6.5 compares the British standard marking system with that of the Norton Abrasive Company's system.

Grain size (grit size)

The number indicating the grain or grit size represents the number of openings per linear 25 mm in the sieve used to size the grains. The larger the grain size number, the finer the grain. Table 6.6 gives a general classification. The sizes listed as *very fine* are referred to as 'flours' and are used for polishing and super-finishing processes.

Table 6.5 Abrasive types (Norton abrasives)

Manufacturer's type code	BS code	Abrasive	Application
A	A	Aluminium oxide	A high strength abrasive for hard, tough materials
32A	A	Aluminium oxide	Cool; fast cutting, for rapid stock removal
38A	A	Aluminium oxide	Light grinding of very hard steels
19A	A	Aluminium oxide	A milder abrasive than 38A used for cylindrical grinding
37C	C	Silicon carbide	For hard, brittle materials of high density such as cast iron
39C	C (green)	Silicon carbide	For very hard, brittle materials such as tungsten carbide

Table 6.6 Grit size

Classification	Grit sizes
Coarse	10, 12, 14, 16, 20, 24
Medium	30, 36, 40, 46, 54, 60
Fine	70, 80, 90, 100, 120, 150, 180
Very fine	220, 240, 280, 320, 400, 500, 600

Table 6.7 Grade

Classification	Letter codes
Very soft	E, F, G
Soft	H, I, J, K
Medium	L, M, N, O
Hard	P, Q, R, S
Very hard	T, U, W, Z

Grade

This indicates the strength of the bond and, therefore, the 'hardness' of the wheel. In a *hard* wheel the bond is strong and securely anchors the grit in place, thus reducing the rate of wear. In a soft wheel the bond is weak and the grit is easily detached, resulting in a high rate of wear.

The bond must be carefully related to the use for which the wheel is intended. Too hard a wheel will result in dull, blunt grains being retained in the periphery of the wheel resulting in the generation of excessive heat at the wheel/work interface leading to 'blueing' and softening of the tool being ground. Too soft a wheel would be uneconomical due to rapid wear and would also result in lack of control of dimensional accuracy in the workpiece when precision grinding. Table 6.7 gives a general classification of the hardness of the wheel using a letter code.

Structure

This indicates the amount of bond between the grains and the closeness of adjacent grains, i.e. in hacksaw blade parlance the '*chip clearance*'. An open structured wheel cuts freely and tends to generate less heat in the cutting zone. Therefore, an open structured wheel has '*free-cutting*' and rapid material removal characteristics. However, it will not produce such a good finish as a closer structured wheel. Table 6.8 gives a general classification of structure

Bond

There is a wide range of bonding materials available and care must be taken to ensure that the bond is suitable for a given application, as the safe use of the grinding wheel is very largely dependent upon the bond selected.

- *Vitrified bond*. This is the most widely used bonding material and is similar to glass in composition. It has a high porosity and strength, producing a wheel suitable for high rates of material removal. Further, it is not adversely affected by water, acids, oils or ordinary temperature conditions.
- *Resinoid bond*. This is used for high-speed wheels where the bursting forces are great. Such wheels are used for portable grinding machines for dressing castings and welded joints. Resinoid bond wheels are also used for the larger sizes of cutting-off wheels. They are strong enough to withstand considerable abuse and are mainly found in fabrication shops, construction sites and foundries.
- *Rubber bond*. This is used where a small amount of flexibility is required in the wheel, such as in thin cutting-off wheels and centreless grinding control wheels.
- *Shellac bond*. This is used for heavy-duty, large-diameter wheels, where a fine finish and cool cutting is required. Such wheels are used for grinding mill rolls.
- *Silicate bond*. This is little used for precision grinding. It is mainly used for finishing cutlery (knives) and edge tools such as carpenters' chisels. The cool cutting characteristics of this bond help to prevent the fine cutting edges of such tools being softened.

Table 6.9 lists the literal codes used to specify the bonding materials discussed above.

Table 6.8 Structure

Classification	Structure numbers
Close spacing	0, 1, 2, 3
Medium spacing	4, 5, 6
Wide spacing	7, 8, 9, 10, 11, 12

Table 6.9 Bond

Classification	BS code
Vitrified bond	V
Resinoid bond	B
Rubber bond	R
Shellac bond	E
Silicate bond	S

6.6 Blanking (stamping) and piercing

Hand and power presses similar to those shown in Fig. 6.52 can be used for:

- Cutting out shaped blanks (blanking operations).
- Cutting holes in the blanks (piercing operations).
- Bending and flow-forming the pierced blank into the finished shape (see Section 7.1.5).

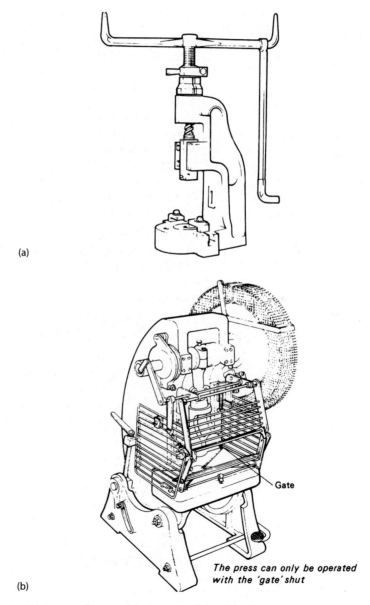

(a)

(b)

The press can only be operated with the 'gate' shut

Gate

Figure 6.52 Types of press. (a) Fly press; (b) power press

6.6.1 Blanking

Figure 6.53(a) shows a typical blanking tool for cutting out circular blanks from strip metal. In a blanking tool the die orifice controls the blank size so, in this tool, the die orifice will be 10 mm diameter and the punch will be 100 mm minus the shear clearance on the diameter. For low-carbon steel this will be about 10% of the

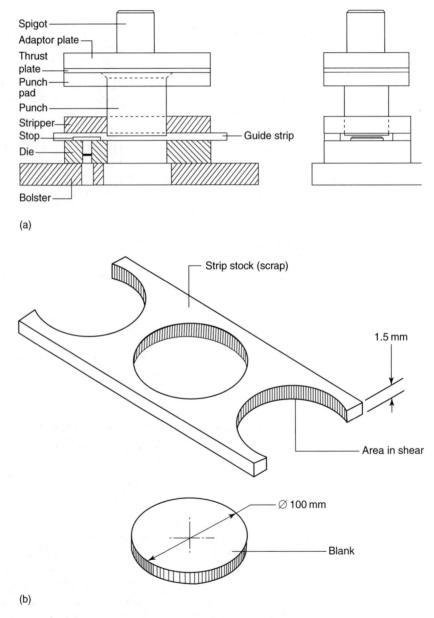

Spigot
Adaptor plate
Thrust plate
Punch pad
Punch
Stripper
Stop
Die
Guide strip
Bolster

(a)

Strip stock (scrap)

1.5 mm

Area in shear

Ø 100 mm

Blank

(b)

Figure 6.53 Blanking. (a) Blanking tool; (b) circular blank produced by tool (a)

Table 6.10 Die clearances

Material	Clearance per side (double the value given for diameters)
Aluminium	$\frac{1}{60}$ material thickness
Brass	$\frac{1}{40}$ material thickness
Copper	$\frac{1}{50}$ material thickness
Steel	$\frac{1}{20}$ material thickness

material thickness. Some typical clearances between the punch and the die are listed in Table 6.10. These need to be doubled for diameters. The die orifice is given a slight taper to allow the blank to drop clear. The stripper guides the punch and keeps it centralized over the die as well as removing the stock material from the punch. The stock material tends to shrink back onto the punch after shearing and requires quite a considerable force to strip it off.

Figure 6.53(b) shows the strip and the blank produced from it. Blanking is a very rapid process with up to several hundred components being produced per minute. A set of press tools can be very expensive and pressing is only used where very large numbers of identical components need to be produced. Example 6.1 shows how the blanking force can be calculated. This is important in ascertaining the size of press in which the blanking too needs to be used.

Example 6.1

Determine the force required to press the 100 mm diameter circular blank shown in Fig. 6.53(b) from strip metal whose ultimate shear stress is 450 N/mm ($\pi = 3.14$).

Solution

$$
\begin{aligned}
\text{Area in shear} &= \text{circumference of blank} \times \text{thickness} \\
&= \pi D \times \text{thickness} \\
&= 3.14 \times 100\text{mm} \times 1.5\text{mm} \\
&= \mathbf{471mm^2}
\end{aligned}
$$

$$
\begin{aligned}
\text{Blanking force} &= \text{area in shear} \times \text{ultimate shear stress of metal} \\
&= 471\text{mm}^2 \times 450\text{N/mm} \\
&= \mathbf{212kN}
\end{aligned}
$$

6.6.2 Piercing

Piercing is the punching of holes in a blank prior to forming to shape. The pierced hole can be any desired shape but, for simplicity, let's consider a circular hole. Figure 6.54 shows a simple piercing tool for use in a hand (fly) press. In a piercing tool the punch controls the hole size so, in this tool, the punch will be the hole diameter and the die orifice will be the hole size plus the shear clearance on the diameter. This will be the same as for blanking. As in blanking, the stripper guides the punch and keeps it centralized

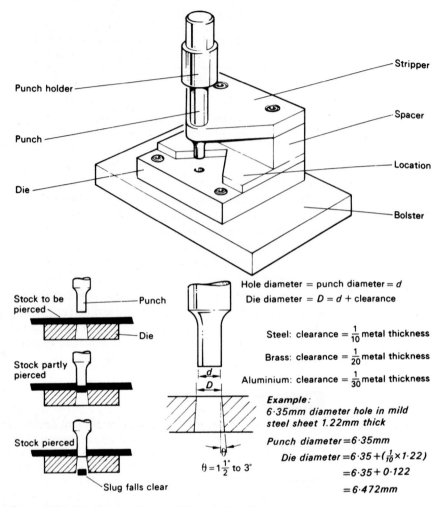

Figure 6.54 Simple piercing tool for use in a fly press

over the die as well as removing the blank from the punch. The stock material tends to shrink back onto the punch after shearing and requires quite a considerable force to strip it off. The piercing force is calculated in the same manner as for blanking.

6.6.3 Universal shearing machine

In most fabrication shops, cutting operations on rolled steel sections are carried out on power-driven machines. Machines are available which perform a combination of cutting operations such as notching, punching and shearing. The shearing operations include not only section shearing but round and square bar cropping and plate shearing as well. Angle section has to be notched in order to permit it to be bent and most of the notches are the vee-notch and the square notch. Figure 6.55 shows a typical universal shearing machine, whilst Fig. 6.56 shows some typical operations performed on such a machine.

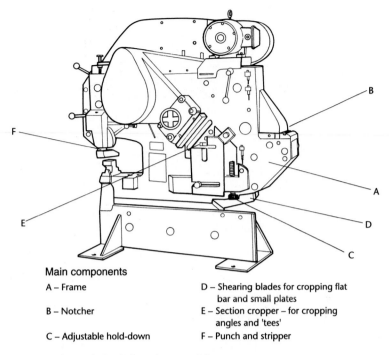

Main components

A – Frame

B – Notcher

C – Adjustable hold-down

D – Shearing blades for cropping flat
  bar and small plates

E – Section cropper – for cropping
  angles and 'tees'

F – Punch and stripper

Figure 6.55 Universal steel shearing machine

6.7 Flame-cutting

Flame-cutting is used for cutting thick plate and structural sections. It can only be used for cutting ferrous metals since the process depends upon a chemical reaction between the hot steel and a stream of high-pressure oxygen. Despite its appearance a set of flame-cutting equipment is *not* the same as welding equipment although it looks similar. The oxygen regulator must be capable of passing a much larger volume of gas and a special torch and nozzle is used. The purity of the oxygen stream is very important and should be 99.5% or better. For every 1% reduction in purity the oxygen consumption increases by 25% and the cutting speed is also reduced by 25%.

A typical cutting nozzle is shown in Fig. 6.57. It provides a ring of flames to preheat the metal being cut to 870–900°C (red heat). When the metal reaches this ignition temperature, a powerful jet of pure oxygen is released into the centre of the preheated area. This initiates an *exothermic reaction* between the hot steel and the oxygen jet. The steel immediately below the oxygen jet is converted into metal oxides or *slag*. These oxides are blown away by the jet of oxygen in a shower of sparks. If the jet is not powerful enough or the speed of cutting is too quick the slag will solidify in the cut and the cut will not be achieved. This oxidizing reaction helps to heat up the metal being cut and the process becomes continuous. The cutting torch is then moved forward along the line of the cut. Figure 6.58 shows a cutting torch and various types of nozzle. The purity of the cutting oxygen stream is very important, and the preheat flame acts a barrier, keeping out atmospheric

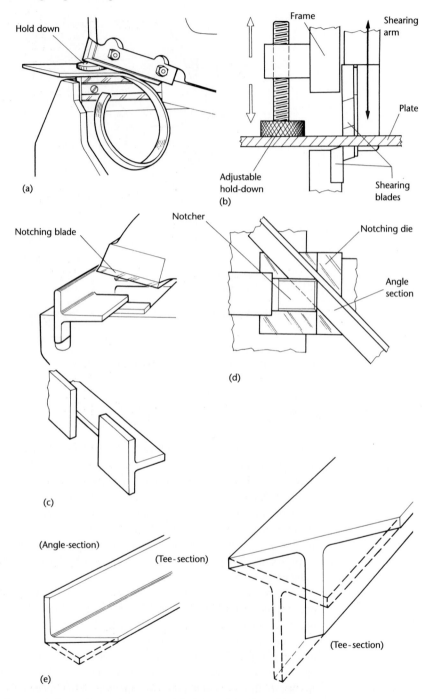

Figure 6.56 Cutting operations: (a) shearing plate; (b) provision for holding plate; (c) notching tee-section; (d) mitring angle flange with the notching tool; (e) typical mitre cutting operations performed on section cropper

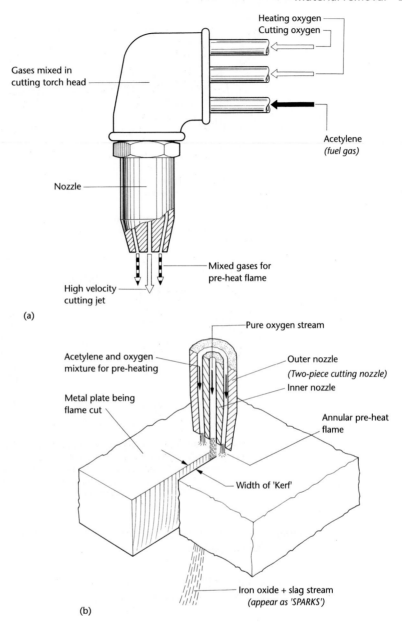

Figure 6.57 The action of oxygen cutting

nitrogen that would react with the oxygen to produce oxides of nitrogen. These oxides will reduce the speed of cutting and increase the oxygen consumption.

Oxygen cutting can be used to cut a wide range of steel types, but the composition of the steel may require some other process such as preheating prior to cutting. It is important that the full composition of the steel is known before cutting commences so that metallurgical or process problems

An exothermic reaction is a chemical reaction that generates heat

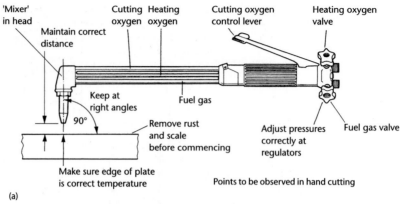

(a)

The size of the torch used depends upon whether it is for light duty or heavy continuous cutting and the volume of oxygen used is much greater than that of fuel gas (measured in LITRES PER HOUR)

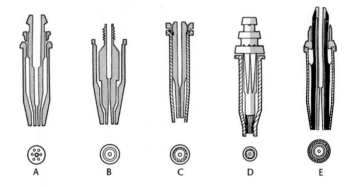

A   One-piece ACETYLENE cutting nozzle – parallel bore, 3 – 9 pre-heat holes, no skirt
B   Two-piece ACETYLENE cutting nozzle – venturi bore, pre-heat annulus, no skirt
C   Two-piece NATURAL GAS nozzle – venturi bore, pre-heat flutes, long skirt
D   Two-piece PROPANE nozzle – parallel bore, pre-heat slots, long skirt
E   Two-piece PROPANE nozzle – parallel bore, pre-heat flutes, long skirt, oxygen curtain

(b)

Figure 6.58 Oxygen cutting-torch details: (a) the cutting torch; (b) cutting nozzle design feature

can be avoided. Table 6.11 lists some of the more common constituents of steels and their effect on the oxygen cutting process.

6.7.1 Fuel gases for cutting

Although the ignition temperature is dependent on the composition of the steel being cut, the choice of fuel gas has an effect on how quickly the flame will raise the material to the ignition temperature. The higher flame temperature of oxy-acetylene (3160°C) compared with oxy-propane (2828°C) will mean that if oxy-acetylene is used then cutting would commence sooner than for oxy-propane. The thicker the material being cut, the more pronounced this effect becomes. However, acetylene is more costly than propane.

Various fuel gases such as acetylene, propane, mixed gases (MAPP) and natural gas may be used in conjunction with oxygen for cutting, and these gases together with their characteristics are listed in Table 6.12.

Table 6.11 Constituents of steel that affect oxygen cutting

Alloying element	Range for good cutting	Comments
Carbon	up to 0.25%	Above 0.25% carbon preheating over a wide area of 250–300°C may be necessary
Manganese	up to 14%	Cutting is difficult above 14% and will need extra pre-heat
Chromium	up to 1.5%, but cutting speeds will be lower	Chromium at 1.5% to 5% requires extra pre-heat. Above 5% process is not recommended
Nickel	up to 3%	Between 3% and 7% requires additional pre-heat. Above 7% process not recommended
Molybdenum	up to 1.5%	Molybdenum at 1.5% to 5% requires extra pre-heat. Above 5% process is not recommended

Table 6.12 Characteristics of fuel gases used when oxygen cutting

	Acetylene C_2H_2	Propane C_3H_8	MAPP C_3H_4
Ratio fuel to oxygen	1:1.1	1:3.75	1:2.5
Max. flame temperature (°C)	3160	2828	2976
Heat distribution to primary flame (kJ/m^3)	18890	10433	15445
Heat distribution to secondary flame (kJ/m^3)	39882	85325	56431
Calorific value of flame (kJ/m^3)	54772	95758	71876

Acetylene

As previously stated, the high flame temperature of acetylene means that the pre-heat times are the shortest of any of the fuel gases. This is particularly important as the thickness of the material being cut increases. Acetylene is also beneficial when bevelling plates, as the cut length is greater than the thickness of the material. Acetylene also tends to produce cuts with much smaller and less pronounced drag lines on the surface. Further, the depth of flame-hardening is less than for other fuel gases. Although the depth and degree of hardening is not important for all applications, if a component is to be welded after cutting, the hard edge can have a harmful effect on the welded joint and possibly lead to cracks forming. Similarly a hard edge can lead to problems with cutting tools if the component is to be machined after cutting. The more localized area of heating associated with oxy-acetylene cutting tends to reduce the level of distortion experienced when cutting. Table 6.13 lists the maximum depth of hardening for a range of pre-heat gases.

Propane

The main advantage of propane is that most of its heat energy is concentrated in the outer or secondary envelope of the flame. While this leads to heating over a much

Table 6.13 Maximum depth of hardening for a range of pre-heat gases

Pre-heat gases	Maximum depth of hardening	
	Mild steel	Alloy steel
Oxy-propane	6.0 mm	3.5 mm
Oxy-MAPP	5.0 mm	2.5 mm
Oxy-acetylene	4.75 mm	2.0 mm

wider area than experienced with gases such as acetylene, it does mean that it is less sensitive to the distance between the end of the nozzle and the workpiece. Thus it is especially useful where operators lack experience and where operators are piercing holes in components because there is less chance of the molten metal splashing up into the nozzle, affecting its performance.

MAPP

As MAPP is a mixture of several fuel gases including propane and methyl acetylene it combines the properties of both gases. It is the dilution of the methyl acetylene with the other gases which reduces its instability, thus making it safer to handle than acetylene alone. The addition of methyl acetylene gives MAPP a higher primary flame temperature and heat distribution than propane, whilst the propane provides a higher secondary flame temperature and energy distribution than acetylene. This even heat distribution makes MAPP less susceptible to changes in stand-off distances between the nozzle and the workpiece. One niche area where MAPP is widely used is underwater cutting, where its stability allows for a greater outlet pressure than would be safe with acetylene alone. Acetylene cannot be used below about 10 m of water, whereas MAPP can be used at much greater operational depths.

Natural gas

This is available as a bulk supply from the local mains distribution system. It is not only the cheapest of the fuel gases that can be used for oxygen cutting but only requires the least amount of oxygen – other than acetylene – for complete combustion with a ratio of 1:1.6. Natural gas has the lowest flame temperature (2786°C) and the lowest calorific value of all the fuel gases. It finds applications in cutting thick materials where the above characteristics result in slow cutting, which can be an advantage since it reduces the flame-hardening effect.

6.7.2 Process comparisons

Oxygen cutting versus plasma cutting:

- Oxygen cutting can cut through steel 2000 mm thick whereas plasma cutting is limited to steel 500 mm thick.
- Oxygen cutting can only cut steels whereas plasma cutting is suitable for all materials.
- Manual oxygen cutting does not require an electrical supply.

- Manual oxygen cutting can be carried out on site.
- Oxygen-cutting nozzles are relatively cheap to replace and last a long time. Plasma-cutting parts are expensive and have a short working life.
- Above 25 mm material thickness, oxygen-cutting speeds are faster than those achievable with the plasma processes. Below 6 mm material thickness plasma cutting is very much faster than oxygen cutting.
- Equipment costs for oxygen cutting are much lower than for plasma cutting.
- Plasma cuts tend to have one straight and one bevel edge. This can lead to difficulties when profile cutting as to the direction in which the job must be cut.
- Oxygen cutting produces a symmetrical cut face so this is not an issue.

Oxygen cutting versus laser cutting:

- Oxygen cutting can cut through steel 2000 mm thick whereas laser cutting is limited to steel about 30 mm thick.
- Oxygen cutting can only cut steels whereas laser cutting can cut most fabrication materials.
- Oxygen cutting does not require an electrical supply, whereas laser cutting machines do require an electrical supply.
- Manual oxygen cutting can be performed on site whereas laser cutting is a factory-based automated production process.
- Oxygen-cutting nozzles are relatively cheap and have a long working life. Laser optics are very expensive and have a shorter working life.
- Above about 25 mm material thickness oxygen cutting is much faster than laser cutting. Below 6 mm material thickness laser cutting is very much faster than oxygen cutting.

6.7.3 Flame-cutting torch attachments

Figure 6.59 shows some useful attachments which, when fitted to the hand-held cutting torch, ensure a steady rate of travel and enable the operator to execute straight lines, bevels and circles with relative ease.

Single cutting support

This simple device may either be a 'spade-type' support or a single 'roller guide' which can be adjusted vertically to give the correct 'stand-off'. Figure 6.59(a) shows a single roller guide supporting the cutting torch at the correct height above the work and guiding it by means of a straight-edge clamped to the workpiece.

Circle cutting device (small diameter)

This is simply a pivot which is attached to the shank of the torch at the required distance from the nozzle according to the radius required. The point of the pivot is located in a centre punch mark as shown in Fig. 6.59(b). Vertical adjustment is provided in order to set the correct 'stand-off'.

Circle cutting device (large diameter)

An example of a large-circle cutting attachment is shown in Fig. 6.59(c). This is similar to the small-diameter-circle cutting device except that it has a graduated radius bar to facilitate setting the pivot position.

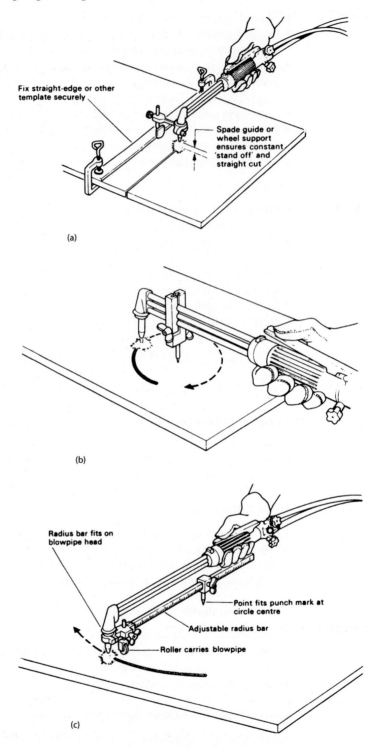

Figure 6.59 Useful cutting torch attachments. (a) Cutting with straight-edge and single support; (b) small circle cutting; (c) large circle cutting

6.7.4 Flame gouging

This process is very similar to flame-cutting except that instead of severing the metal, a groove is gouged out of the surface of the plate. The principle of operation is the same as that used in oxy-fuel gas-cutting processes except that a special type of nozzle is used in a standard cutting torch. When gouging, the metal is preheated and then the cutting oxygen is turned on as shown in Fig. 6.60. Unlike the flame-cutting process where the flame and cutting oxygen is at 90° to the surface of the plate being cut, when gouging, the flame and cutting oxygen is at 5° to 15° to the surface of the plate. Gouging is a useful process for removing defective welds for maintenance purposes and also for removing local defects in a plate prior to re-welding.

6.7.5 Machine cutting (oxy-fuel gas)

Oxy-fuel gas cutting machines consist of one or more cutting torches and a means for supporting and propelling them in the required direction with high precision. There are many machines available, ranging from simple light-weight portable devices to floor-mounted static types having multiple heads to maximize production and controlled by templates or by computer numerical control systems. A typical portable cutting machine is shown in Fig. 6.61. It is basically a self-propelled light-weight tractor or 'mouse' which carries the cutting equipment. The machine is electrically propelled and will make runs of any length. For straight-line work the tractor runs along an extruded aluminium track or an inverted length of channel iron clamped to the plate being cut. For circle cutting, the tractor is placed directly onto the plate being cut and is controlled by a radius bar as shown.

Floor-mounted profiling machines often have several oxy-fuel gas cutting heads and can cut multiple blanks from plate in one operation. Older-type machines use a template to control the profile being cut but more modern machines are computer controlled (CNC). The computer program is far less costly and quicker to produce than a traditional template.

When using a template-type profiling machine, the template must make allowance for:

1. The 'kerf' of the cut. This is the actual width of the metal removed in the cutting process. It may vary between 1.5 and 2 times the diameter of the cutting oxygen orifice of the cutting nozzle used.
2. The diameter of the tracing roller.

6.7.6 Health and safety when oxy-fuel gas cutting

In oxy-fuel gas cutting several types of safety hazard are present. These include:

- Particulate and gaseous fumes.
- Burns caused by the high levels of heat produced by the process.
- Burns caused by metal ejected from the process.
- Potential asphyxiation if work is carried out in a confined space.
- Oxygen enrichment in a confined space.
- Fire hazards due to the flammability of the fuel gases employed.

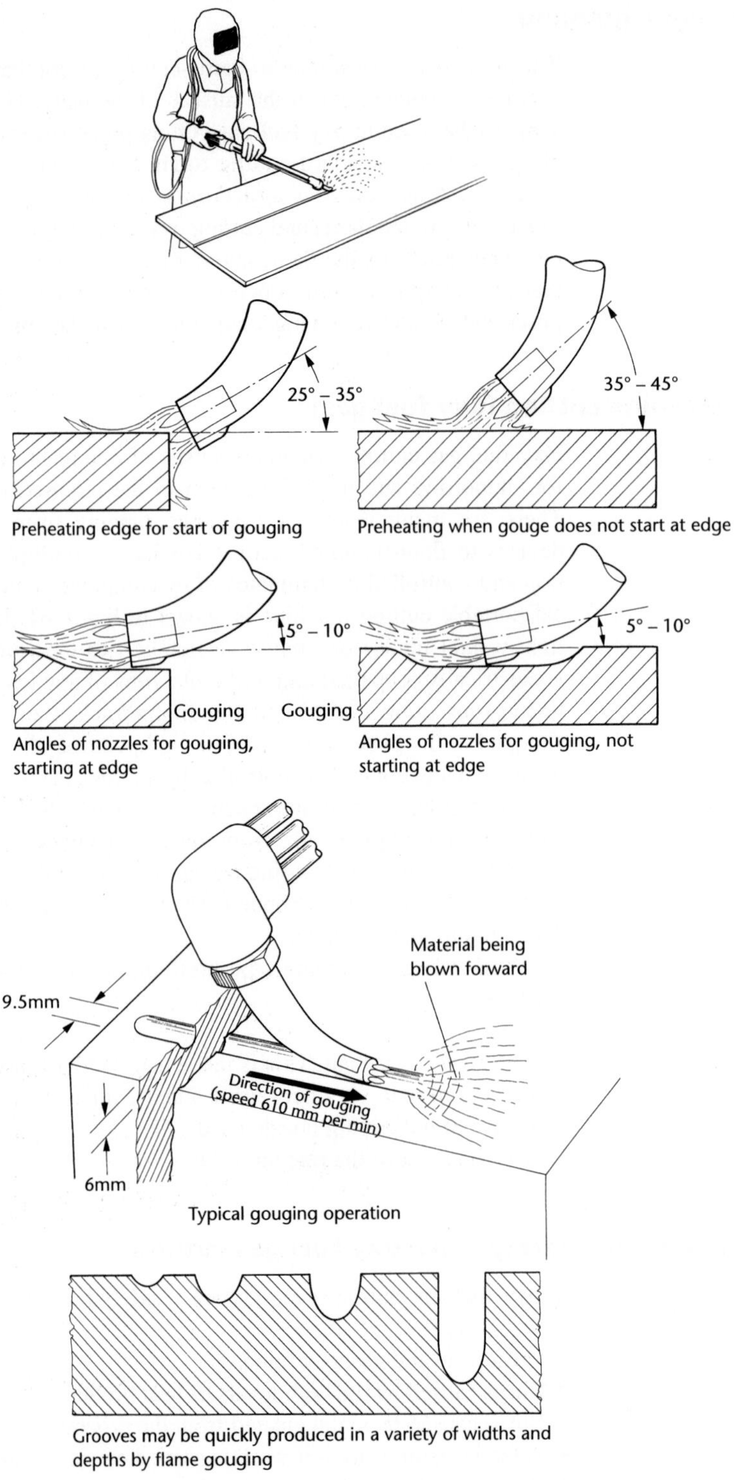

Preheating edge for start of gouging

Preheating when gouge does not start at edge

25° – 35°

35° – 45°

Angles of nozzles for gouging, starting at edge

Angles of nozzles for gouging, not starting at edge

5° – 10°

5° – 10°

Gouging

Gouging

9.5mm

6mm

Material being blown forward

Direction of gouging
(speed 610 mm per min)

Typical gouging operation

Grooves may be quickly produced in a variety of widths and depths by flame gouging

Figure 6.60 Flame gouging

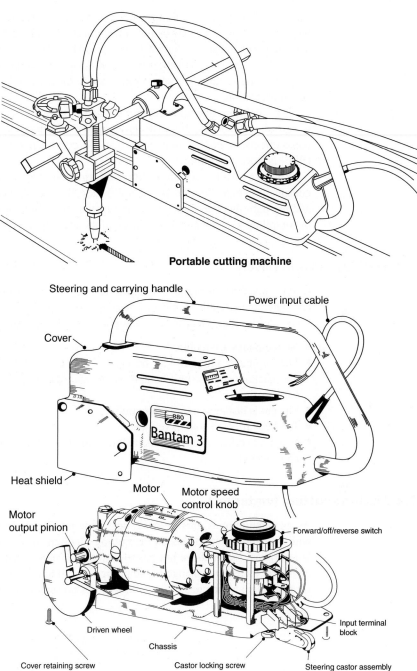

Portable cutting machine

Figure 6.61 Portable type oxygen cutting machine

The main safety issue when oxygen cutting is the production of fumes; especially when cutting paint-coated or galvanized surfaces. These fumes can be hazardous, causing problems such as *metal fume fever*. Therefore, it is necessary to ensure there is an adequate fume extraction in operation which is capable of removing the hazards. In some instances, it may be necessary for the operator to use a respirator for added protection.

It is important that all necessary precautions are taken to ensure that each hazard is considered individually (risk assessment), and that action is taken to minimize these effects. It is also important that the correct personal protective equipment (PPE) is worn at all times.

Readers wishing to obtain further information about oxy-fuel gas cutting should consult the Trade File 5.6 published by the British Oxygen Company. The author and the publishers are indebted to the British Oxygen Company for their assistance in compiling the above sections on oxy-fuel gas cutting.

6.7.7 Machine cutting (laser)

Cutting machines using an industrial laser in place of an oxy-acetylene flame and computer numerically controlled systems are rapidly replacing the older type of flame-cutting equipment for many applications. The advantages of this process are:

- No templates are required.
- The computer program can be created 'off-line', saved to disk and inserted in the machine control unit as and when required, thus saving on 'down-time' and 'setting-time'.
- The cut is neater and more accurate.
- The heat affected zone is reduced.
- Materials other than ferrous metals can be cut (i.e. non-ferrous metals and non-metals).

6.7.8 Machine cutting (water jet)

As its name implies, this process uses a fine jet of very-high-pressure water mixed with fine abrasive particles as the cutting medium. It has all the advantages of the laser process listed above but has the added advantage that no heat is involved and, therefore, there is no change to the properties of the material. The cut edge has a finish equivalent to fine machining and requires no subsequent finishing.

Exercises

6.1 Cutting tool principles

a) Sketch a single-point metal-cutting tool and label it to show the rake angle, clearance angle and the wedge (tool) angle.

b) List the factors that influence the choice of rake angle.

c) With the aid of sketches show how sheet metal and tin plate can be held in a vice so that it can be sawn and filed without vibrating.

d) Show with the aid of a sketch what a vice shoe is, how it is used, and its purpose.

e) With the aid of sketches show how the basic cutting angles are applied to the teeth of a hacksaw blade and explain the purpose of the 'set' and chip clearance of the blade.

f) With the aid of sketches show how the basic cutting angles are applied to a cold chisel and how these are influenced by the angle of inclination of the chisel.

g) With the aid of sketches show how sheet metal may be cut between the chisel and the vice jaws to create a shearing action.

h) Explain how a file is specified and give typical applications of (i) a half-round file, (ii) a three square file, (iii) a ward file, (iv) a round file and (v) a flat file.

i) Name the faults you would look for before using the following items of equipment:
i) hammer
ii) chisel
iii) file
iv) spanner
v) screwdriver.

6.2 Hole cutting equipment

a) Explain why a drill may cut a hole that is larger than the nominal size of the drill but never smaller.

b) i) With the aid of a sketch show how the basic cutting angles are applied to a twist drill.
ii) State the purpose of the flutes of a twist drill.

c) Explain why a twist drill is unsuitable for cutting large diameter holes in sheet metal and thin plate.

d) Compare the advantages and limitations of a trepanning cutter and a hole saw for cutting large diameter holes in sheet metal and thin plate.

e) State what is meant by the term 'sensitive feed' as applied to drilling machines.

f) Compare the relative advantages and limitations of a bench drilling machine, a pillar drilling machine and a radial arm drilling machine. Give examples where each type would be used.

6.3 Metal-cutting by shearing

a) Give examples showing where the following types of shear would be used to cut sheet metal: snips, bench shear, guillotine shear.

b) With the aid of sketches explain why a small clearance gap should be left between the blades of a guillotine shear, state the size of this clearance and state the approximate shear angle between the blades.

c) With the aid of sketches show the principles of the rotary shear and how this principle can be employed in both a straight line and a circle cutting machine.

d) Sketch a simple blanking tool for use in a stamping press and also a simple piercing tool for a rectangular component 50 mm by 75 mm with corner radii of 5 mm. There is a hole 25 mm diameter to be pierced in the centre of the blank. The mild steel strip from which this component is to be cut is 60 mm wide by 1.0 mm thick and is supplied in 1.5 m length strips.

e) i) Calculate the clearance between the punch and the die for the blanking tool and state whether this is applied to the punch or the die orifice.
ii) Given that the shear strength of the mild steel strip is $450\,\text{N/mm}^2$, calculate the force required to pierce the hole through the blank.

6.4 Portable power tools

a) In the interests of safety what visual inspections of an item of portable equipment should be carried out before accepting it for use?

b) State how the hazard of electric shock can be minimized when using portable power tools.

c) With the aid of sketches describe the difference between portable electric shears and portable electric nibblers. List the relative advantages and limitations of portable shearing and nibbling machines.

d) State the safety precautions that must be taken when using portable grinding machines. Under what conditions must the maximum safe operating speed be marked on the grinding wheel and on the grinding machine?

e) With the aids of sketches describe the difference between straight-type and angle-type portable grinding machines and give examples where each type would be used.

f) A grinding wheel is marked 38A40-K4R. What does this mean?

6.5 Flame-cutting

a) State the main differences between a set of oxy-acetylene flame-cutting equipment and a set of oxy-acetylene welding equipment.

b) Explain the basic principles of flame-cutting and explain why this technique can only be applied to ferrous metals.

c) List the essential protective clothing and equipment that should be worn when flame-cutting. What additional safety precautions should be taken?

d) With the aid of diagrams show how the cutting torch can be guided when cutting straight lines and cutting circles.

e) Describe an item of equipment that may be used to automate the cutting process.

f) Compare the advantages and limitations of: oxy-acetylene cutting, laser cutting and water-jet cutting.

7 Sheet and plate metalwork

7.1 Sheet and plate metalwork (introduction)

Sheet metalwork is the manipulation of sheet metal of 3.5 mm in thickness, or less, using mainly hand tools or portable power tools in order to manufacture a range of diverse products. Plate metalwork is the manipulation of metal plate over 3.5 mm in thickness using mainly power tools. The fabricated products for both sheet metalwork and for thin plate metalwork are produced from flat blanks that have been marked out as discussed in Chapter 5 and then cut out (contoured) prior to forming to shape. The forming operations used range from simple bending and rolling operations to more complex flow-forming operations such as spinning and presswork as shown in Fig. 7.1.

7.1.1 Forming by bending (folding)

The terms folding and bending are loosely used in the sheet-metal industry and largely interchangeable in common parlance. To be precise, the term 'folding' refers to sharp corners with a minimum bend radius. The term 'bending' refers to deflections of relatively large corner radii. Folding and bending involve the deformation of material along a straight line in two dimensions only.

When a bending force is applied to a workpiece under free bending conditions, the initial bending is elastic in character. This is because the stresses that are developed in the opposite faces of the material are not sufficiently high to exceed the *yield strength* of the material. The stresses developed on the outside of the bend tend to stretch the metal and are, therefore, *tensile stresses*. The stresses developed on the inside of the bend tend to shorten the metal and are, therefore, *compressive stresses*. The movement or strain which takes place in the metal as a result of the initial bending force is elastic only and, upon removal of the force, the workpiece springs back to its original shape.

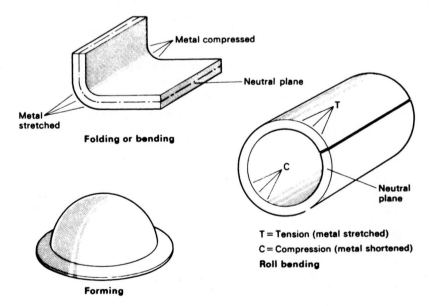

Figure 7.1 Comparison of common cold-forming applications

As the bending force is gradually increased these stresses, both tensile and compressive, produced in the outermost regions of the material, will eventually exceed the yield strength of the material. Once the yield strength of the material has been exceeded, the movement (strain) which occurs in the material becomes *plastic* and the material takes on a *permanent set*. This permanent strain occurs only in the outermost regions furthest from the *neutral plane* (*neutral axis*). The neutral plane is an imaginary plane situated between the tension side and the compression side of the bend of the material where the metal is neither stretched or shortened but maintains its original length. Its position will vary slightly due to the differing properties of different materials, their thickness and their physical condition. Therefore, there is a zone adjacent to the neutral plane where the strain remains elastic.

On release of the bending force the material adjacent to the neutral plane will try to give up its elastic strain energy and straighten the material out. However, the greater portion of the material which has suffered plastic deformation will resist this release of elastic strain energy and the material will remain bent. Nevertheless, there will be some slight recovery of shape and this is known as '*spring-back*'. To allow for this spring-back a degree of '*over-bend*' is required. Figure 7.2 shows the effects of a bending force on a material.

In reality, 'free-bending' conditions rarely, if ever, occur intentionally. Folding or bending usually occurs in press tools (pressure bending) or in folding machines. The principle of pressure bending is shown in Fig. 7.3.

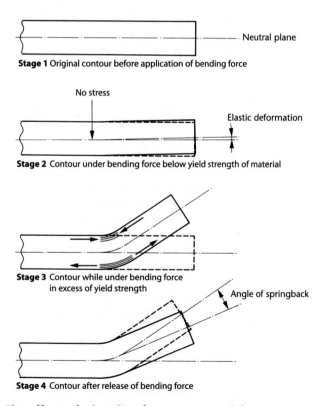

Stage 1 Original contour before application of bending force

Stage 2 Contour under bending force below yield strength of material

Stage 3 Contour while under bending force in excess of yield strength

Stage 4 Contour after release of bending force

Figure 7.2 The effects of a bending force on a material

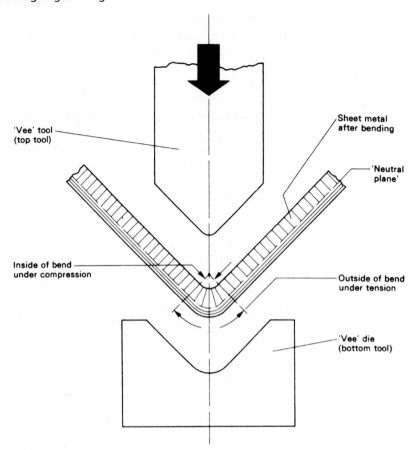

Figure 7.3 Bending action – pressure bending

7.1.2 Spring-back

Spring-back has already been mentioned in the previous section. We will now consider it in more detail. When bending a material an unbalanced system of varying stresses occurs in the region of the bend. When the bending operation is complete and the bending force is removed, this unbalanced system of stresses tends to return to a state of equilibrium. The material tries to spring back and any part of the elastic stress which remains in the material becomes *residual stress* in the bend zone. The amount of spring-back to be expected will vary because of the differing composition and mechanical properties of the materials used in fabrication processes. Some materials, because of their composition, can withstand more severe cold-working than others.

The severity of bending a specific material can withstand without cracking depends upon the *ratio of bend radius to material thickness*.

- A 'tight' (small) radius causes greater cold-deformation than a more generous bend in a material of the same thickness.
- A thicker material develops more strain-hardening (work-hardening) than is experienced in a thinner material bent to the same inside radius.

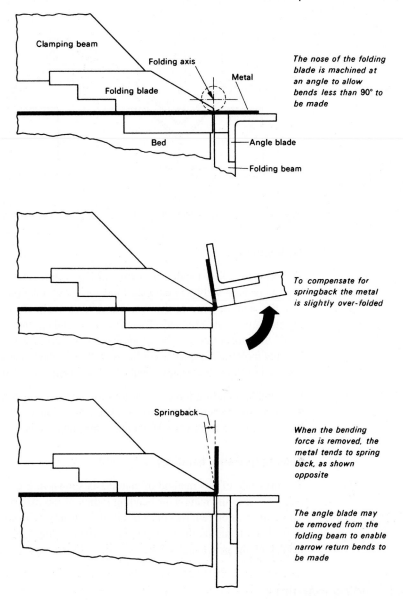

Figure 7.4 Allowing for springback on a folding machine

The '*condition*' of the material being bent will influence the amount of spring-back likely to occur. For example, an aluminium alloy that has been cold-rolled to the *half-hard condition* will exhibit greater spring-back than the same alloy in the *fully annealed condition* for the same degree of bending.

7.1.3 Compensating for spring-back

Figure 7.4 shows how the clamping beam of a folding machine is specifically designed to compensate for spring-back, whilst Fig. 7.5 shows two methods of compensating for spring-back when using a *press-brake* or a 'vee' tool in a fly press.

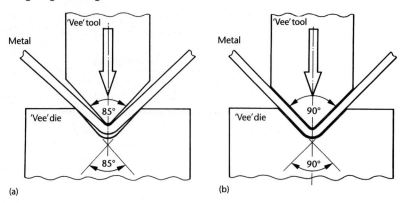

Figure 7.5 Two methods of pressure bending. (a) Air bending; (b) coining

Air-bending

The principle of air-bending is shown in Fig. 7.5(a). This allows for various angles of bend to be achieved by *three-point loading*. These three points are the *two edges* of the 'vee-die' and the *nose* of the 'vee-punch' (top tool). During air-bending, the material retains some of its elasticity. Therefore, the bending angle must be over-closed (over-bend) to compensate for the spring-back that occurs when the tools are opened. The bending tools are designed accordingly, with both the top and bottom tools having 'vees' that are less than 90°, usually about 85°. The advantages of air-bending are:

- A smaller bending force is required for any given material.
- The ability to bend heavy (thick) sheets and plates.
- The ability to form various angles in the same tools.

Coining (pressure bending)

The principle of this method of bending is shown in Fig. 7.5(b). The nose of the vee-punch crushes the natural air-bending radius of the material on the inside of the bend. This compression removes most or all of the elasticity from the bent material, resulting in the bend retaining the exact angles of the bending tools. Therefore, when coining a bend, both the punch and the die have an included angle of 90°.

7.1.4 Folding machines

Manually operated folding machines are usually used for folding tinplate and thin sheet metal up to 1.62 mm in thickness. An example of a manually operated folding machine is shown in Fig. 7.6. The smallest width of bending is 8 to 10 times the material thickness and the minimum inside corner radius of the bend is 1.5 times the metal thickness.

The procedure for bending sheet metal in a folding machine is as follows:

1. *Clamping*. In clamping, the amount of lift of the clamping beam is important. It should be sufficient to allow the fitting and use of special clamping blades (fingers) and to give adequate clearance for previous folds.

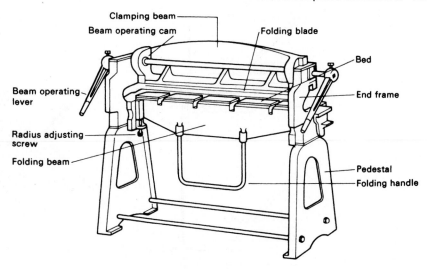

Figure 7.6 Manual folding machine

2. *Folding*. Care must be taken to see that the folding beam will clear the work, particularly when making second or third folds. Some folding machines are designed to fold radii above the minimum, either by fitting a radius bar or by adjustment of the folding beam.

3. *Removal of work*. Care must be taken when folding to ensure that the work can be easily removed on completion of the final bend. The sequence of folding must be carefully planned. The lift of the clamping beam is important when removing the work. Some folding machines, known as *universal folders*, have swing beams. The work may be folded completely around the beam, which is swung out to allow removal of the work.

Some of the above points are shown in Fig. 7.7.

Figure 7.7(a) shows a section through a 'box and pan' folding machine. It is fitted with a standard bed bar and fingers. The sheet metal is shown in position after completion of a right-angle bend when using a standard-angle folding bar.

Figure 7.7(b) shows a small radius bend being made. This time there is a gap between the nose of the folding blade and the face of the folding bar, thus air bending is taking place. This allows a larger radius to be formed.

Figure 7.7(c) shows a series of small return bends being made on this machine using a specially stepped bed bar. Such a bar is very useful for moulded work. The clamping beam lifts high enough to allow that part of the metal on the inside of the beam to be withdrawn over the bar. In this example a narrow blade has been substituted for the standard folding bar. This presents a smaller face width to the folding beam.

Figure 7.7(d) shows the use of radius fingers in conjunction with the standard folding bar. This allows radius bends up to a maximum of 25 mm radius to be made. The radius fingers may be positioned where required on the clamping beam to allow short lengths to be folded.

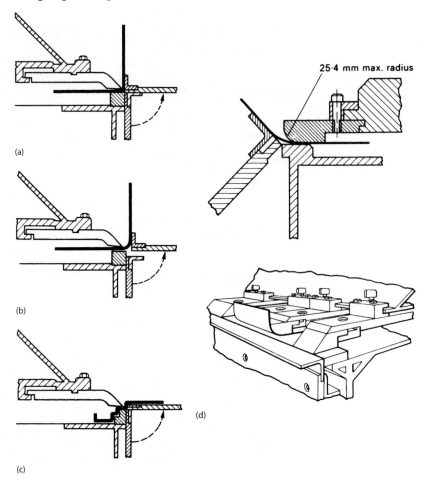

Figure 7.7 Use of folding machines. (a) Folding with standard bed bar; (b) small radius bending; (c) making small reverse bends; (d) use of radius fingers

Figure 7.8(a) shows the variety and combination of bends that can be produced on a standard folding machine, whilst Fig. 7.8(b) shows the use of a mandrel providing the folding machine is fitted with trunnion arms to carry the mandrel. This is only possible if the lift of the clamping beam is adequate. A machine with a clamping beam lift of between 175 and 200 mm will allow a mandrel of 152 mm maximum diameter to be used.

The folding of shallow boxes and pans can also be performed on a universal folding machine, provided there is sufficient lift of the clamping beam to allow an angle clamping blade to be attached to the clamping beam as shown in Fig. 7.8(c).

7.1.5 Bending in press tools

Pressure bending (coining) has already been introduced in Section 7.1.3. This is a common presswork operation for the batch production of small clips and brackets.

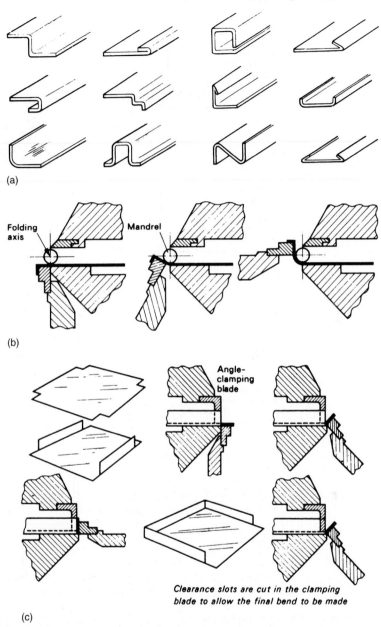

(a)

(b)

(c)

Figure 7.8 Further examples of the use of folding machines. (a) Examples of work produced on a folding machine; (b) use of a mandrel in a folding machine; (c) use of an angel-clamping blade in a folding machine

A typical vee-bending tool suitable for a fly press (hand press) is shown in Fig. 7.9(a) and a typical U-bending tool is shown in Fig. 7.9(b). For larger components and thicker materials, similar tools can be used in power presses. Dedicated angle bending machines are manufactured and an example is shown in Fig. 7.9(c).

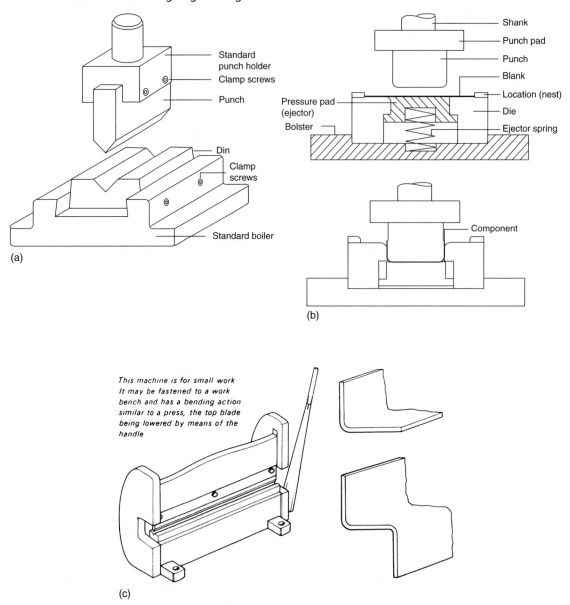

Figure 7.9 Pressure bending. (a) Simple vee-bend tool; (b) U-bending; (c) angle-bending machine

7.1.6 The press-brake

A typical *press-brake* is, in effect, a power press with a very wide but narrow ram and bed. It may be mechanically or hydraulically operated. Press-brakes are designed to bend to a rated *capacity* based on a *die ratio* of 8:1 which is accepted as the ideal bending condition. The meaning of die ratio for a vee-bending tool is explained in Fig. 7.10, whilst Table 7.1 shows the effect of the die ratio on the bending pressure

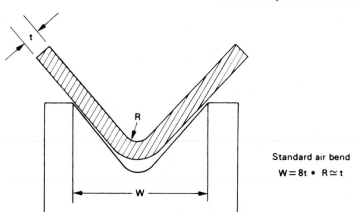

Figure 7.10 Die ratio

Table 7.1 Comparison of 'Vee' die ratios

Metal thickness		Force tonnes/metre (die ratio)*		
s.w.g.	mm	8:1	2:1	16:1
20	0.9	6.8	4.1	3.0
18	1.2	9.1	5.8	4.1
16	1.62	12.2	7.5	5.4
14	2.0	14.9	9.5	6.8
12	2.64	19.6	12.2	8.8
	3.2	23.7	14.6	10.5
	4.8	35.2	22.0	15.9
	6.4	47.4	29.5	21.3
	8.0	58.9	36.6	26.8
	9.5	70.8	44.0	31.8
	11.0	82.6	51.5	37.3
	12.7	94.5	58.9	42.7
	14.3	115.3	66.0	48.1
	15.9	118.2	73.5	53.2
	17.5	129.7	80.9	58.6
	19.0	141.9	88.1	64.0
	20.4	153.5	95.5	69.1
	22.2	165.2	102.6	74.5
	23.8	177.1	110.1	79.9
	25.4	189.3	117.5	85.3

*Required to produce 90° 'air bends' in mild steel (Tensile strength 450 N/m² using a die ratio).

required when bending mild steel. Table 7.2 gives the multiplying factors required when bending metals other than mild steel.

For thin material the die ratio may be reduced to 6:1 but the bending pressure will increase. For thick material the die ratio may have to be increased to 10:1 or even 12:1 to keep the bending pressure within the capacity of the machine.

Let's now consider some typical press-brake operations as shown in Fig. 7.11.

Table 7.2 Bending forces required for metals other than mild steel

Material	Multiplied by
Stainless steel	1.5
Aluminium – soft temper	0.25
Aluminium – hard temper	0.4
Aluminium alloy – heat treated	1.2
Brass – soft temper	0.8

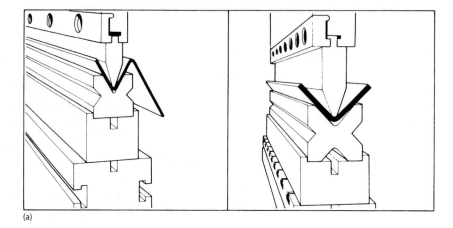

(a)

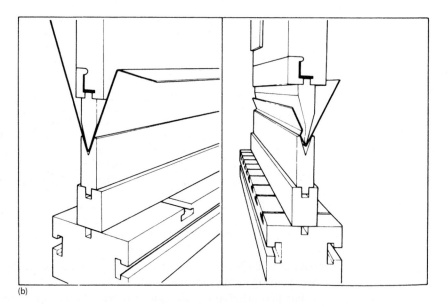

(b)

Figure 7.11 Versatility of pressure bending. (a) Four-way dies; (b) acute angle tools; (c) goose-neck punches; (d) flattening; (e) radius bending; (f) channel forming; (g) box making; (h) beading (see stiffening of sheet metal)

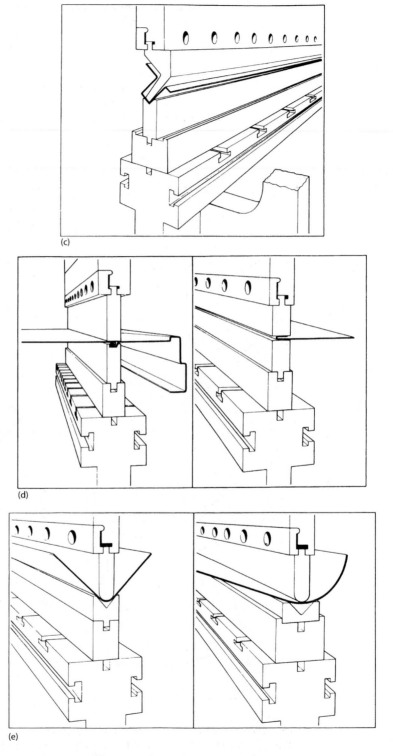

Figure 7.11 (Continued)

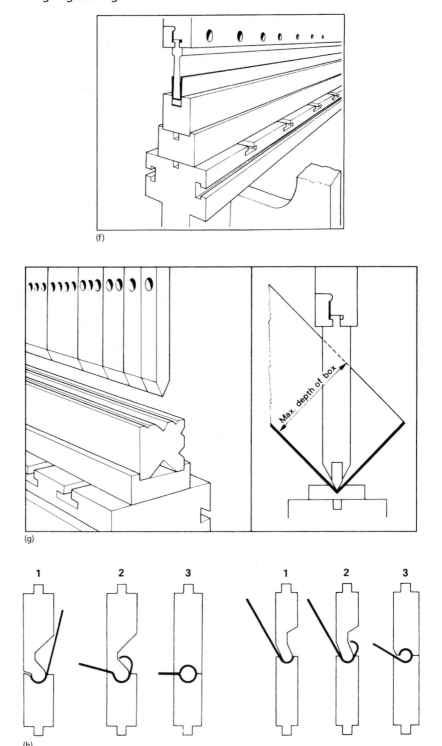

(f)

(g)

(h)

Figure 7.11 (Continued)

Interchangeable four-way dies – Figure 7.11(a)

These are used for air-bending medium and heavy plate. The bottom tool (die) has four 'vee' openings depending upon the die ratio required. The 'vee' openings have an included angle of 85° to provide *over-bend* to allow for *spring-back*. The punch for use with four-way dies has an included nose angle of 60°.

Acute-angle dies – Figure 7.11(b)

Acute-angle dies have many uses and, if used in conjunction with flattening dies, a variety of seams and hems may be produced in sheet metal. These tools are available for any angle but if the angle is less than 35° the sheet-metal component tends to stick in the lower tool (die). Acute-angle dies may be set to air-bend a 90° angle by *raising* the ram and punch.

Goose-neck punch – Figure 7.11(c)

When making a number of bends in the same component, clearance for previous bends has to be considered. *Goose-neck* punches are specially designed for this purpose. These tools are very versatile, enabling a variety of sheet-metal sections to be formed.

Flattening (planishing) tools – Figure 7.11(d)

Flattening tools of various forms may be used either in pairs for flattening a returned edge or hem on the edge of sheet metal, or in conjunction with a formed male or female die. Figure 7.11(d) shows a flat male tool (punch) and a formed female tool (die) closing a countersunk seam in a sheet-metal fabrication.

Radius bending – Figure 7.11(e)

A radius bend is best formed in a pair of suitable tools. The radius on the male punch is usually slightly less than that required in order to allow for 'spring-back' in the material. Large radii, particularly in plate, can be produced by simply adjusting the height of the ram and progressively feeding the sheet or plate through the tools.

Channel dies – Figure 7.11(f)

Channel dies are made with *pressure pads* so that the blank material is held against the punch during the forming operation to avoid the blank slipping out of position at the start of the operation. A channel section in heavy-gauge metal is best made in a 'vee' die with a 'goose neck'-type male tool (punch).

Box-making – Figure 7.11(g)

Male punches for box-making must be as deep as possible. Most standard machines are fitted with box tools which can make boxes of any depth up to a maximum of 170 mm deep. For deeper boxes machines with greater die space than standard are required. For each extra 25 mm of die space (*daylight*) the depth of the box increases by 17 mm. The punches are made in segments of standard widths to suit any size of box.

Beading – Figure 7.11(h)

This shows the three operations required to form a beaded edge. Sheet-metal edges are dangerously sharp and, for safety reasons as well as adding strength and stiffness to the product, have to be beaded, wired (beaded around a wire core) or folded to make them safe. Figure 7.11(h) shows the three operations required to form a beaded edge (hem).

7.1.7 Bend allowance for sheet metal

When sheet metals are bent through an angle, the metal on or adjacent to the outside surfaces becomes *stretched*, whilst the metal on or adjacent to the inside surfaces of the bends becomes *compressed*. It is necessary to make allowance for these effects when developing a template or when marking out a blank sheet prior to bending. The enlarged cross-section of a 90° bend shown in Fig. 7.15 emphasizes the importance of the 'neutral line'.

> This neutral line is an imaginary curved line lying somewhere inside the metal in the bend. Its position and length does not alter from the original flat length during bending

Because there is a slight difference between the amount of *compressive strain* and the amount of *tensile strain*, the *neutral line* does not lie on the centre line of the metal but lies in a position nearer to the inside of the bend as shown in Fig. 7.12.

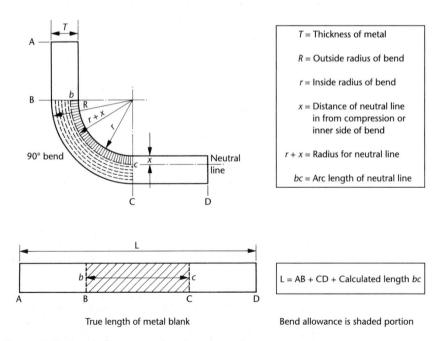

T = Thickness of metal

R = Outside radius of bend

r = Inside radius of bend

x = Distance of neutral line in from compression or inner side of bend

$r + x$ = Radius for neutral line

bc = Arc length of neutral line

L = AB + CD + Calculated length bc

True length of metal blank

Bend allowance is shaded portion

Figure 7.12 Bend allowances for sheet metal

For the purpose of calculating the allowance for a bend in sheet metal, the neutral line curve is regarded as the arc of a circle whose radius is equal to the sum of the inside bend radius plus the distance of the neutral line in from the inside of the bend. The precise position of the neutral line inside the bend depends upon a number of factors which include:

- The properties of the metal.
- The thickness of the metal.
- The inside radius of the bend.

Table 7.3 lists the approximate positions of the neutral line for some common materials and should be read in conjunction with Fig. 7.12. For general sheet metalwork the values for the radius of the neutral line may be used (where precision is unimportant). These are shown in Table 7.4.

Generally, the position of the neutral line is 0.4 times the thickness of the metal in from the inside of the bend. Therefore, the radius used for calculating the bend allowance is equal to the sum of the inside bend radius plus 0.4 times the thickness of the metal. Furthermore, the bend radius is rarely less than twice the metal thickness and rarely more than four times the metal thickness. Therefore, for all practical purposes, when calculating the required length of a thin sheet-metal blank when forming cylindrical or part cylindrical work, the mean circumference is used. That is, the neutral line is assumed to be the central axis of the metal thickness. It is only when working with thin plate and thick plate that the neutral line needs to be calculated more accurately. The terminology used when bending metal is as follows:

- *Bend radius* – the inside radius of the bend.
- *Outside bend radius* –the inside radius of the bend plus the metal thickness.
- *Bend allowance* – the length of the metal required to produce only the radius portion of the bend.

Table 7.3 Neutral line data for bending sheet metal

Material	Average value of ratio x/T
Mild steel	0.433
Half-hard aluminium	0.442
Heat-treatable aluminium alloys	0.348
Stainless steel	0.360

Table 7.4

Thickness of material		Approximate value of neutral line radius
s.w.g.	mm	
30 to 19	0.315 to 1.016	One-third metal thickness plus inside bend radius
18 to 11	1.219 to 2.346	Two-fifths metal thickness plus inside bend radius
10 to 1	3.251 to 7.620	One-half metal thickness plus inside bend radius

Example 7.1

Calculation – centre line bend allowance

Calculate the length of the blank required to form the 'U' clip shown in Fig. 7.13. The position of the neutral line = $0.5T$ (centre line), where $T = 12.7$ mm.

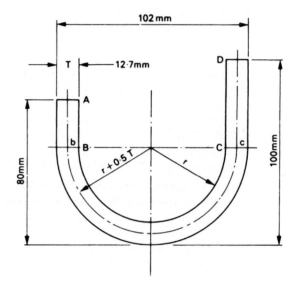

Figure 7.13 Calculation – centre line bend allowance

Solution
The length (L) of the blank is equal to the sum of the straight arm lengths 'AB' and 'CD' plus the mean line (radius) length 'bc'. Thus

$$L = AB + CD + bc$$

where bc represents a semi-circular arc whose mean radius R is equal to the inside radius r plus half the metal thickness T.

The outside diameter of the diameter of the semicircle is given as 102 mm.

$$= 49$$

Therefore, the inside diameter of the semicircle = $102 - (2T)$
$$= 102 - 25.4 = 76.6 \text{ mm}$$

Therefore, the inside radius $r = (76.6)/2 = \textbf{38.3 mm}$
And the mean radius $R = 38.3 + (0.5 \times 12.7)$
$$= 38.3 + 6.35 = \textbf{44.65 mm}$$

Length of arms:

$$AB = 80 - (102)/2 = 80 - 51 = \mathbf{29\,mm}$$
$$CD = 100 - (102)/2 = 100 - 51 = \mathbf{49\,mm}$$

Bend allowance:

$$bc = \pi R = 3.142 \times 44.65 = \mathbf{140.3\,mm}$$

Total length of blank:

$$AB + CD + bc = 29 + 49 + 140.3 = \mathbf{218.3\,mm}$$

Example 7.2

Calculation – neutral line bend allowance

Calculate the length of the blank required to form the bracket shown in Fig. 7.14, using the neutral line value of $0.4T$ from the inside bend radius, and given that the metal thickness $(T) = 6.35\,mm$ and the inside bend radius $(r) = 2T$.

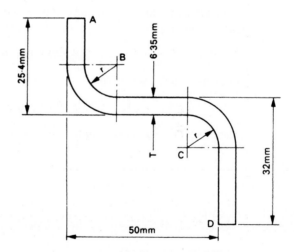

Figure 7.14 Calculation – neutral line bend allowance

Solution
Length of flats:

$$
\begin{aligned}
AB &= 25.4 - (r + T) = 25.4 - (2T + T) \\
&= 25.4 - 3T \\
&= 25.4 - (3 \times 6.35) \\
&= 25.4 - 19.05 = \mathbf{6.35\,mm}
\end{aligned}
$$

Similarly:
$$CD = 32 - 3T = 32 - 19.05 = \mathbf{12.95\,mm}$$

But
$$\begin{aligned}
BC &= 50 - (r + T + r)\\
&= 50 - (2r + T)\\
&= 50 - (4T + T)\\
&= 50 - 5T\\
&= 50 - (5 \times 6.35) = \mathbf{18.25\,mm}
\end{aligned}$$

Total length of flats:
$$6.35 + 12.95 + 18.25 = \mathbf{37.55\,mm}$$

The bend allowance radius:
$$\begin{aligned}
R &= r + 0.4T\\
&= 2T + 0.4T\\
&= 2.4T
\end{aligned}$$

Therefore
$$R = 2.4 \times 6.35 = \mathbf{15.24\,mm}$$

Bend allowance for bends B and C ($90° + 90° = 180°$)
Bend allowance:

$$2\{\theta \times R \times (2\pi/360°)\}$$

where $\theta = 180° - 90° = 90°$

$$= 2 \times 90 \times 15.24 \times 0.0175 \approx \mathbf{48\,mm}$$

Therefore, the total blank length = total length of flats + total bend allowance
$$= 37.55 + 48 = \mathbf{85.55\,mm}$$

7.2 Roll-bending sheet metal and plate

Roll-bending sheet metal and plate is used for:

- Producing cylindrical components.
- Producing conical components.
- Wiring cylindrical edges.

7.2.1 Roll-bending machines

Bending rolls for sheet metal and plate are made in a variety of sizes. Those intended for thin sheet metal and wiring beaded edges are usually manually operated, whilst those for plate work are always power-driven. Bending rolls for sheet metal are

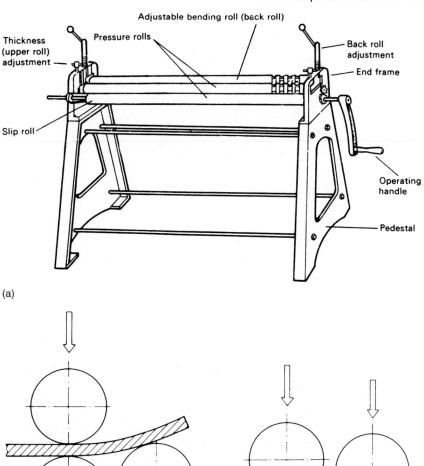

Figure 7.15 Roll bending sheet metal. (a) Sheet-metal bending rolls; (b) roll-up type; (c) roll-down type

known as *pinch-type* machines, whilst those intended for plate work *are pyramid-type* machines. Also available for heavy-duty plate work are universal machines, which may be used for both *pinch* and *pyramid* rolling.

Figure 7.15(a) shows a set of manually operated sheet-metal bending rolls. The grooves at the right-hand end of the rolls are for wire-beading the edges of cylindrical or conical components. This is a *pinch*-type machine. It has two front rollers that are geared together and lightly grip (*pinch*) the sheet, propelling it through the machine.

There is a third 'free' roller at the rear of the machine to 'set' the sheet to the required radius. This third roller may be below the sheet or above the sheet.

Roll-up-type machine

This has the 'free' roller *below* the sheet as shown in Fig. 7.15(b). Roll-up machines have adjustment on the top or bottom pinch roll to compensate for various material thicknesses and adjustment in the upward direction on the back (free) roller to adjust the set of the sheet to the required radius. As a general rule, the minimum diameter that can be rolled on a pinch-type, roll-up machine is about 1.5 to 2 times the diameter of the rear roller.

Roll-down-type machine

This has the free roller *above* the sheet as shown in Fig. 7.15(c). Roll-down machines have adjustment in a vertical direction on the top and bottom pinch rollers to compensate for various material thicknesses and adjustment in a downward direction on the back roller. This type of machine will not roll more curvature than will pass beneath the pedestal frame.

7.2.2 Rolling plate

Rolling machines used for plate work are very much more robust than those used for sheet metal and tin plate. They are always power driven. Whereas heavy-duty, motorized pinch-type rolls are suitable for thin plate work, machines intended for thick plate work are of the *pyramid type*. These have three rolls arranged in a pyramid formation as shown in Fig. 7.16(a).

Most plate-rolling machines are provided with longitudinal grooves along the lower rolls to assist in gripping and driving the plate. These grooves can also be used for the initial alignment of the plate. The top roll is adjustable up or down and may be 'slipped' to allow removal of the work when rolling is complete to form a cylinder (see Section 7.2.3).

Figure 7.19(b) shows the rolls set for 'pinch-bending'. The main advantage of the three-roll pyramid type machine is that for heavy plate the bottom roll centres are wide apart. This reduces the load on the top roller. Since the bottom rollers are mounted in inclined slideways, the bottom roll centres are automatically reduced as the rollers are adjusted upwards for work on thin plate and small diameters.

Figure 7.17(a) shows the layout of a 'four-in-one' universal pyramid/pinch type rolling machine. These machines are capable of performing all the roll-bending operations normally carried out in fabrication workshops. In the hands of a skilled operator it is a universal machine for all types of roll-bending in both thin and thick plate work. Figure 7.17(b) shows the sequence of operations for rolling a section of steel pipeline on a 'four-in-one' rolling machine.

7.2.3 Slip-rolls

When rolling complete cylinders, the finished cylinder is left surrounding the rear roll, so provision has to be made for its removal. Rolls with this provision are referred to as *slip-rolls*. With most sheet-metal rolling machines the roll around

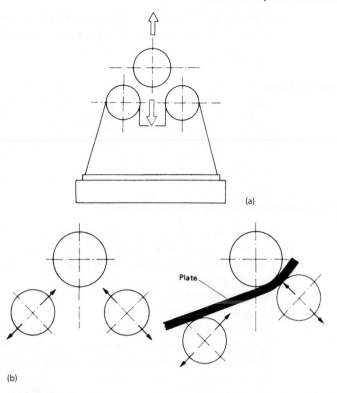

Figure 7.16 Pyramid-type rolling machine. (a) Pyramid-type rolls (standard design); (b) pyramid-type rolls with adjustable bottom rollers

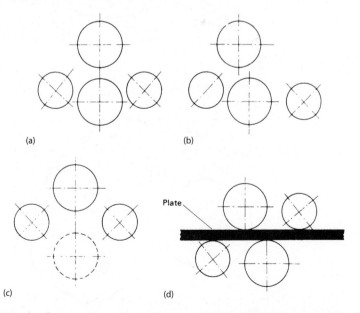

Figure 7.17 The four-in-one universal pyramid/pinch-type rolling machine showing alternative settings of the rollers. (a) Four-roll pinch; (b) three-roll or offset pinch; (c) inclined-roll pyramid bender; (d) four-roll flattening machine

which the cylinder is formed is made to slip out sideways so that the cylinder can be removed. The slip-roll on heavy-duty, power-driven plate-rolling machines usually slips upwards for removal of the cylinder.

7.2.4 Cone rolling

It is possible to roll conical components on both hand- and power-operated rolling machines providing the included angle of the cone is relatively small. This is done by adjusting the rear (curving) roller so that it is at an angle in the horizontal plane to the pinch rollers or, in the case of pyramid rolling machines, so that the curving roller is at an angle to the other two rolls.

7.2.5 Ring-bending rolls for angle sections

Ring-bending rolls may be hand-operated by suitable gearing or power-operated. They are used for the cold-bending of channel, angle and tee-section stock to produce bar rings. The axes of the rollers may be either horizontal or vertical as shown in Fig. 7.18.

Ring-bending rolling machines consist of three rollers arranged in triangular formation (similar to pyramid rolls). Each roller can be split into two sections to accept the flat flanges of angle sections or channels as they are being bent. When bending an outside ring, the flat flange of the angle lies in the slots between the two bottom rollers. These slots are adjusted so as to prevent the flat flange from puckering during the roll-bending operation. For an inside ring the flat flange of the angle lies in the slot of the single

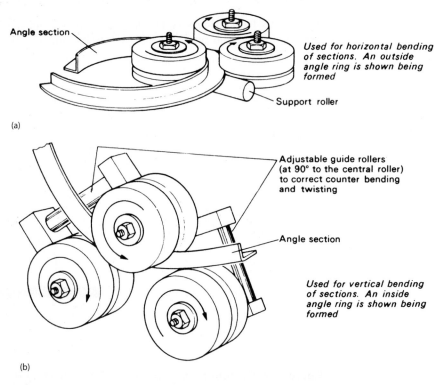

Figure 7.18 Angle ring-bending rolls. (a) Vertical roller arrangement; (b) horizontal roller arrangement

central roller. Pressure is exerted, during the rolling operation, by a screw arrangement which moves the single central roller towards the gap between the other two rollers.

7.3 Flow-forming sheet metal

So far, we have only considered the bending and rolling of sheet metal and plate in two planes. Flow-forming is the shaping or bending of metal in three planes (three-dimensional). This is much more difficult than manipulating metal in two planes since, in three planes, some part of the metal must be *stretched* or *shrunk* or both.

7.3.1 Flanging sheet metal

Let's consider a *flange* that is to be 'thrown' on a curved surface such as a cylinder as shown in Fig. 7.19(a). It can be seen that the edge of the flange, after externally flanging, has a greater circumference than it had before the flange was thrown. In this case the metal has been *stretched*. Now let's consider a flange to be worked up around the edge of a flat metal disc as shown in Fig. 7.19(b). It can be seen that, in this instance, the edge of the disc after flanging has a smaller circumference than it had before flanging. In this case the metal has been *shrunk* (*compressed*). Shrinking or compressing the metal will *increase* its thickness.

Stretching the metal will **reduce** its thickness

Figure 7.20 shows how an angle section can be curved by increasing or decreasing the surface area of one flange. In practice metal is not normally removed by the simple expedient of cutting 'vee-slots'. The surface area is reduced by shrinking (compressing) the metal. This is more difficult and requires greater skill than when producing an externally curved flange because it is much easier to stretch metal by thinning it, than to compress metal by thickening it.

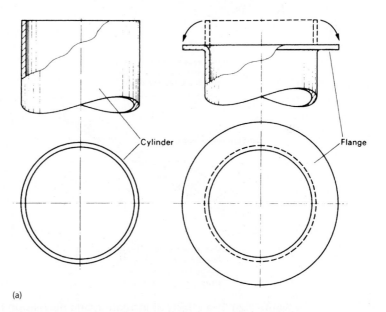

Cylinder

Flange

(a)

Figure 7.19 Comparison of flanging methods. (a) Left: cylinder before flanging, right: cylinder after flanging; (b) left: disc before flanging, right: disc after flanging

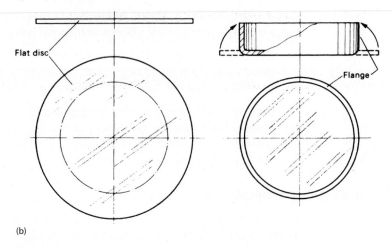

(b)

Figure 7.19 (Continued)

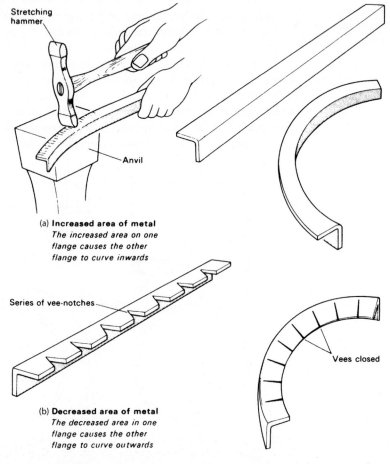

Figure 7.20 The effects of increasing and decreasing the surface area of one flange of an angle strip

7.3.2 The principles of flow-forming sheet metal by hand

A craftsperson needs to posses a thorough knowledge of the properties of the materials which he or she has to use. This enables the craftsperson to understand and even forecast the behaviour of materials which are subjected to applied forces and so be in *control of the desired direction of flow* during flow-forming operations. During flow-forming the metal will tend to *work-harden*. The degree of work-hardening will depend upon the composition of the metal and the severity of the forming operation to which it is subjected. It may be necessary to re-anneal the work several times during a flow-forming operation to prevent it cracking and splitting.

For successful flow-forming the metal must be in the *annealed* (soft) condition

The techniques used for forming work by hand are similar for most materials. The main differences are concerned with:

1. The *force* with which the metal is struck.
2. *The direction* in which the force (blow) is applied.

If a piece of aluminium sheet and a piece of low-carbon steel sheet of the same thickness are struck with blows of equal force, the aluminium, being the softer and more malleable, will deform to a much greater extent than the steel. Since the flow-forming

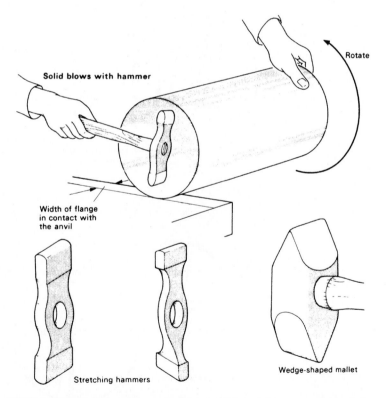

Figure 7.21 Increasing circumference by stretching. (a) Increased area of metal. The increased area on one flange causes the other flange to curve inwards; (b) decreased area of metal. The decreased area in one flange causes the other flange to curve outwards

of sheet metal is essentially a 'hammering' processes, it is most important that we consider the types of blow which can be struck on sheet metal and that each type of blow has its own field of application for any given purpose.

Solid blow

When metal is struck solidly over a steel anvil or 'head', the solid blow will stretch the metal as the typical application in Figs. 7.21 and 7.22 shows. For very soft materials a wooden mallet is less likely to bruise the metal.

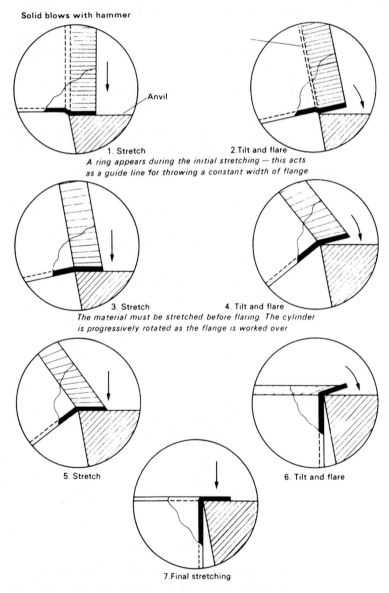

Figure 7.22 Sequence of operations for producing an external flange on a cylindrical body

Elastic blow

Where either the head or the tool (or both) are made of a resilient material such as wood an elastic blow is delivered. An elastic blow will form sheet metal without unduly stretching it and can be used to advantage to thicken the metal when shrinking it. The use of an elastic blow is shown in Fig. 7.23.

Floating blow

Where the head of the anvil is not directly under the hammer a floating blow is delivered. The floating blow is one which is used to control the direction in which the

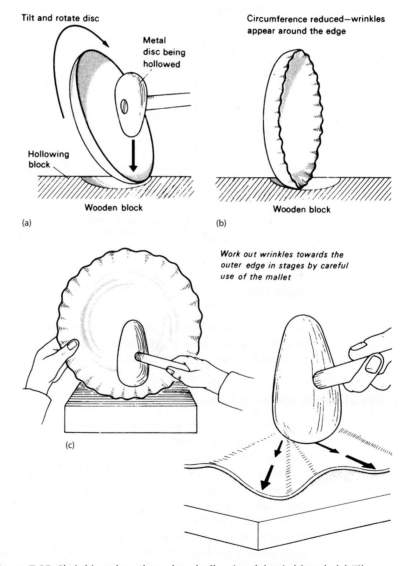

Figure 7.23 Shrinking the edge when hollowing (elastic blows). (a) Tilt to steeper angle; (b) edge caused to wrinkle; (c) use elastic blows

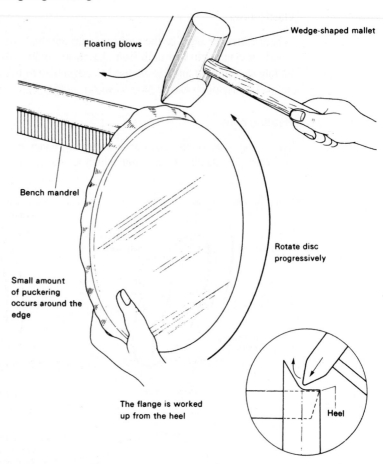

Figure 7.24 Raising a flange on a metal disc

metal is required to flow during the forming process. It is delivered while the metal is held over a suitable head or stake, hitting it 'off the solid' so as to form an indentation at the point of impact. Figure 7.24 shows the use of floating blows.

7.3.3 Use of bench tools for forming sheet metal

Craftpersons frequently find it necessary, when suitable machines are not available, to resort to the use of various types of metal anvils and heads when forming sheet-metal articles. These devices are commonly referred to as 'stakes' and are designed to perform many types of operations for which machines are not readily available or readily adaptable. Good-quality stakes are made from malleable cast iron or cast steel. They are all sold by weight because of their variety of size and shape.

A 'stake' used for sheet metalwork consists basically of a shank supporting a head or horn. The shanks are generally tapered to a standard size and shape at their lower end and are designed to fit into a bench socket. The heads and horns are available in a great variety of shapes and sizes with their working faces machined and polished. Figure 7.25 shows some of the more common types that are available.

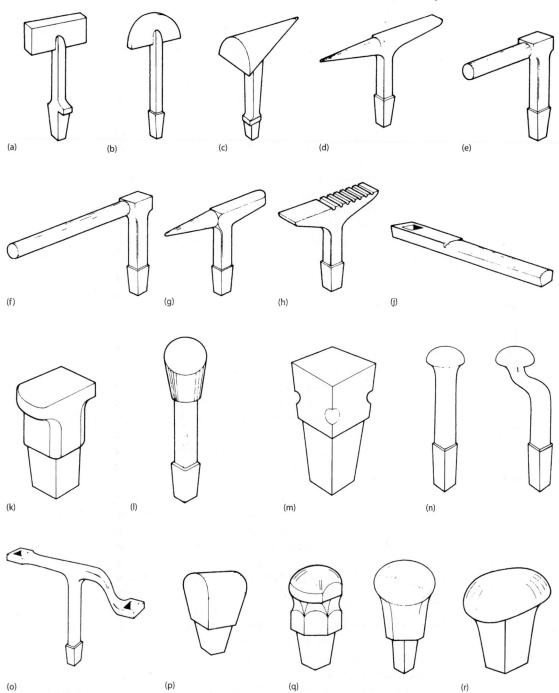

Figure 7.25 Typical bench stakes. (a) Hatchet stake; (b) half-moon stake; (c) funnel stake; (d) beak or brick-iron; (e) side stake; (f) pipe stake; (g) extinguisher stake; (h) creasing iron; (j) bench mandrel; (k) planishing anvil; (l) round bottom stake; (m) canister stake; (n) convex-head stakes; (o) horse; (p) long-head stake; (q) round-head stakes; (r) oval-head stake

- *Hatchet stake* – Fig. 7.25(a). The hatchet stake has a sharp, straight edge bevelled on one side. It is very useful for making sharp bends, folding the edges of sheet metal, forming boxes and pans by hand, and 'tucking-in' wired edges and seaming.
- *Half-moon stake* – Fig. 7.25(b). The half-moon stake has a sharp edge in the form of an arc of a circle bevelled on one side. It is used for throwing up flanges on metal discs, or profiled blanks, preparatory to wiring and seaming. It is also used for 'tucking-in' fired edges on curved work.
- *Funnel stake* – Fig. 7.25(c). As the name implies, this stake is used when shaping and seaming funnels and tapered articles with part conical corners such as 'square-to-round' transformers.
- *Beak- or bick-iron* – Fig. 7.25(d). This stake has two horns, one of which is tapered and the other is a rectangular anvil. The thick, tapered horn or 'beak' is used when making spouts and sharp tapering articles. The anvil may be used for squaring corners, seaming and light riveting.
- *Side stake* – Fig. 7.25(e). A side stake has only one horn which is not tapered. It is more robust than a bick-iron and can withstand considerable hammering. Its main uses are forming, riveting and seaming pipe work. It is also used when forming tapered work of short proportions.
- *Pipe stake* – Fig. 7.25(f). A pipe stake is an elongated version of the side stake and, because of the overhang, is less robust. As its name implies it is used when forming and seaming sheet-metal pipes.
- *Extinguisher stake* – Fig. 7.25(g). This is very similar to a bick-iron. It has a round and tapered horn at one end and a rectangular-shaped horn at the other. Some extinguisher stakes contain a number of grooving slots on the working surface of the rectangular horn. These are useful when creasing metal and bending wire. The tapered horn is used when forming, riveting, or seaming small tapered articles. It is also useful when forming wrinkles or puckers prior to 'raising'.
- *Creasing iron* – Fig. 7.25(h). This has two rectangular shaped horns, one of which is plain. The other horn has a series of grooving slots of various sizes. The grooves are used when 'sinking' a bead on a straight edge of a flat sheet (i.e. reversing wired edges).
- *Bench mandrel* – Fig. 7.25(j). This is firmly fixed to the bench by means of strap clamps which may be quickly released, allowing the mandrel to be reversed or adjusted for length of overhang. The mandrel is double-ended – the rounded end is used for riveting and seaming pipes, whilst the flat end is used for seaming the corners of pans, boxes, square or rectangular ducting and riveting. It also has a square tapered hole in the flat end for receiving the shanks of other stakes and heads. Bench mandrels are available in four sizes ranging from 20 kg to 114 kg.
- *Planishing mandrel* – Fig. 7.25(k). Planishing mandrels are available in a variety of shapes and sizes. The one shown is called a 'Tinsmith's anvil' and is used when planishing flat surfaces in all types of work. The working surface is highly polished.
- *Round bottom stake* – Fig. 7.25(l). These stakes are available in various diameters and have flat working surfaces. They are used when forming the bases of cylindrical work and for squaring knocked-up seams.
- *Canister stake* – Fig. 7.25(m). This stake has square and flat working surfaces. Its main use is for working in the corners and squaring up the seams when working with square or rectangular products.

- *Convex-head stakes* – Fig. 7.25(n). These are used when forming or shaping double-contoured and spherical work. They are usually available in two patterns – with a straight shank and with an off-set (cranked) shank.
- *Horse* – Fig. 7.25(o). This adaptable stake is really a double-ended support. At the end of each arm (one of which is cranked downwards for clearance) there is a square, tapered hole for the reception of a wide variety of heads. Four typical heads will now be shown.
- *Long-head* – Fig. 7.25(p). This is used when making knocked-up joints on cylindrical articles, and also when flanging.
- *Round-heads* – Fig. 7.25(q). Two types of round head are shown and these are used when 'raising'.
- *Oval head* – Fig. 7.25(r). This is oval in shape as shown and has a slightly convex working surface. It sometimes has a straight edge at one end.

The condition of the stake has much to do with the workmanship of the finished articles. Therefore, great care must be taken when using them. If a stake has been roughened by centre-punch marks or is chisel-marked, such marks will be impressed upon the surface of the workpiece and spoil its appearance. Therefore, a stake should never be used to back up work directly when centre-punching or cutting with a cold chisel.

A mallet should be used wherever possible when forming sheet metal. When a hammer has to be used, care must be taken to avoid hitting the metal at an angle so as to leave 'half-moon' impressions on the surface of the stake. Bench tools that have been abused and damaged must be reconditioned immediately. Regular maintenance will avoid marking the surface of the workpiece. Such marks cannot be removed.

7.3.4 'Hollowing' and 'raising'

Hollowing and raising are both methods employed for the purpose of flow-forming sheet metal. Basically:

- *Hollowing* is employed when the desired shape is only slightly domed or hollowed. The sheet metal is beaten into a small indentation; therefore, the metal being formed is stretched and its *original thickness is reduced*.
- *Raising* is always employed where shapes of much greater depth are required. It is a process whereby sheet metal is beaten and induced to flow into the required shape by the application of 'floating' blows struck whilst the metal is slightly off the head or former shape being used. The metal formed by raising is compressed and has its *original thickness increased*.

Figures 7.26 to 7.31 inclusive show the comparison between the two techniques.

7.3.5 Panel beating

Panel beating was originally much in demand when car bodies where built as 'one-offs' by highly skilled coach builders. Nowadays car bodies and sub-frames are integrated fabrications assembled by robots from pressed steel panels. Panel beating is now largely relegated to firms restoring vintage and veteran cars and aircraft, firms producing prototypes, firms rebuilding cars damaged in accidents and similar specialist work.

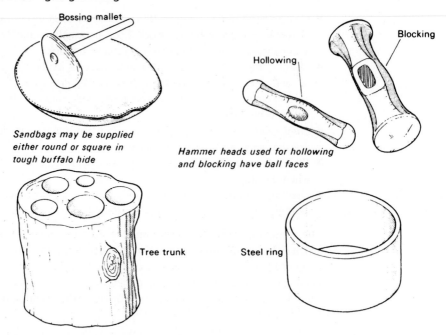

Figure 7.26 The basic tools for hollowing

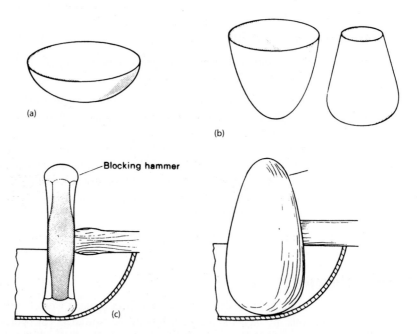

Figure 7.27 Comparison of hollowing and raising. (a) Shallow depth – this would be formed by hollowing; (b) too deep for hollowing – this would be formed by raising; (c) depth limitations

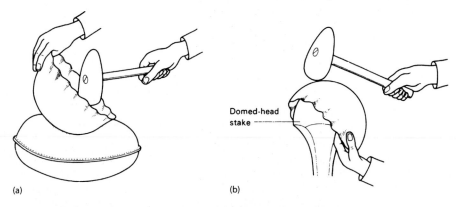

(a)   (b)

Figure 7.28 The two basic methods of forming a bowl. (a) Hollowing a hemispherical bowl on a sandbag; (b) shaping a hemispherical bowl by raising – the metal is made to flow over a solid steel head

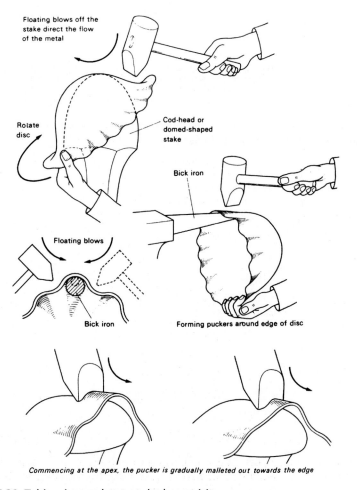

Floating blows off the stake direct the flow of the metal

Rotate disc

Cod-head or domed-shaped stake

Bick iron

Floating blows

Bick iron

Forming puckers around edge of disc

Commencing at the apex, the pucker is gradually malleted out towards the edge

Figure 7.29 Taking in surplus metal when raising

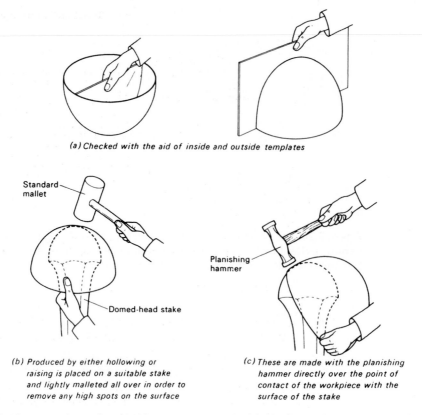

(a) Checked with the aid of inside and outside templates

Standard mallet

Domed-head stake

(b) Produced by either hollowing or raising is placed on a suitable stake and lightly malleted all over in order to remove any high spots on the surface

Planishing hammer

(c) These are made with the planishing hammer directly over the point of contact of the workpiece with the surface of the stake

Figure 7.30 Finishing processes for double curvature work. (a) Checking the contour; (b) double-curvature work; (c) overlapping blows

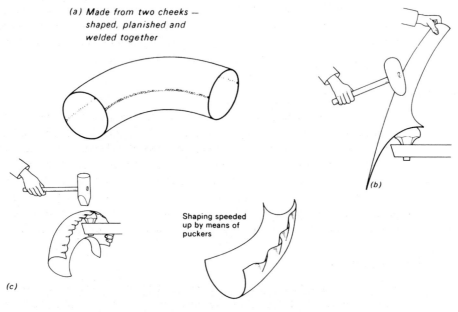

(a) Made from two cheeks — shaped, planished and welded together

(b)

Shaping speeded up by means of puckers

(c)

Figure 7.31 Pipe bend fabricated from sheet metal. (a) Pipe bend; (b) working up the throat of one cheek; (c) working up the back of one cheek

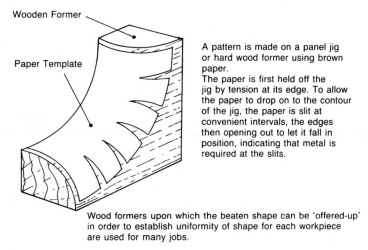

A pattern is made on a panel jig or hard wood former using brown paper.
The paper is first held off the jig by tension at its edge. To allow the paper to drop on to the contour of the jig, the paper is slit at convenient intervals, the edges then opening out to let it fall in position, indicating that metal is required at the slits.

Wood formers upon which the beaten shape can be 'offered-up' in order to establish uniformity of shape for each workpiece are used for many jobs.

Figure 7.32 Split and weld pattern (panel beating)

The skilled panel beater is a craftsperson who, to a great extent, relies on a good eye for line and form. It is a specialized skill which can only be cultivated by years of experience combined with dexterity in the use of hand tools.

In general, most panel-beating work is carried out on deep-drawing quality steel or aluminium alloy. Although aluminium alloy is softer and more malleable than steel, it is more easily overstretched and great care is required when forming it. Providing they have skills and lightness of touch, many panel beaters prefer working with aluminium alloys which they find responds more readily than steel and is much lighter to handle. Whichever metal is used – steel or aluminium – the techniques used are the same.

Basically, double curvature work is produced by hollowing or raising to the required shape, followed by planishing to achieve a smooth surface finish. In beating certain complex shapes, hollowing and raising are often combined. The planishing operation, which 'fixes' the metal to shape and gives a perfectly smooth surface, demands particularly clean, smooth tools whether a hammer or wheeling is used.

7.3.6 Split and weld method

This process, as shown in Fig. 7.32, is simpler, less laborious and quicker than other methods of panel beating. It is clear from the paper template being fitted over a wooden former that material is required where the slits open out. Stretching the metal at these points is time-consuming and it is often quicker and cheaper to weld in V-shaped metal gussets. Conversely, when beating certain forms of double curvature where 'shrinking' is necessary, it is much quicker to 'lose' the surplus metal by simply cutting V-shaped pieces out of it. The workpiece is beaten to shape so that the cut edges meet and then weld them together. An example is shown in Fig. 7.38.

7.3.7 The wheeling machine

The wheeling machine is used where sheet metal has to be shaped into double curvature forms. Although their most frequent use is to smooth metal panels that have been roughly beaten to shape, they can be used for 'crushing' welded seams and shaping panels of shallow curvature without any preparatory beating.

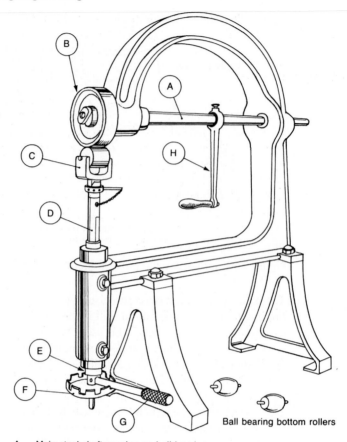

Ball bearing bottom rollers

A — Main steel shaft running on ball bearings.
B — Flanged top roller (fixed).
C — Detachable bottom roller supported in carrier that has quick release action.
D — Perpendicular pillar supporting bottom roller.
E — Screwed shaft for pillar adjustment.
F — Flanged turret wheel enabling bottom roller assembly to be raised or
    lowered, during operation, by turning the wheel with the foot lever 'G'.
H — Handle freely rotating on shaft, can, when necessary, be tightened and
    used to rotate top roller when wheeling work of small dimensions.

Figure 7.33 Standard type wheel machine

Figure 7.33 shows a wheeling machine and names its parts. The machine has two wheels or rollers, the upper wheel being almost flat and the lower wheel being convex with both wheels meeting at a common centre. The upper wheel is fixed on its spindle and the lower wheel runs freely and is carried on a vertical pillar which can be adjusted up or down by a screw mechanism to regulate the pressure on the sheet metal. Interchangeable bottom wheels (rollers) are available in a number of shapes to suit work with various curvatures. The wheels (rollers) are made from hardened and tempered steel and polished. They run on anti-friction bearings. A quick-release mechanism is provided for the lower wheel support column to facilitate the insertion and removal of the workpiece without altering the pressure setting. Wheeling machines are available in various sizes, some with additional clearance for large work such as aircraft cowlings; however, the majority conform to the design shown in Fig. 7.33.

7.3.8 Wheeling processes

The skill in using a wheeling machine lies in acquiring a 'feel' for the work as it passes back and forth between the rollers. Panels of moderate curvature can be produced by wheeling alone. The sheet is placed between the wheels and, depending upon the thickness, temper and composition of the material, the pressure is applied accordingly. Repetition of the passing movement between the wheels in both directions, as shown in Fig. 7.34, stretches the metal so that it takes on a convex curvature. It is very important that each backward and forward movement should be accompanied by alteration of direction so that the wheels make contact with the panel in a different place at each pass while ensuring complete coverage of the entire surface. The panel should not be pinched between the wheels too tightly and, by 'feel', the operator should allow the panel to follow its own shape during wheeling. The movement is varied until the desired shape is obtained, with those parts of the panel which only require a slight curvature receiving less wheeling that those parts that require greater curvature. Wheeling pressure is applied as required by raising the lower wheel (roller) and, after a few passes, the necessary curvature will be accomplished.

When working with soft metals such as aluminium, which is very easily scratched, it is essential that the surfaces of the tools, rollers and workpiece are perfectly clean. Care must be taken not to put excessive pressure on the workpiece because three times as much 'lift' is obtained compared with harder metals such as steel for the same setting. Figure 7.34 shows the basic wheeling process and the means of correcting a common error (corrugation).

Panels of shallow curvature

Panels with very little curvature require the minimum of pressure and should be wheeled to the required shape a gently as possible. Excessive working will result in a fault known as 'corrugating' as shown in Fig. 7.34(c). This type of fault is not easily corrected by wheeling as this tends to make the corrugations worse. It may be necessary to reset the panel using a mallet and panel head. However, slight corrugations can be removed by wheeling across the wheel tracks responsible for the fault as shown in Fig. 7.34(d).

Panels of varying contour

The contours of some panels vary from one point to another, with the result that the curve is greater in some places and less in others. In order to produce this variable curvature it is necessary to wheel over the heavily curved parts more often than the surrounding areas as shown in Fig. 7.34(e). A better finish is obtained by lightly wheeling across the panel in a diagonal direction as indicated by the dotted lines.

Panels of deep curvature

Panels with very full contours can be subjected to considerably greater wheeling pressure without danger of corrugating the surface. When wheeling panels of very full shape, the passage of the wheel over work should be controlled so that the start and

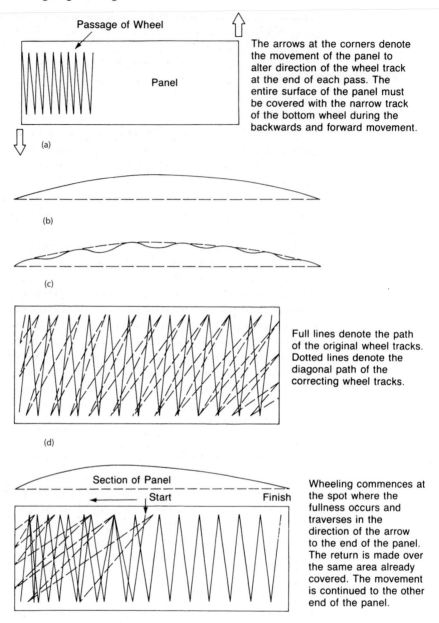

Passage of Wheel

Panel

(a)

The arrows at the corners denote the movement of the panel to alter direction of the wheel track at the end of each pass. The entire surface of the panel must be covered with the narrow track of the bottom wheel during the backwards and forward movement.

(b)

(c)

Full lines denote the path of the original wheel tracks. Dotted lines denote the diagonal path of the correcting wheel tracks.

(d)

Section of Panel

Start   Finish

Wheeling commences at the spot where the fullness occurs and traverses in the direction of the arrow to the end of the panel. The return is made over the same area already covered. The movement is continued to the other end of the panel.

(e)

Figure 7.34 The technique of wheeling. (a) Correct method of wheeling; (b) section of a panel with a clean contour; (c) section of a panel with corrugation faults; (d) correcting a corrugated panel; (e) wheeling a panel with a varying contour

finish points of each pass do not occur in the same position, the panel being moved so that the raising of the shape is performed evenly. Sometimes too much shape is wheeled into the panel. This excess shape can be corrected by turning the panel upside down and wheeling the outside edges. With panels of only very slight curvature, this reverse wheeling can be performed over the entire panel.

Large panels

An assistant may be required to hold one side when wheeling large panels. Success when wheeling one panel with two operators depends upon each person doing their own pulling. On no account should either operator push the panel whilst the other operator is pulling. Since no two operators have the same 'pull', it is important that they change sides half-way through the job to avoid giving the panel uneven curvature. Any roughness when pulling or pushing will cause corrugations and unevenness of shape.

As previously stated, wheeling is widely used for finishing panels that have be formed by the 'cut-and-weld' method. The approximate shape and the protrusion of the weld-bead can be smoothed out and brought to the required contour on the wheel after preliminary dressing with a mallet over a suitable panel head. Wheeling not only crushes and smoothes over the weld which will disappear after dressing with a portable sander but, as a result of the cold-working imposed by wheeling, also partially restores the temper of the metal in the weld zone. Metal adjacent to the welded seam will become soft due to the heat of the welding process. Figure 7.35 shows the stages in forming a large roof panel for a prototype motor vehicle.

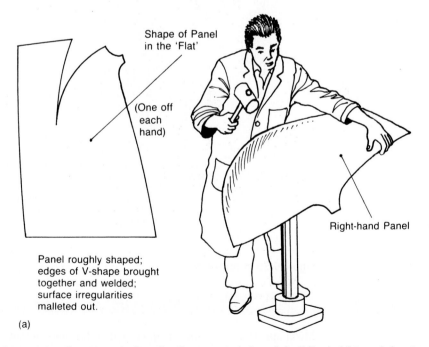

Shape of Panel in the 'Flat'

(One off each hand)

Right-hand Panel

Panel roughly shaped; edges of V-shape brought together and welded; surface irregularities malleted out.

(a)

Figure 7.35 The stages in forming large panels (coach building). (a) Panel shaping; (b) planishing on the wheel; (c) use of a jig

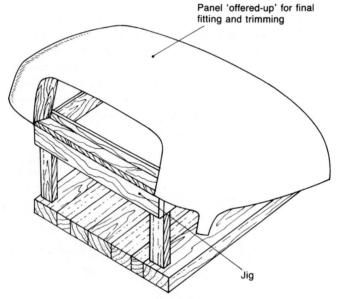

Left-hand Panel

(b)

Panel 'offered-up' for final
fitting and trimming

Jig

Frequent use of jigs during the fabrication of the component
panels, and for final fitting and trimming, ensure the symmetry
of the final composite panel.
The composite panel is fabricated by welding. It consists basically
of a central panel and two end panels. The central panel is shaped
entirely by wheeling.

(c)

Figure 7.35 (Continued)

7.4 The principles of metal spinning

Metal spinning is another method of flow-forming suitable for producing surfaces
of revolution. Most sheet metals can be cold-formed by spinning circular blanks in a
spinning lathe and applying pressure to make the circular blank flow over a rotating

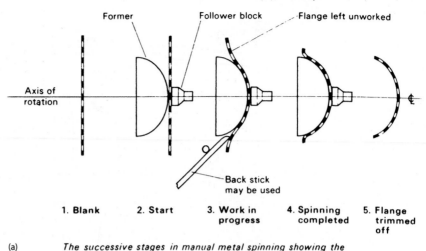

Axis of rotation

Former   Follower block   Flange left unworked

Back stick
may be used

1. Blank   2. Start   3. Work in   4. Spinning   5. Flange
                            progress   completed   trimmed
                                                            off

(a)   *The successive stages in manual metal spinning showing the*
             spreading or drawing action which is characteristic of the process

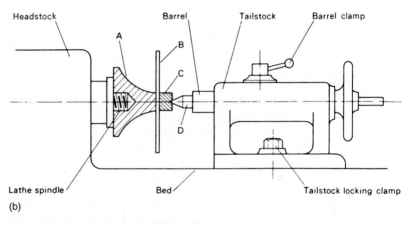

Headstock   Barrel   Tailstock   Barrel clamp

A

B

C

D

Lathe spindle   Bed   Tailstock locking clamp
(b)

Figure 7.36 The basic principles of metal spinning. (a) Metal spinning; (b) the basic features of the spinning lathe

former. The flat blank undergoes *plastic flow* during the spinning process and the pressure is applied manually through the *leverage* of spinning tools. This calls for considerable skill acquired through training and experience. The ease with which a metal can be spun depends upon the individual properties of the metal being used. The way in which the change of shape is accomplished is shown in Fig. 7.36(a). The spinning tool is not shown, but the 'back-stick' is introduced as a means of preventing the blank from collapsing.

The spinning lathe is much simpler than the engineer's centre lathe. It consists of a headstock with a solid spindle with a screwed nose to carry the former and a tailstock with a rotating live centre. The tool is supported on a tee-rest similar to a woodturning lathe. The formers are made from hardwood such as *mahogany* or *lignum vitae* for short runs or steel and cast iron for long runs. The basic features of a spinning lathe are shown in Fig. 7.36(b).

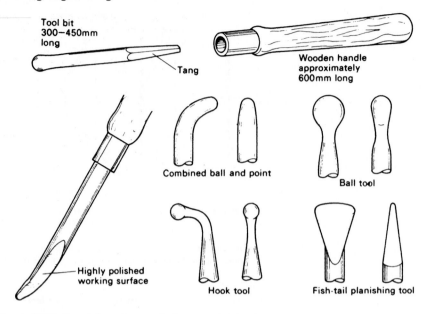

Figure 7.37 Hand forming tools for metal spinning

7.4.1 The spinning process

The hand-spinning process which is most commonly used is performed with the aid of a number of uniquely designed tools whose hardened working surfaces are shaped and polished according to the nature of the work being spun. Some typical hand-spinning tools are shown in Fig. 7.37. The tools are not standardized and many craftpersons choose to make their own tools.

Hand-spinning tools consist of two parts:

- *The tool bit*. This is approximately 300 to 450 mm long and usually forged to shape from high-speed steel round bar hardened and tempered. Opposite the working end is a 'tang' which fits into a long wooden handle.
- *The wooden handle*. This is approximately 600 mm long. The tool bit, when securely fitted, projects from the handle for about 200 mm. Therefore, the average overall length of a hand-spinning tool is between 750 and 850 mm.

The most common spinning tools consist of a combination ball and point. Its range of usefulness is large on account of the variety of shapes that may be utilized by rotating the tool in different directions.

- The *ball tool* is used for finishing curves.
- The *hook tool* is shaped for inside work.
- The *fish-tail planishing tool* is commonly used for finishing work. It can also be used for sharpening any radii in the contour.

The majority of spinning operations involve starting the work and bringing it approximately to the shape of the former, after which '*smoothing*' or '*planishing*' tools are used to remove the spinning marks imparted during the initial forming and producing a smooth finish. These hand tools are used in conjunction with a *tee-rest*

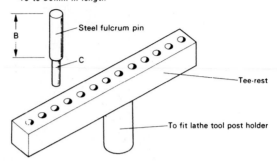

'B' represents the working portion of the fulcrum pin.
This varies in length between 75 and 100mm.
'C' represents the portion of the pin which is a free
fit in the holes provided in the tee-rest.
It is smaller diameter than 'B' and is approximately
40 to 50mm in length

Simple tee-rest and fulcrum pin

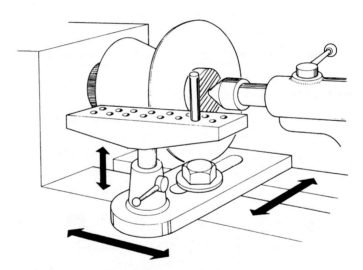

Figure 7.38 Adjustment for variations of fulcrum pin position

and *fulcrum pin*. The manner in which the fulcrum pin is advanced as the spinning proceeds is extremely simple and is shown in Figs 7.38 to 7.41 inclusive. The tee-rest is supported in the tool-rest holder, which is clamped to the bed of the lathe. The tool rest provides a wide range of adjustments in six directions and a further fine adjustment can be made by releasing the clamp bolt and swivelling the tee-rest. All these features are shown in Fig. 7.41. The action of the spinning tool is shown in Fig. 7.39.

When commencing spinning operations, the initial strokes are made outwards towards the edge of the circular blank being spun. In order to speed up the process and to avoid thinning the metal unduly, strokes in the opposite direction are also made, i.e. inward strokes. In both cases it is important not to dwell in any one position on the workpiece so as to cause excessive local work-hardening. The tee-rest and the position of the fulcrum pin are reset as the work progresses and both the forming tool and the

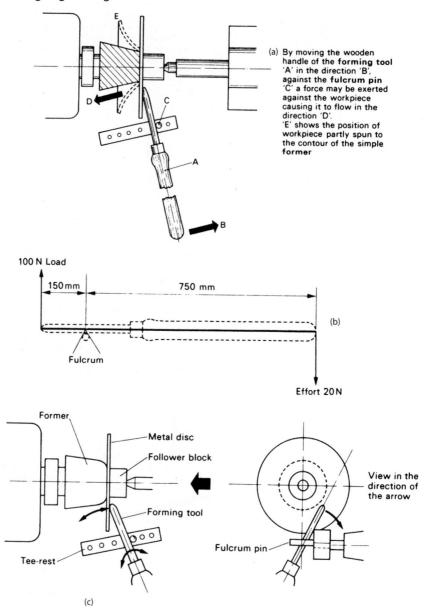

(a) By moving the wooden handle of the **forming tool** 'A' in the direction 'B', against the **fulcrum pin** 'C' a force may be exerted against the workpiece causing it to flow in the direction 'D'.
'E' shows the position of workpiece partly spun to the contour of the simple former

Figure 7.39 The action of hand spinning. (a) Exerting a force against a workpiece; (b) moments acting on the spinning tool; (c) three-dimensional movement of the forming tool

outer surface of the metal blank have to be frequently lubricated. Figure 7.40 shows the metal spinning process.

Back-sticks, as their name implies, are always positioned at the back of the blank being spun immediately opposite the forming tool as shown in Fig. 7.41. They are used to prevent wrinkles forming and, in the case of thin metal, they prevent it collapsing. Pressure is applied in the direction indicated and the work revolves between the back-stick and the forming tool, two fulcrum pins being used in this instance.

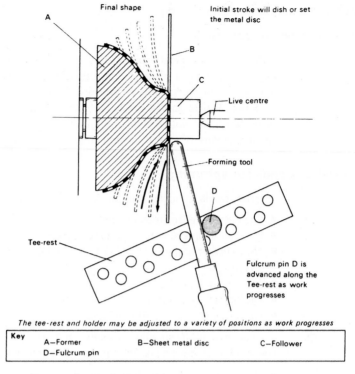

Figure 7.40 The metal spinning process

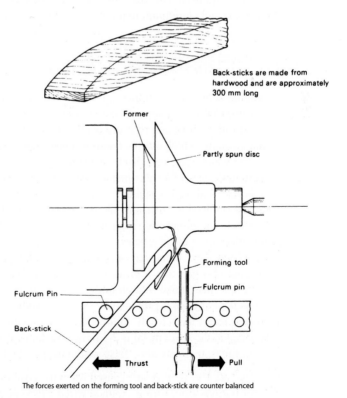

Figure 7.41 The use of the back-stick

7.4.2 Lubrication for spinning

Lubrication is essential in order to minimize the friction between the work and the tool to prevent excessive heating, scratching of the work or damage to the tool. A lubricant must be applied frequently to the surface of the work and tools during the spinning operation. It is important to use the correct type of lubricant. It should adhere to the metal blank and not be thrown off by the high rotational speed involved. For hand spinning, *tallow* or *industrial soap* or a *mixture of both* are used as lubricants.

7.4.3 Spindle speeds for spinning

Spindle speeds for metal spinning are fairly critical, and they will depend upon:

- The ductility of the metal being spun.
- Whether the metal is ferrous (hard to spin) or non-ferrous (easy to spin).
- The diameter of the blank being spun (the larger the diameter, the higher the surface speed).
- The thickness of the metal being spun.
- The shape of the former.
- The shape of the spinning tool used.

The drive to the spindle is usually via a two-speed electric motor and a three-step cone pulley, giving six possible speeds. As a general rule:

- Mild steel requires the slowest spindle speed.
- Brass requires about twice the speed of mild steel.
- Copper and aluminium requires a speed only slightly higher than brass.

7.5 Swaging sheet metal

Swaging is an operation that is used to raise a moulding (*swage*) on the surface of a sheet-metal component. A swage is produced by means of a pair of special contoured rollers. Swaging rolls are available in a large variety of contours to fit a swaging machine, which may be hand- or power-operated.

The wired edges can also be made using simple bench tools primarily to make the sharp, raw edges of a sheet-metal article safe and also to add strength and stiffness to the article. Although swaging has many similar functions to that of wired beads, it is not just confined to stiffening edges but may be used some distance from the edge of the sheet. The projecting shape of the swage above the surface imparts considerable strength to sheet-metal articles.

Wiring sheet metal not only makes the edges of sheet metal articles rigid and safe, but also provides a pleasing appearance. Wiring or 'beading' is a process for forming a sheet metal fold around a wire of suitable diameter. Much of the strength of this type of edge is provided by the wire. Additional strength is obtained from the stressed metal that closely follows the exact contour of the wire. The allowance to be added to the sheet metal is 2.5 times the diameter of the wire. In addition to 'true' wired edges, there are also 'false' wired edges which may be one of two types:

1. *Applied*. The applied type is used when the position or metal thickness is unsuitable for normal 'true' wiring. Applied wired edges are attached and fastened in position by a return flange, riveting, spot-welding or soldering.

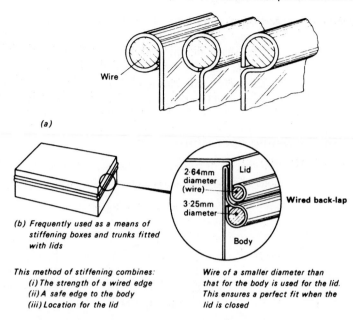

(a)

2·64mm
diameter
(wire)

3·25mm
diameter

Lid

Wired back-lap

Body

(b) Frequently used as a means of
stiffening boxes and trunks fitted
with lids

This method of stiffening combines:
  (i) The strength of a wired edge
  (ii) A safe edge to the body
  (iii) Location for the lid

Wire of a smaller diameter than
that for the body is used for the lid.
This ensures a perfect fit when the
lid is closed

The edge of the metal is hidden on the inside in the case of the 'back-lapped wired edge'

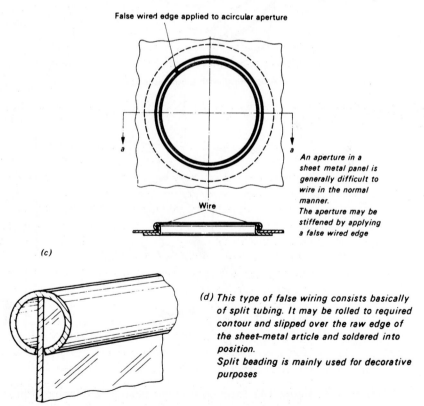

False wired edge applied to acircular aperture

a

a

Wire

An aperture in a
sheet metal panel is
generally difficult to
wire in the normal
manner.
The aperture may be
stiffened by applying
a false wired edge

(c)

(d) This type of false wiring consists basically
of split tubing. It may be rolled to required
contour and slipped over the raw edge of
the sheet-metal article and soldered into
position.
Split beading is mainly used for decorative
purposes

Figure 7.42 Types of wired edge; (a) three common types of wired edges;
(b) back-lapped wired edges; (c) cross-section *a – a* of applied false wired edge;
(d) split beading

2. *Hollow*. This type of 'hollow bead' is usually produced by folding the edges around a wire core that is then withdrawn. Although hollow beads are rigid due their shape and the work-hardening that the process induces in the metal, because hollow edges do not contain a wire they will not withstand an impact blow and are relatively easily damaged. Figure 7.42 shows some types of wired edges, whilst Figure 7.43 shows how a hollow bead may be produced on a spinning lathe.

7.5.1 Further swaging operations

The 'ogee', or return curve swage, is frequently used to strengthen the centre portions of cylindrical containers and drums because of its high resistance to internally

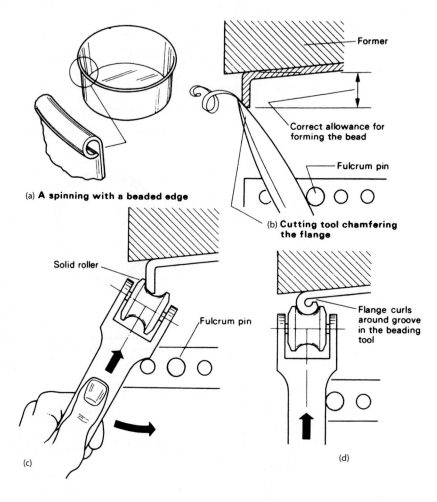

(a) **A spinning with a beaded edge**

(b) **Cutting tool chamfering the flange**

(c)

(d)

Figure 7.43 Forming a hollow bead on a spinning lathe. (a) A spinning with a beaded edge; (b) cutting tool chamfering the flange; (c) beading wheel advanced at an angle and then gradually brought square; (d) to the spinning axis for final closing of the bead

and externally applied forces in service. Examples of the combination of strength and decoration associated with swaging are to be found in the design of circular sheet-metal objects such as drums, dustbins, waste-paper baskets, buckets and water tanks. The maximum thickness of sheet metal that can be swaged is 1.62 mm. Figure 7.44 shows some aspects of swaging.

The swage is a very important part of the automobile industry to add strength and stiffness to large body panels and also to prevent them 'drumming', thus reducing the noise level in the passenger compartment. The body panels are formed in

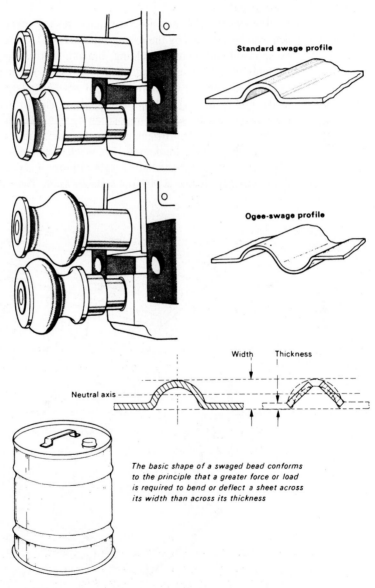

Standard swage profile

Ogee-swage profile

Width   Thickness

Neutral axis

The basic shape of a swaged bead conforms to the principle that a greater force or load is required to bend or deflect a sheet across its width than across its thickness

Figure 7.44 The swaging of sheet metal

large power presses, and the correct contour and the swage is formed at the same time as an integral part of the pressing operation.

7.5.2 The use of stiffeners

Large panels may be reinforced by the application of stiffeners. Usually panels are stiffened by the fact that they are fastened to some sort of rigid framework. These frameworks are usually fabricated from metal sections that are inherently strong and rigid because of their form. Sheet-metal sections may be roll-formed for the purpose of providing internal and external stiffening of large components of cylindrical or circular shape.

The edges of fabrications constructed in sheet metal which is too thick to wire or hem can be stiffened by the use of a flat bar or D-shaped section. It may be attached by spot-welding, manual welding, brazing or riveting. One of the more common methods of achieving strength and stiffness is the use of 'angle-section frames'. Figure 7.45 shows various methods of stiffening large panels. All four edges are made rigid by folding. A 'top-hat' section is used to stiffen the centre section of the panel and is usually secured in position by spot-welding.

Another method of stiffening large sheet fabrications, also shown in Fig. 7.48, is to attach them to a rigid framework. The welded section is fabricated from a 'P-section', which has a very high strength/weight ratio for a sheet-metal section. All four edges of the panel are folded at 90° to a suitable width. The panel is then placed in position

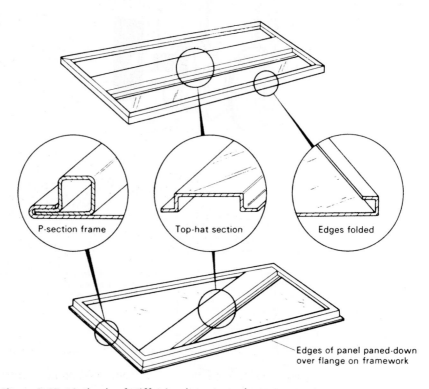

P-section frame   Top-hat section   Edges folded

Edges of panel paned-down over flange on framework

Figure 7.45 Methods of stiffening large panels

over the frame and the edges are paned down over the flange of the 'P-section'. The centre of the panel can be stiffened by means of a diagonal 'top-hat' section.

Figure 7.46 shows the use of angle stiffeners. Welded angle frames are widely used as a means of stiffening and supporting rectangular ducting for high-velocity systems. They also serve as a jointing media when assembling sections together by bolting as shown in Fig. 7.46(a). The angle section is riveted to the duct and sections of duct are then bolted together – flange to flange – with a suitable gasket between the angle flanges. The large sizes of square or rectangular ducting tend to 'drum' as a result of turbulence in the air passing through them. To overcome this drumming it is advisable

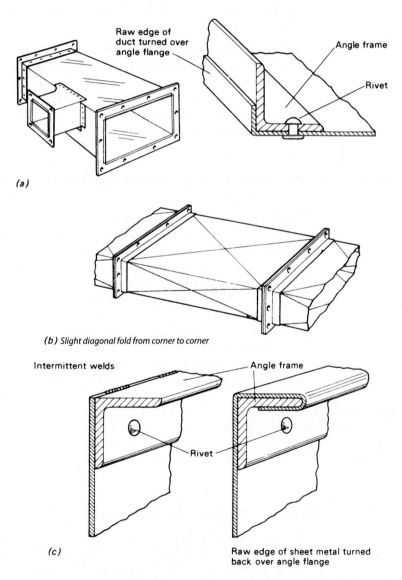

Figure 7.46 The use of angle stiffeners. (a) Section of rectangular ductwork; (b) diamond-break stiffening of duct walls; (c) welded or riveted

to provide adequate stiffening to the walls of the duct. This may be achieved by the use of swaging, but more often a diamond break is used as shown in Fig. 7.46(b).

Simple welded angle frames may be used as a means of supporting and stiffening the open ends of tanks or bins fabricated from sheet metal. Two methods of attaching angle frame stiffeners are shown in Fig. 7.46(c).

7.6 Basic fabrication procedure

The production of fabricated components and structures involves five principal operational stages:

1. Measuring, marking out and the production of templates (see Chapter 5).
2. Cutting and preparing blanks from stock material (see Chapter 6).
3. Forming blanks to make the required article (see Chapter 7).
4. Joining and assembly (see Chapters 10, 11 and 12).
5. Surface finishing – galvanizing, electroplating and polishing, vitreous enamelling and painting. These topics are beyond the scope of this book.

7.6.1 The need for planning the sequence of operations

Sheet-metal articles may be made from one or more pieces of metal by the use of hand tools, machine tools or a combination of both. Most components and assembles involve a sequence of operations that must be planned in order to produce the finished article as economically as possible. The sequence of operations may vary slightly between individual craftpersons and the equipment available in the fabrication workshop. If a sheet-metal article requires a number of joints to be made during the course of its fabrication, the sequence of operations must be planned to ensure that access is possible.

Usually, operations on sheet metal and plate commence with cutting out the blank. This often involves the use of a guillotine shear. It is considered good practice to make a 'trim cut' on standard sized sheets or plates to ensure one straight edge is available to use as a datum or service edge from which other measurements and cuts are made.

Economy in the use of materials should be the prime consideration of all craftpersons engaged in fabrication

This reduces the size of the flat material, usually in length, and this must be taken into consideration when marking out blank sizes for cutting to ensure maximum yield and minimum waste. Intelligent marking-out on the flat material is essential in order to avoid unnecessary waste in the form of 'off-cut material' or 'scrap'.

7.6.2 Transferring paper or metal templates onto sheet metal

The first operation is to transfer the pattern or template outline onto the sheet metal. When using patterns or templates it is important to position them in the proper manner on the stock sheet in order to avoid unnecessary scrap. Templates must be held in position in order to restrain any tendency to move during the marking-out operation. This initial marking-out operation is shown in Figure 7.47.

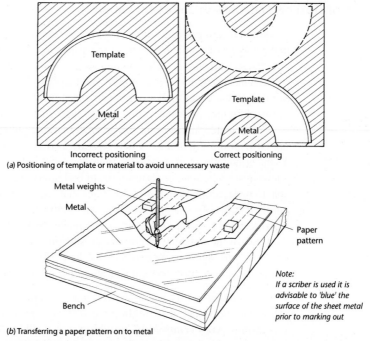

Incorrect positioning   Correct positioning

(*a*) Positioning of template or material to avoid unnecessary waste

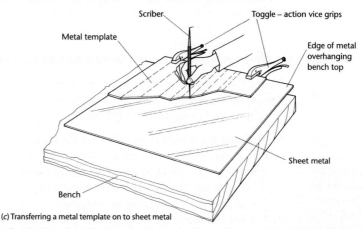

(*b*) Transferring a paper pattern on to metal

Note:
If a scriber is used it is advisable to 'blue' the surface of the sheet metal prior to marking out

(i)   Place the sheet of metal to be used on the surface of the bench and position the paper to avoid waste
(ii) To prevent the paper from creeping hold it in position with metal weights
(iii) With a hard sharp pencil or scriber scribe the outline of the pattern on the sheet metal
(iv) Remove the weights and the pattern and cut the metal to the outline scribed upon it using universal hand shears removing all burrs with a suitable file

(*c*) Transferring a metal template on to sheet metal

(i)   Place the sheet of metal to be used on the bench with one of its squared sides slightly overhanging
(ii) Position the metal template in position as shown, and clamp it securely with vice grips to restrain any movement
(iii) Scribe the outline of the template on the sheet metal, using a sharp scriber
(iv) Release vice grips to remove template and cut the sheet metal to the outline scribed upon it with a suitable pair of hand shears. Remove all burrs with a file

Figure 7.47 Methods of transferring patterns to metal

7.6.3 Use of notched corners in sheet-metal work

A common requirement for light sheet-metal fabrications is the *notched corner*, as used in the making of simple folded trays and boxes. Notching is an essential operation where thin gauge fabrications include wired edges, self-secured joints, lap joints and welded corner seams. The term 'notching' is used to describe, in simple form, the removal of metal from the edges and/or corners of sheet-metal blanks or patterns prior to carrying out any forming operations in order to facilitate such operations.

Considerable thought must be given to marking out for notching since good notching is of prime importance where the finished article is to have a neat appearance. There is nothing more unsightly than overlaps, bulges and gaps resulting from not allowing for notching, or resulting from bad notching, on the initial blank. It is advisable to make rough sketches of junctions that require notching before starting to mark out the pattern. This simple exercise will enable the craftsperson to determine where metal removal (notching) is necessary to make the article.

Usually, the notching of corners on sheet-metal blanks that are to be formed into boxes is controlled by the depth of the box. Unfortunately, a large amount of scrap often results from this notching operation. For example:

- A rectangular tray is required to be made by square notching the corners, folding the sides up square and welding the vertical corner seams. The depth of the notch is 20 mm. **Area removed by notching = 4 × 20 × 20 = 1600 mm^2.**
- A rectangular box made in the same way but 140 mm deep will have a **total area removed by notching = 4 × 140 × 140 = 78400 mm^2.** By comparison, it can be seen that although the box is only seven times deeper than the tray the amount of scrap metal produced by notching is 49 times greater.

7.6.4 Fabricating a sheet-metal pan with wired edges and riveted corners

Figure 7.48 shows the details and blank layout for making the sheet-metal pan from thin-gauge sheet metal. Table 7.5 lists the tools and equipment required.

The following sequence of operations is required to make the pan:

> Remember that handling freshly cut sheet-metal can be dangerous so always remove sharp edges and burrs after any cutting operation

1. Mark out the overall dimensions for the blank on a suitable sheet of metal of the correct thickness. If a standard size sheet is used, square the sheet by making a trim-cut on the guillotine shear and, using this cut edge as a datum edge, mark out the overall length and width of the required blank. Check the rectangular outline for squareness by measuring the diagonals.
2. Cut out the blank on the guillotine and remove all sharp edges or burrs with a suitable file.
3. Mark off the allowances for the riveted flanges. These are normally positioned on each end of the long sides of the pan.
4. Mark the centre lines for the rivets. *Operations 3 and 4 may be performed after square notching the corners if preferred.*
5. Notch the corners. Care must be taken, when cutting with snips, not to 'over cut' the corners. The notching of corners can be performed much quicker and more conveniently on a notching machine. Most machines are capable of making square notches up to 102 mm in depth in sheet metal up to 1.6 mm thick.

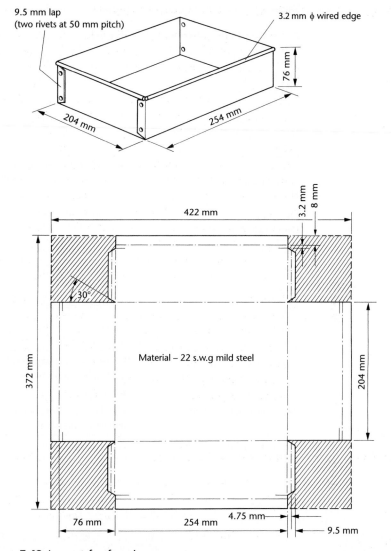

Figure 7.48 Layout for forming a pan

6. Mark out the clearance notches at the end of the rivet flanges. These are normally made at an angle of 30° as shown in Fig. 7.48, and may be marked with the aid of a bevel gauge.

7. Mark the positions for the rivets on the flanges and centre punch.

8. Drill or punch the rivet holes. Figure 7.49shows a 'tinman's hand lever punch', which is used for punching holes in thin sheet metal. Note the conical nipple for aligning the punch in the centre punch mark.

9. Bend the long sides up first in a folding machine to an angle of about 45° and flatten back as shown in Fig. 7.50(a). This provides a crease line for final bending by hand over a hatchet stake after the short sides (ends) have been folded up.

Table 7.5 Tools and equipment required to make a rectangular pan

Tools and equipment required	Remarks
Steel rule Straight edge Flat steel square Dividers Jenny odd-legs Bevel Scriber	These are used for marking-out the blank
Centre punch Nipple punch Hammer	Required for marking the positions of the rivet holes. The hammer is also used for riveting
Guillotine Universal snips	For cutting out the blank and notching the corners
Rivet set	For the riveting operation
Folding machine	For bending up the sides
Bench stakes Mallet	These are required for completing the bending operations by hand and for wiring the edge
Cutting pliers	For cutting the wire and holding it in position during the wiring operation
Bench vice	For bending the wire if a frame is used, and for holding the hatched stake when throwing the flange off for the wired edge
Tinman's hand-lever punch	May be used for punching the rivet holes in the corner flanges on the blank
Drilling machine or portable drill	For drilling the rivet holes on assembly

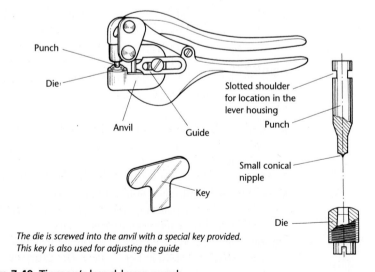

The die is screwed into the anvil with a special key provided.
This key is also used for adjusting the guide

Figure 7.49 Tinman's hand-lever punch

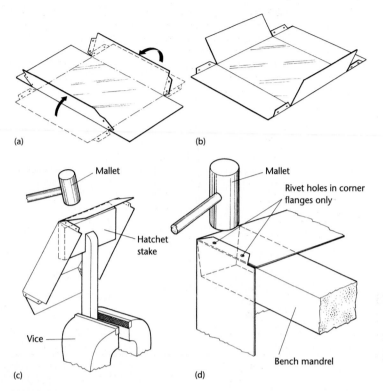

Figure 7.50 Bending the sides of the pan. (a) First bend; (b) second bend; (c) final bend; (d) bending the laps

10. Bend up the ends in the folding machine to 90°. This operation will also bend up the rivet flanges as shown in Fig. 7.50(b).
11. Complete the bending operations over a hatchet stake as shown in Fig. 7.50(c).
12. During operation 11, the flanges have to be knocked back slightly to accommodate the ends. This is rectified by bending the laps over a suitable stake with a mallet as shown in Fig. 7.50(d).
13. Support each corner, in turn, on a suitable anvil or bench stake and mark the centres of the rivet holes through the holes in the flanges with a suitable size nipple punch.
14. Drill or punch the top hole in each corner first and insert a rivet to maintain alignment before drilling or punching the bottom holes.
15. Deburr the holes ready for riveting.
16. The corners are riveted as shown in Fig. 7.51. Further information on making riveted joints is given in Chapter 9.
17. To wire the edge, bend the wire frame in a bench vice and apply it to the pan. Tuck in the bead using a chisel stake and mallet as shown in Fig. 7.51.

Typical faults that can occur when wiring a straight edge are shown in Fig. 7.52.

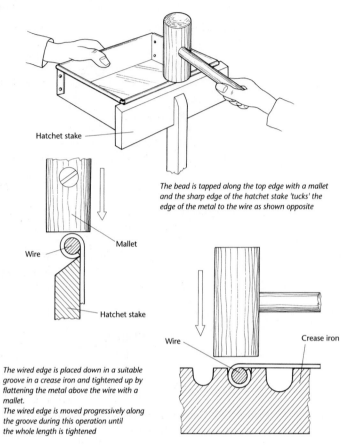

Hatchet stake

Mallet

Wire

Hatchet stake

The bead is tapped along the top edge with a mallet and the sharp edge of the hatchet stake 'tucks' the edge of the metal to the wire as shown opposite

Wire

Crease iron

The wired edge is placed down in a suitable groove in a crease iron and tightened up by flattening the metal above the wire with a mallet.
The wired edge is moved progressively along the groove during this operation until the whole length is tightened

Figure 7.51 Wiring the edge of the pan

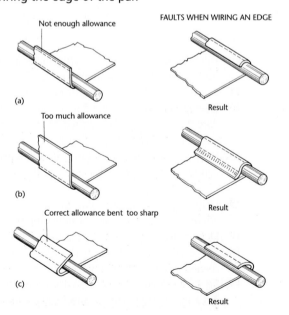

FAULTS WHEN WIRING AN EDGE

Not enough allowance

(a)

Result

Too much allowance

(b)

Result

Correct allowance bent too sharp

(c)

Result

Figure 7.52 Typical faults when wiring as straight edge

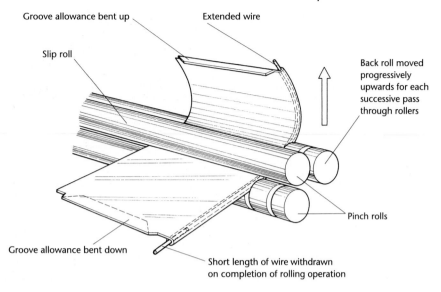

Groove allowance bent up

Extended wire

Slip roll

Back roll moved progressively upwards for each successive pass through rollers

Pinch rolls

Groove allowance bent down

Short length of wire withdrawn on completion of rolling operation

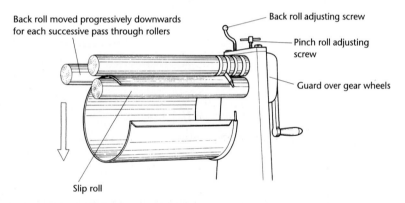

Back roll moved progressively downwards for each successive pass through rollers

Back roll adjusting screw

Pinch roll adjusting screw

Guard over gear wheels

Slip roll

Figure 7.53 Rolling after wiring in the flat

7.6.5 Wiring cylinders and cones

There are two methods of wiring the edge of a curved surface:

- Wiring in the flat *before* forming by rolling.
- Wiring the edge *after* forming by rolling.

Wiring before rolling

Figure 7.53 shows how a cylinder is rolled after it has been wired in the flat. *Slip-rolls* are used for rolling cylinders after wiring. The cylinder is rolled over the *slip-roll* so that the cylinder can be removed after rolling is complete. Before inserting the metal, care must be taken to ensure that:

- The wired edge rests in the correct groove in one of the 'pinch rolls'.
- The machine is checked to determine whether it 'rolls-up' or 'rolls-down'.

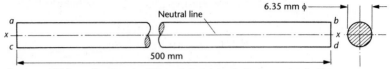

If a piece of 6.35 mm diameter wire 500 mm in length is rolled to form a ring it will have a mean diameter of 159 mm. Likewise, when sheet of 1.6 mm thick metal of the same length is rolled to form a cylinder, it will have a mean diameter of 159 mm

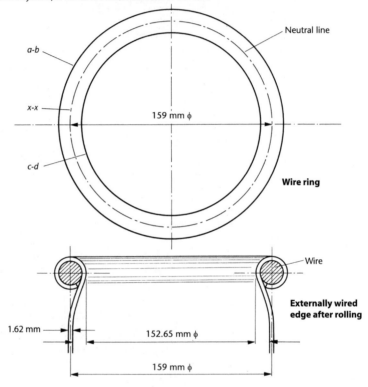

If the sheet is wired along on edge (i.e. the 500 mm length) and then rolled to form a cylinder, the above effect will result. It will be seen that both sheet metal and wire forms assume a common **mean diameter** thus causing a considerable constriction as illustrated above

Figure 7.54 The effect of wiring before rolling

The rolling operation reduces the internal diameter of the cylinder in the vicinity of the wired edge as shown in Fig. 7.54 if the edge is wired before rolling.

The rolling operation

On *internally* wired edges, rolling after wiring produces the *opposite effect* to that obtained on *externally* wired cylinders. The wired edge tends to become slightly larger in diameter than that of the cylinder.

Wiring after rolling

Figure 7.55 shows the sequence of operations for wiring the edge of a cylinder after it has been rolled. Rolling can be employed as a final operation to true up the wired edge.

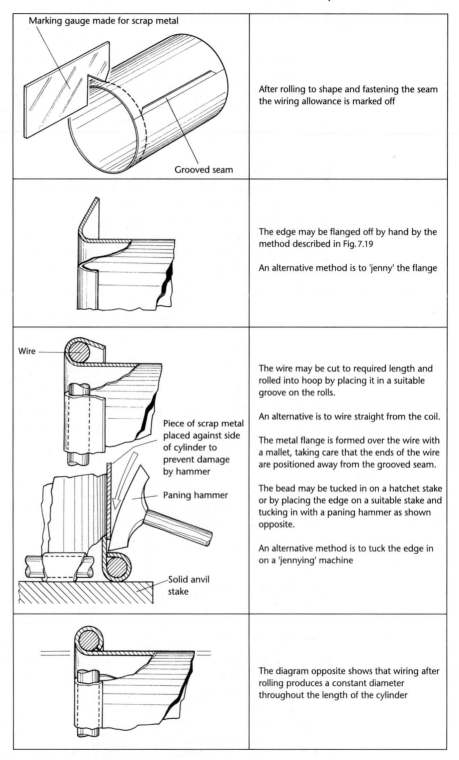

Marking gauge made for scrap metal / Grooved seam	After rolling to shape and fastening the seam the wiring allowance is marked off
	The edge may be flanged off by hand by the method described in Fig. 7.19 An alternative method is to 'jenny' the flange
Wire / Piece of scrap metal placed against side of cylinder to prevent damage by hammer / Paning hammer / Solid anvil stake	The wire may be cut to required length and rolled into hoop by placing it in a suitable groove on the rolls. An alternative is to wire straight from the coil. The metal flange is formed over the wire with a mallet, taking care that the ends of the wire are positioned away from the grooved seam. The bead may be tucked in on a hatchet stake or by placing the edge on a suitable stake and tucking in with a paning hammer as shown opposite. An alternative method is to tuck the edge in on a 'jennying' machine
	The diagram opposite shows that wiring after rolling produces a constant diameter throughout the length of the cylinder

Figure 7.55 Wiring a cylinder after rolling

7.6.6 Fabricating a domestic tin-plate funnel

The funnel shown in Fig. 7.56 is made from tin plate in two parts, a *body* and a *thimble*. Both are frustums of right cones and their patterns are developed by the radial line method previously described in Section 4.15.2. Bend lines are marked out in pencil to avoid damaging the tin plating and allowing the funnel to rust in service. Lines showing where the metal is to be cut can be scribed as usual. The developments for both the body and the thimble are quadrants with the appropriate allowances for the seams and for notching made as usual.

Breaking the grain

Before commencing work on the sheet of tin plate the first operation should be to '*break the grain*' of the metal in order to prevent ridges forming in the metal. This is done by rolling the piece of metal backwards and forwards a few times though bending rolls set to impart a shallow curvature. The direction of bending is reversed each time. This will ensure that the parts of the funnel will have a smooth surface, free from ridges, when they are formed. Breaking the grain is always sound practice prior to rolling, particularly on metals that have been cold-reduced. Once the breaking operation is complete, the metal should be rolled out flat by suitable adjustment of the rolls in readiness for cutting out the blank and forming to shape.

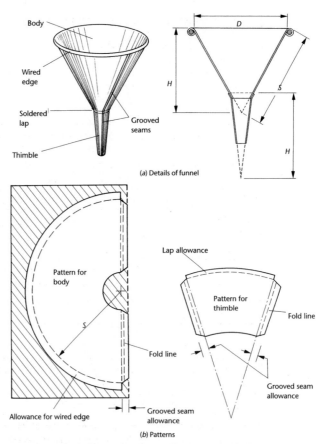

Figure 7.56 Details and patterns for fabricating a domestic funnel

Making the funnel

Table 7.6 lists the tools and equipment needed to make the funnel, whilst Fig. 7.57 shows some of the operations involved.

1. Cut out the blanks from tin plate that has had its grain broken, as previously described, using bench shears and snips.
2. Fold the locks for the grooved seam, one up and one down as shown at A and B in Fig. 7.57(a) on both the body and the thimble using a bench folding machine.
3. The body may be formed to shape over a funnel stake. This operation consists of bending the body of the funnel by hand with a sliding motion over the stake. A mallet is used for final forming to shape after grooving the seams as shown in Fig. 7.57(b). For further information on making self-secured grooved joints see Section 9.4.
4. The thimble is formed in a similar manner over a long tapering bick-iron as shown in Fig. 7.60(c). It is formed roughly to shape, the grooved seams are interlocked, and the thimble is driven hard onto the bick-iron to hold the seam tight until it is completed by grooving.
5. The body and thimble are joined by soft-soldering as shown in Fig. 7.60(d). For further information on soldering see Section 10.1. The grooved seams are sealed by soldering on the inside.

Table 7.6 Tools and equipment required to make a funnel

Tools and equipment required	Remarks
Steel rule Scriber Divider	These are used for marking out
Guillotine Bench shears Universal snips	For cutting out the blanks and developed shapes. Straight snips may be used for straight line and external curve cutting, but bent pattern snips will be required for cutting internal curves. Universal snips will perform both these operations
Rolling machine	for BREAKING-IN the TINPLATE
Flat bed folding machine	For forming the locks for the grooved seams. An alternative method of performing this operation is to throw the edges over a hatchet stake using a mallet
Funnel stake Bick iron Mallet Stretching hammer	Used for forming the body of the funnel Used for forming the thimble For stretching lap circumference of thimble
Grooving tool Hammer	For fastening the seams
Soldering stove Soldering iron	For soft soldering the inside of the seams and the lap joint between body and thimble
Cutting pliers	For cutting wire for the wired edge
Jennying machine	Two operations are required: 1. Turning the wire allowance 2. 'Tucking' the edge over the wire

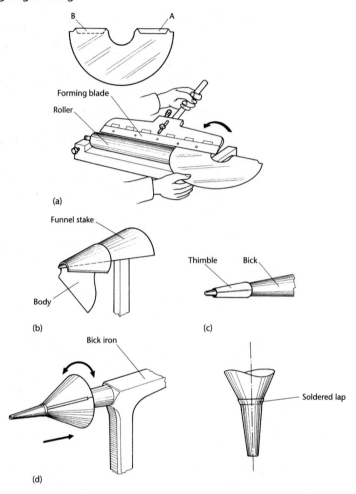

Figure 7.57 Operations for forming a funnel. (a) Folding the locks; (b) forming the body; (c) forming the thimble; (d) finishing the funnel

6. Finally the top edge of the body has to be wired. Cut a length of wire and roll it into a hoop on the bending rolls. The wire allowance is flanged up on a jennying machine (see Section 7.5.5). The wire is fitted in place and the metal is closed over the wire with a suitable pair of rolls on the jenny, making sure that the butt joint of the hoop is positioned away from the grooved seam of the body.

7.6.7 Forming operations using a universal jennying machine

Figure 7.58 shows some examples of the types of wheels or rolls that are used on a universal jennying machine together with the various operations which they perform.

Figure 7.59 shows some typical operations using a universal jennying machine for preparing edges.

■ Figure 7.59(a) shows how the rolls can be used to turn up a narrow edge on circular and irregular components ready for wiring.

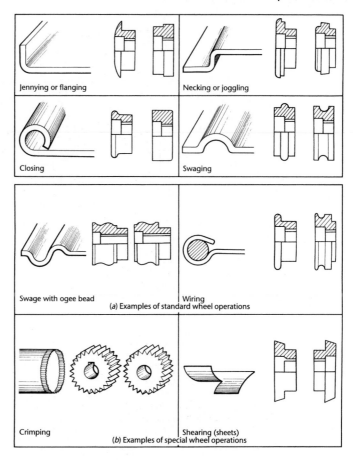

Figure 7.58 Wheels and rollers for jennying machines

- Figure 7.59(b) shows how the sharp edge of the upper roll can be used for tucking-in wired edges and also for turning single edges on curved work and discs.
- The vee-shaped rolls shown in Fig. 7.59(c)are ideal for turning up a double edge on elbows for paned-down or knocked-up seams.
- For a wired edge the allowance must equal 2.5 times the diameter of the wire. This measurement is taken from the face of the 'gauge' to the centre of the upper roll as shown in Fig. 7.59(d). The gauge is adjusted, usually by a knurled screw, at the side of the machine. After setting to the required measurement the gauge is locked in position.

Caution: When using sharp edged rolls for flanging operations, do not over-tighten the top roll. If the top roll is too tight the metal will be sheared. The top roll should be adjusted so as to afford only a light grip on the metal between the rolls.

When flanging a disc, a small piece of thin, scrap metal folded as shown Fig. 7.60 should be used to prevent injury to the operator's hand. As an extra precaution remove all burrs from the metal blank before commencing the operation.

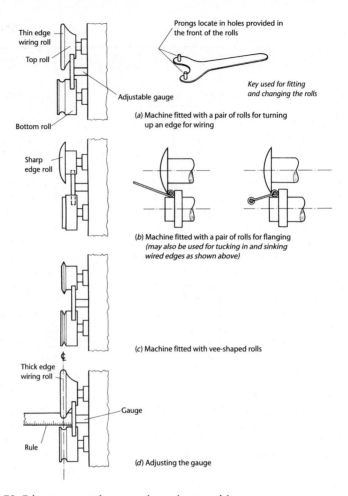

Figure 7.59 Edge preparation on a jennying machine

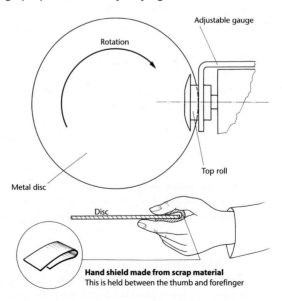

Figure 7.60 Safety precaution when flanging a disc with the jenny

Exercises

7.1 Forming sheet metal and thin plate by folding and bending
  a) Explain what is meant by 'spring-back' and explain how this effect may be overcome.
  b) With the aid of sketches explain the difference between 'air-bending' and 'pressure-bending'.
  c) With the aid of sketches illustrate the principle of operation of a manually operated folding machine.
  d) Sketch a typical U-bending tool suitable for use in a fly press and label its principal parts.
  e) Describe in what ways a press-brake differs from a power press and show a typical example of its use.
  f) With the aid of sketches explain what is meant by 'the neutral line' when bending sheet metal and thin plate, and explain the difference between neutral line allowance and centre line allowance when calculating the sized of a blank.
  g) For the component shown in Fig. 7.61, calculate:
    i) The neutral line blank length
    ii) The centre line blank length.

7.2 Roll-bending sheet metal and plate
  a) With the aid of sketches explain the difference between 'pinch rolls' and 'pyramid rolls'.
  b) With the aid of sketches explain the difference between 'roll-up'-type machines and 'roll-down'-type machines.
  c) Explain:
    i) the purpose of 'slip-rolls'
    ii) the difference between cylinder rolling and cone rolling.

  d) Describe what is meant by the 'flow-forming' of sheet metal.
  e) With the aid of sketches show how a flange may be 'thrown' on the end of a sheet-metal cylinder.
  f) When using a hammer or a mallet, describe the differences between:
    i) a solid blow
    ii) an elastic blow
    iii) a floating blow.
  g) With the aid of sketches explain how a hemisphere 75 mm diameter can be raised from a flat copper blank 1.00 mm thick using hand tools only. List the tools required and describe any inter-stage heat treatment that may be necessary as the metal work-hardens.
  h) Describe the difference between 'hollowing' and 'raising'.
  i) Briefly describe why the 'split and weld' technique is used when panel beating.
  j) Describe the essential requirements of a wheeling machine and the purpose of its use.

7.3 Metal spinning
  a) Describe the essential differences between an engineer's centre lathe and a sheet metalworker's spinning lathe.
  b) With the aid of sketches explain the basic principles of metal spinning.
  c) Explain what is meant by a 'back stick', why it is required and how it is used.

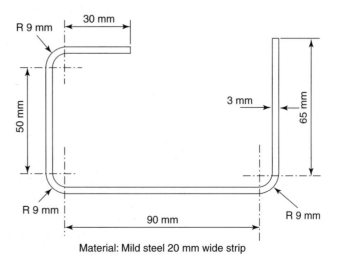

R 9 mm   30 mm

50 mm

R 9 mm

3 mm

65 mm

90 mm

R 9 mm

Material: Mild steel 20 mm wide strip

Figure 7.61

d) Why must the blank be lubricated when spinning and what special property must the lubricant possess?

7.4 Swaging and stiffening processes

a) With the aid of sketches describe the essential principles of swaging and why this process is often required in sheet metalwork.

b) State the reasons for wiring the edge of a sheet-metal component and, with the aid of sketches, show how a straight edge may be wired.

c) With the aid of sketches show how the ends of a cylindrical sheet-metal component may be wired.

d) Explain why stiffening is required for large sheet-metal surfaces and, with the aid of sketches, show TWO examples of applied stiffeners.

7.5 Fabrication planning and procedures

a) Describe how the outline of a template may be transferred to the sheet-metal stock from which a blank is to be cut.

b) Explain why notched corners are often required when making sheet-metal products.

c) Sketch a blank layout and write out a planning sheet and a list of tools required to make the box shown in Fig. 7.62.

d) Describe the procedure for making the box shown in Fig. 7.62.

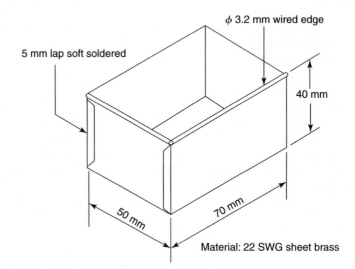

ϕ 3.2 mm wired edge

5 mm lap soft soldered

40 mm

50 mm

70 mm

Material: 22 SWG sheet brass

Figure 7.62

8

Structural steelwork and pipework

When you have read this chapter, you should be able to:

- Identify the more common types of rolled steel sections
- Identify structural connections and assemblies
- Identify the types of stanchions and stanchion splices
- Identify the methods of connecting beams to stanchions
- Identify the types and uses of trusses, frames and beams
- Set out pipework bends
- Understand the principles of pipe bending
- Understand the safety precautions to be observed when loading pipes ready for bending
- Understand the methods of bending pipes by hand and by machine
- Identify the types of pipe and their applications
- Identify the types of pipe joints and their applications
- Understand the use of pipe-threading equipment
- Identify the contents of pipe systems by colour coding

8.1 Rolled steel sections

The materials most commonly used in structural steel work are rolled steel sections made from low-carbon (mild) steel. These are made to British Standard sizes, some of which are shown in Fig. 8.1 and whose corresponding dimensions are shown in Table 8.1. For efficiency and ease of handling, a structural member should use as little material as possible whilst carrying as much load as possible. The 'I-section', which is extensively used in structural steel fabrication work, is designed to give an efficient strength/weight ratio, and the 'universal sections' provide even better efficiency.

8.2 Typical structural steel connections and assemblies

In structural steelwork, the connections of the members are made by one of the following methods:

- Use of hot-forged (black) bolts.
- Use of high-strength friction-grip bolts.
- Riveting.
- Welding.

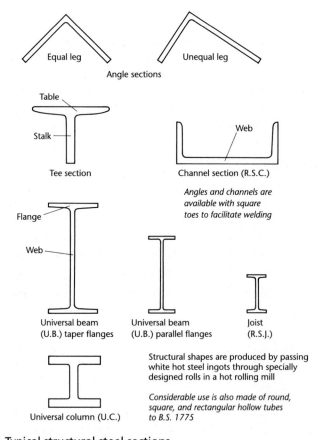

Figure 8.1 Typical structural steel sections

Table 8.1 Standard sizes hot-rolled sections for structural work

Rolled steel section	Range of sizes (dimensions in mm)				Flanges
	From		To		
	Web	Flange	Web	Flange	
Universal beams (UB)	609.6 × 304.8		920.5 × 420.5		Slight taper (2° 52′)
	203.2 × 133.4		617 × 230.1		Parallel
Universal columns (UC)	152.4 × 152.4		474.7 × 424.1		Parallel
Joists (RSJ)	76.2 × 50.8		203.2 × 101.6		5° Taper
Channels (RSC)	76.2 × 38.1		431.8 × 101.6		5° Taper
Equal angles	25.4 × 25.4 × 3.175		203.2 × 203.2 × 25.4		
Unequal angles	50.8 × 38.1 × 4.762		228.6 × 101.6 × 22.13		
	Table	Stalk	Table	Stalk	
Tees (from Universal beams)	133.4 × 101.6		305.5 × 459.2		Parallel or slight taper
Tees (from Universal columns)	152.4 × 76.2		395.0 × 190.5		Parallel
Tees (short stalk)	38.1 × 38.1 × 6.35		152.4 × 152.4 × 15.88		Slight taper (½°)
Tees (long stalk)	25.4 × 76.2		127 × 254		8° Taper

All Universal beams and Universal columns are specified by their serial size and mass per metre.

Essentially a steel structure is designed so that the sizes of the various members are determined by their ability to withstand the effects of applied loads in service. The size of the member must always be as near to the design size as possible, and *always greater* than the design size for safety. The dimensions between the various members or components by which the forces and any moments of forces are transmitted are designed to comply with a rigid code of safe practice.

8.2.1 Stanchion bases

In structural steel work stanchions or columns are vertical members. Stanchion bases transmit the downward thrust on the stanchion to the foundation upon which the stanchion and its load are supported. The stanchion base also spreads the load so that the stanchion does not cut into and/or crush the foundation, which is usually concrete. There are two types of stanchion base:

1. Bases for stanchions transmitting only a *direct load* to the foundations.
2. Bases where the stanchion has to transmit a considerable bending moment to the foundations in addition to the direct, vertical load.

Some examples of stanchion bases are shown in Fig. 8.2. Only the top face of the base slab has to be machined to ensure close contact with the end of the stanchion. Slabs that are less than 50 mm thick do not normally need to be machined. The foot of the stanchion must be machined square to the axis of the stanchion to ensure it stands vertically. This ensures that the load is carried directly onto the base plate or slab and into the foundations. If the foot of the stanchion or column is not square to the base plate, then the total load on the stanchion would be carried through the bolt or rivet shanks where these are used to secure the assembly.

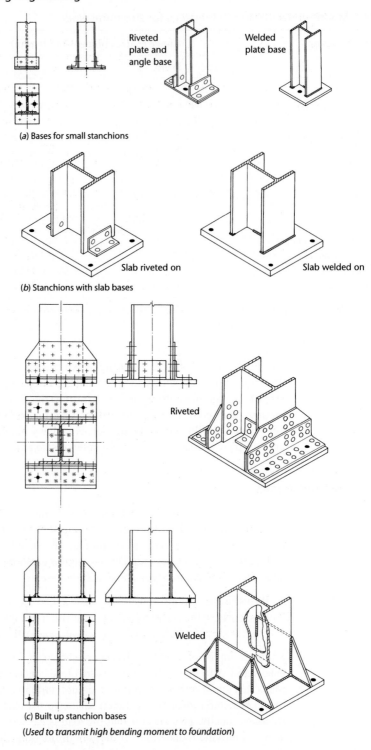

(a) Bases for small stanchions

Riveted plate and angle base

Welded plate base

(b) Stanchions with slab bases

Slab riveted on

Slab welded on

Riveted

Welded

(c) Built up stanchion bases

(Used to transmit high bending moment to foundation)

Figure 8.2 Stanchion bases

8.2.2 Stanchion splices

Splices in stanchions or columns should be arranged at a position above the adjacent floor level so that the joint, including any splice plates, are well clear of any beam (horizontal member) connection. Splices should never be made on a connection, otherwise the bolts or rivets making the joint would be subjected to double loading. Figure 8.3 shows some typical stanchion splices.

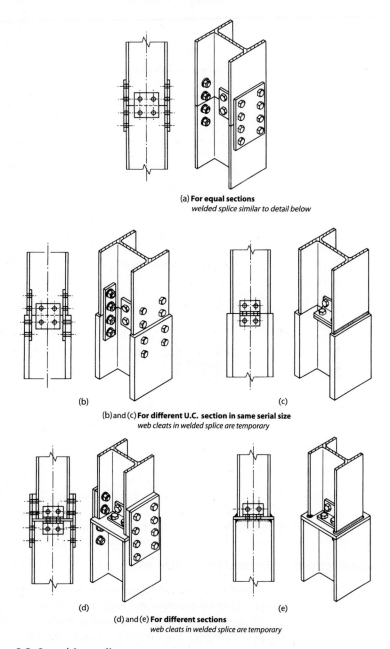

(a) **For equal sections**
welded splice similar to detail below

(b)

(c)

(b) and (c) **For different U.C. section in same serial size**
web cleats in welded splice are temporary

(d)

(e)

(d) and (e) **For different sections**
web cleats in welded splice are temporary

Figure 8.3 Stanchion splices

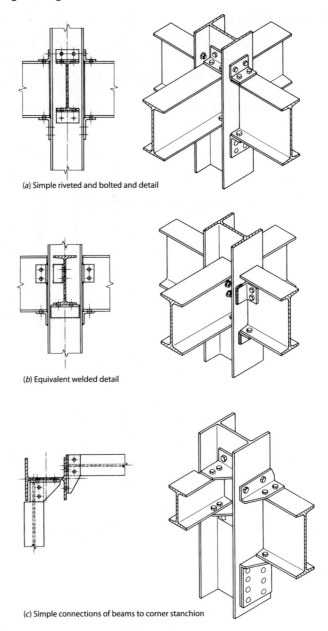

(a) Simple riveted and bolted and detail

(b) Equivalent welded detail

(c) Simple connections of beams to corner stanchion

Figure 8.4 Connection of beams to splices

8.2.3 Connections of beams to stanchions

Usually, beams are horizontal members that are connected at right angles to the vertical stanchions or columns and support the floors of buildings. However, they may be inclined or curved (arched) for special applications. We are only concerned with right-angled connections. Figure 8.4 shows some simple beam-to-stanchion connections using seating cleats and their fixings.

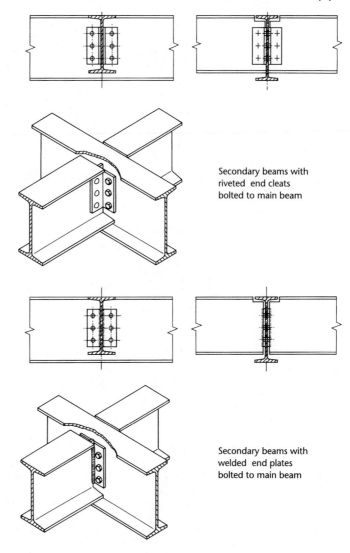

Figure 8.5 Beam to beam construction

8.2.4 Beam-to-beam connections

Where beam-to-beam connections are made, web cleats or welded-on end plates are used to transmit the load from the secondary beams to the main beams. These types of connections are shown in Fig. 8.5.

8.3 Trusses and lattice frames

Figure 8.6 shows details of both riveted and welded roof trusses. It can be seen that these are plane frames consisting of sloping rafters which meet at the ridge. A main

(*a*) Riveted roof truss

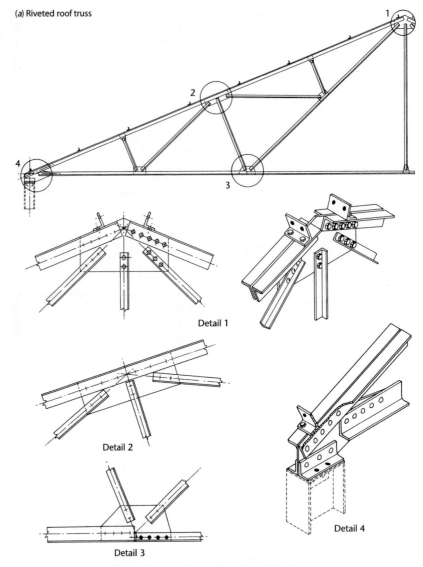

Detail 1

Detail 2

Detail 3

Detail 4

The gusset plate are designed to withstand the forces
applied by the members which they connect together

Figure 8.6 Roof trusses

horizontal tie connects the lower ends (feet) of the rafters, and the internal bracing members. Trusses are used to support the roof in conjunction with purlins. These purlins are secondary members laid longitudinally across the rafters on which the roof covering is laid.

Figure 8.7 shows details of lattice girders – also be sometimes called trusses – which are plane frames of open web construction. They usually have parallel chords or booms which are connected with internal web bracing members. There are two

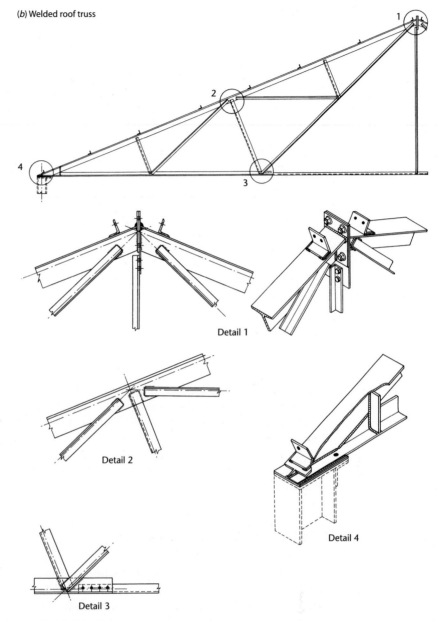

(b) Welded roof truss

Detail 1

Detail 2

Detail 4

Detail 3

Figure 8.6 (Continued)

basic types of lattice girder, the 'N' type and the 'Warren' type, both of which are shown in Fig. 8.7.

As with roof trusses, the framing of a lattice girder should be *triangulated*, taking into account the span and the spacing of the applied loads. The booms are divided into panels of equal length and, as far as possible, the panel points are

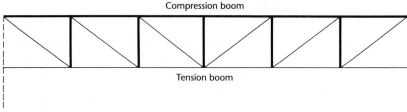

Compression boom

Tension boom

The 'N-type' lattice girder

The diagonal bracing members are arranged so that they act as ties.
If reversed they would become struts and the shorter vertical members would be ties.

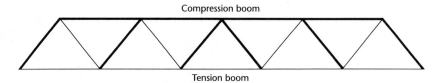

Compression boom

Tension boom

The 'Warren-type' lattice girder

Note: The thick lines in the diagrams represent 'struts'

Figure 8.7 Lattice girders

arranged to coincide with the applied loads. Lattice girders may be used in flat roof construction or in conjunction with trusses or trussed rafters.

Because of the greater forces usually supported by the members of lattice girders, connections are usually riveted or welded into position. Site connections are usually made with high-strength, friction-grip bolts. These have largely replaced rivets for site work. Figure 8.8 shows how lattice girders can be fabricated from standard rolled sections.

8.4 Web stiffeners

Web stiffeners are required when a beam or plated structure is subjected to a twisting force (torsion) or a sideways thrust. The need for web stiffeners, or gussets, increases as the depth of the beam web increases as shown in Fig. 8.9.

Web stiffeners may be welded or riveted into position. When fabricating stiffeners that are to be welded into position, it is important that the stiffener is an exact fit in the beam. The slope of the tapered flanges should be copied exactly. With triangular-shaped gussets 'feather edges' must be avoided, and the sharp edges must be removed, for otherwise the strength of the assembly may be reduced rather than increased. Another important reason for cutting off the corners of webs or gussets is to provide clearance for fillet welds or bend radii. This allows for ease of assembly and permits welding through the gap with the web, or gusset, in position. It also avoids costly fitting operations at the corners. Various applications of web stiffeners and gussets are shown in Figs 8.10 and 8.11.

(a) Riveted lattice girder

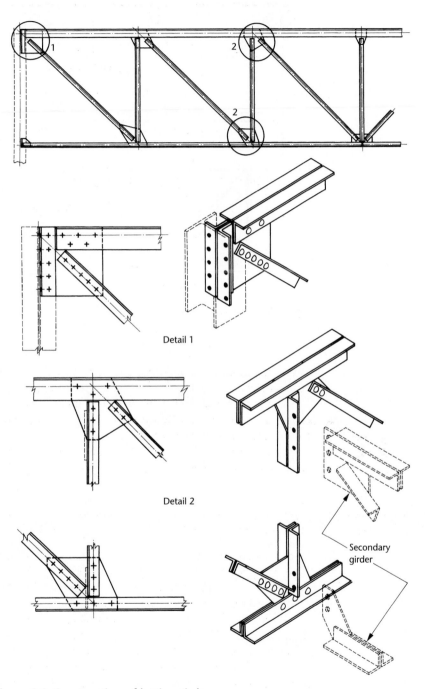

Detail 1

Detail 2

Secondary
girder

Figure 8.8 Construction of lattice girders

(b) Welded lattice girder

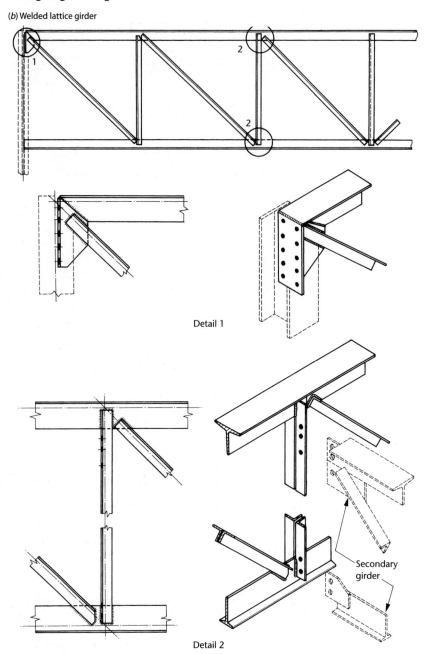

Detail 1

Secondary girder

Detail 2

Figure 8.8 (Continued)

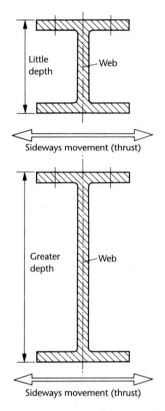

The smaller sizes of universal beam sections will resist a twisting force within its web which has relatively little depth, presenting a short lever arm

The larger sizes of universal beam sections are unable to resist a twisting force within its web as effectively as the smaller sizes

This is because of the much greater depth presenting a very much longer lever arm

In addition, the thickness of the web of universal beam sections does not increase proportionally with the depth of the beam

Figure 8.9 Need for web stiffeners

8.5 Fabricated lightweight beams

Bar frames or *open web frames* are of ultra-light-weight construction and are particularly useful for use as long-span roof purlins. The flanges are usually formed of angle sections and the web diagonals are made from round or square sections which are welded to the flanges as shown in Fig. 8.12.

8.6 Castellated beams

The castellation system of construction is a simple method of increasing the strength/weight ratio of a beam with very little waste. Figure 8.13 shows this method of construction. By flame-cutting and arc-welding the depth of any beam can be increased by as much as 50%, thus increasing its load capacity for large spans. Tapered beams and ridged beams may also be produced by this method.

Handling large beams of any type on site demands the use of mechanical lifting devices – usually a mobile crane or a tower crane. The safe use of cranes was introduced in Chapter 1. Bolted and riveted connections are discussed in Chapter 9 and welded joints are discussed in Chapter 10.

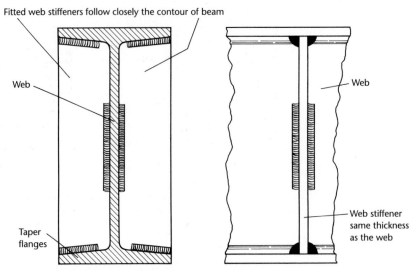

Section of a beam showing fitted web stiffeners (welded)

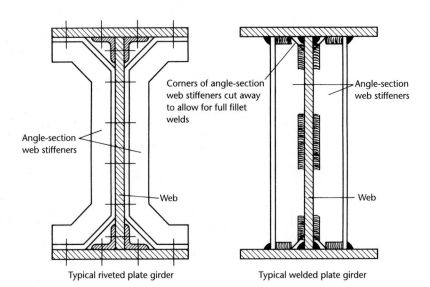

Typical riveted plate girder

Typical welded plate girder

Figure 8.10 Web stiffeners

Gussets for single fillet welded flanges

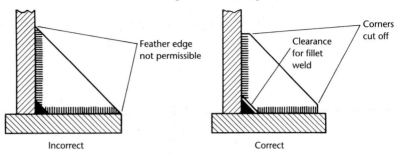

One reason for stiffening a flange is when a single fillet
weld has to be used, the other side being inaccessible
for welding, and if otherwise this single weld would be
subject to bending

Welded connection between two I-section members

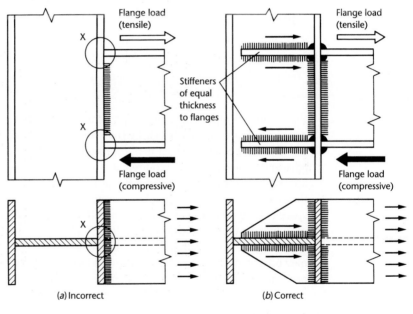

Intense local stresses in welds and web of
vertical member at X, owing to lack of
stiffness of flange in resisting horizontal
flange loads

Provision of suitable stiffeners ensures
proper transfer of horizontal loads to
web of vertical member

Figure 8.11 Applications of stiffeners

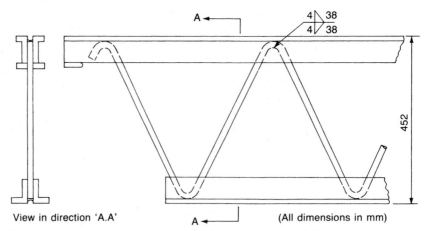

View in direction 'A.A'   A ◄━━━┙   (All dimensions in mm)

This simple lightweight beam is fabricated by welding together 64 mm × 45 mm × 8 mm M.S. angle sections and 16 mm diameter M.S. bar.
The web diagonals are produced by ordinary workshop methods such as localised heat at the bends and pulling the bars round pins.

Figure 8.12 Typical fabricated bar frame or beam

8.7 Pipework (setting out bends)

Pipe bends are represented on working drawings by centre lines only. In practice, the actual bends are marked out (lofted) on the workshop floor for large installations or on sheet fibre board or sheet metal for small installations. When two pipes running adjacent to each other are required, it is usual to make the outer bend with a standard machine former which has a throat radius equal to three times the diameter of the pipe. The inner bend, which is sharper, is made by hand. It is important that the space between the pipes is adequate to allow for compression joints and the means of tightening them up, also for installing 'saddles' for holding the pipes in position. Soldered joints require rather less room between the pipes. Figure 8.14(a) shows the method of setting out right-angle bends. Figure 8.14(b) shows an example of 'offset' bends to enable pipes to change direction to avoid an obstacle such as a girder.

8.7.1 Setting out pipes for machine bending

Figure 8.15(a) shows alternative methods of specifying the measurements for pipe bending, whilst Figs 8.15(b), 8.15(c) and 8.15(d) show the setting-up procedure for machine-bending pipes to the required measurements.

Inside measurements

To bend the pipe to the inside measurements, mark the pipe with the dimension 'x' as shown in Fig. 8.15(b). Place a set-square against the mark and carefully insert the pipe in the bending machine so that the *square touches the inside of the groove* in the former and bend as usual.

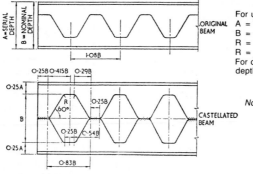

For universal beams and columns:
A = Serial depth of original beam.
B = Nominal depth of original beam.
R = 12 mm for original beams 200 mm and over.
R = 9 mm for original beams under 200 mm.
For other sections A and B are equal to the original depth.

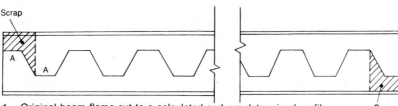

Note: A castellated beam will inherit a greater STRENGTH/WEIGHT RATIO than the original section from which it is fabricated. To enable suitable connections to be made, it is sometimes necessary to fill some holes by welding in shaped plates.

(*a*) **Details for standard castellation**

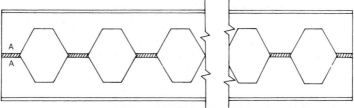

1. Original beam flame cut to a calculated and pre-determined profile.
    For mass production of castellated beams, the profile cutting of the web is performed by multi-head machines using a template.

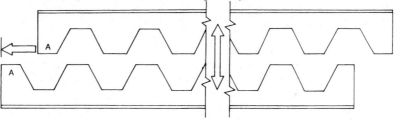

2. Cut beam separated into two components for fitting A to A.

3. The two components welded together to form castellation.
    Note: The depth of a castellated beam is approximately one and a half times that of the original section.

(*b*) **Stages in forming castellated beams**

Figure 8.13 Castellated beams

(a) The setting out of right-angle bends

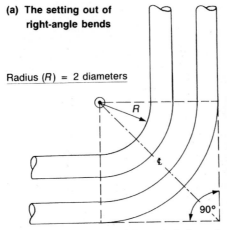

Radius (R) = 2 diameters

The inner pipe is drawn first using a 'square' to make sure it is at right angles. The outer pipe is then drawn leaving sufficient distance between to allow for screwing up any fixing clips. The inner bend shown is drawn on a radius equal to two diameters to the throat or two and a half to the centre of the pipe. The arcs for each bend are struck from a common point (marked with a small circle) on the line bisecting the angle.

The dotted lines forming a right angle indicate the use of the square to obtain the right-angle bend.

(b) The setting out of offset bends

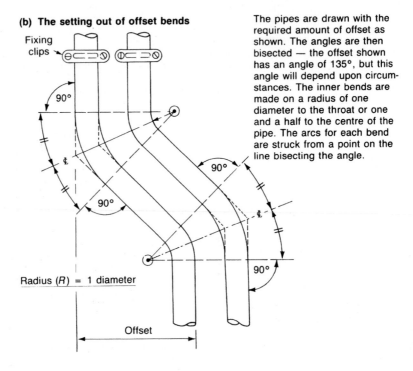

Fixing clips

Radius (R) = 1 diameter

Offset

The pipes are drawn with the required amount of offset as shown. The angles are then bisected — the offset shown has an angle of 135°, but this angle will depend upon circumstances. The inner bends are made on a radius of one diameter to the throat or one and a half to the centre of the pipe. The arcs for each bend are struck from a point on the line bisecting the angle.

Figure 8.14 Setting out pipe bends

Outside measurements

To bend the pipe to the outside measurements, mark the pipe with the dimension 'y' as shown in Fig. 8.15(c). Place a set-square against the mark and carefully insert the pipe in the bending machine so that the *square touches the outside edge of the former* and bend as usual.

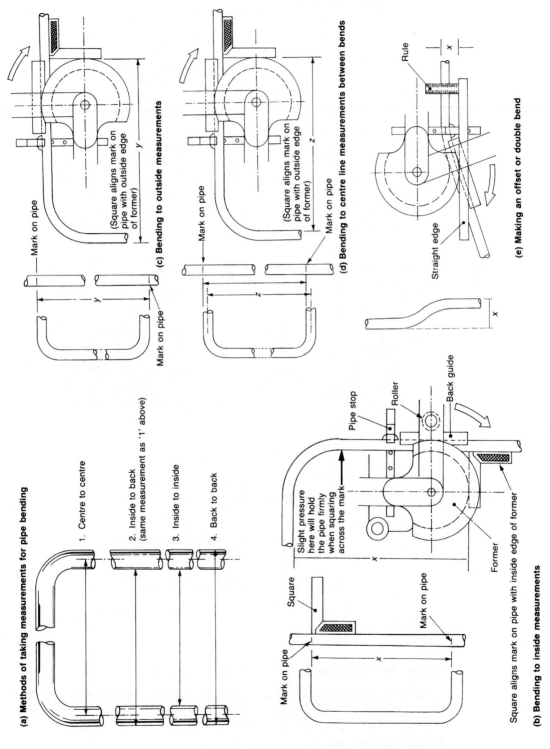

Figure 8.15 Methods of making offsets and double bends by machine

Centre-line measurements

To bend the pipe to the centre-line measurements, mark the pipe with the dimension 'z' as shown in Fig. 8.15(d). Place a set-square on the mark and carefully insert the pipe in the bending machine so that the *square touches the outside edge of the former* as when bending to outside measurements and bend as usual.

Off-set measurements

The set-up for making an 'off-set' or 'double bend' to required measurements is shown in Fig. 8.15(e). After making the first bend to the required bevel, the pipe is set in the machine as shown, taking care to ensure the second bend will be made parallel to the first bend on the same (horizontal) plane. A straight-edge is then placed against the outside edge of the former and parallel to the pipe. When set to the dimension 'x' bend as usual.

8.8 Pipe bending

It is essential to appreciate the deformation – as shown in Fig. 8.16 – which occurs during the pipe-bending process to properly understand the use of the tools used and

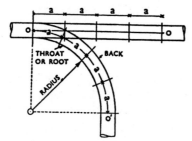

The walls of a straight pipe are parallel and must remain parallel after bending, if its true round section is to be maintained in the bend.
The original length of the pipe (0–0) remains unaltered after bending only along the centre line (0–0).

The inside or throat of the bend is shortened and compressed and the outside or back is lengthened or stretched.

(a) A true bend — each division represents a 'throw' in making the bend

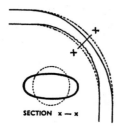

The shortening and lengthening will tend to produce a flattening of the back and an inwards kinking of the throat with a spreading of the sides of the bend. The collapse of the pipe in the bend will occur unless precautions are taking to prevent it.
Pipes with thick walls have less tendency to collapse than thin-walled pipes of the same diameter.

(b) Deformation of a bend in unsupported pipes — ovality

Note: Bend allowances for pipework are determined by the bend radius and the pipe diameter, using a MEAN RADIUS, in the same way as bend allowances for platework are determined by bend radius and plate thickness (see Table 10.4).

Figure 8.16 Pipe bending – true and deformed bends

Table 8.2 Pipe loading for manual bending

Type of loading	Remarks
Steel springs	Only easy bends should be attempted with spring loading due to the possible difficulty of removal. The minimum throat radii are approximately 3 diameters for pipes up to 25 mm and 4 diameters for 32 mm and 55 mm pipes
Lead	Owing to the physical power required and the difficulty of melting out the lead, the bending of pipes over 38 mm diameter is uneconomical; for smaller diameter pipes it is a safe and efficient loading. Sharp bends can be made successfully within radii as small as 2 diameters for pipes up to 25 mm and 2½ diameters for 32 mm and 38 mm pipes
Pitch or resin or a mixture of both	Bending is easier than with lead loading, and is preferred for pipes over 38 mm diameter because less force is required, the pipe is lighter to handle, and loading and unloading is easier. Pipes up to 152 mm or more in diameter can be bent. The minimum throat radii are 2 diameters for 32 mm and 38 mm pipes, and 4 diameters for 50 mm and 100 mm pipes
Sand	Consists of filling the pipe with dry sand, well rammed or 'tamped' to consolidate it. Cold or hot bending can be employed. Bending the pipe at 'red heat' is generally the best method and is essential for sharp bends or bends in pipes having a greater diameter than 50 mm. Bends can be made hot in pipes having diameters up to 152 mm, and at radii as small as 2½ diameters for 25 mm to 50 mm pipes, and 4 diameters for 64 mm and 100 mm pipes present no difficulty
Fusible alloy	*This low melting point alloy can be maintained in its molten state at temperatures lower than the boiling point of water, and quickly solidifies at room temperature.* After bending, the filler will quickly run out when the pipe is dipped into a tank of boiling water, leaving the interior of the pipe perfectly clean

the various techniques employed. Because of their relatively thin walls, light-gauge pipes need greater care if satisfactory bends are to be made by manual methods. In order to prevent 'ovality' or collapse, it is necessary to fill or 'load' the pipe with some material that will support the pipe walls. The materials most commonly used are listed in Table 8.2.

Before loading with lead, pitch, resin or low-melting-point alloy, the pipe must be sealed at one end with a wooden plug. When sand is used the pipe must be plugged at both ends in order to retain the compacted loading.

The use of flexible steel springs for pipe bending is probably the most common and practical method employed when bending thin-walled copper pipes by hand. It is recommended that bending springs should be lightly greased before use to facilitate their removal after bending. Figure 8.17 shows the method of supporting pipes using steel bending springs.

8.8.1 Pipe loading – safety

Lead loading

Note: that lead loading is no longer encouraged for environmental and health reasons. On no account can lead loading be used on any pipes used for domestic water systems or pipes used in the food industry. It is still permissible for central heating systems and gas pipes.

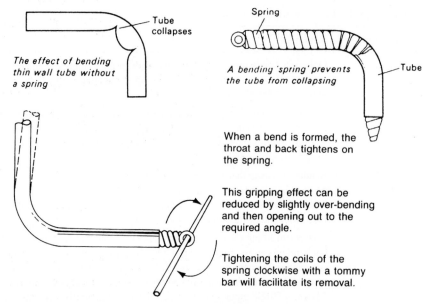

The effect of bending
thin wall tube without
a spring

Tube
collapses

Spring

A bending 'spring' prevents
the tube from collapsing

Tube

When a bend is formed, the
throat and back tightens on
the spring.

This gripping effect can be
reduced by slightly over-bending
and then opening out to the
required angle.

Tightening the coils of the
spring clockwise with a tommy
bar will facilitate its removal.

Use of an extension rod hooked into the eyelet of the spring will enable
it to be inserted and withdrawn when a bend is made in the middle of
a length of pipe.

Figure 8.17 Use of springs for bending

The pipe should be securely fixed or held in tongs since it will become very hot when the molten lead is poured into it. Great care must be taken to ensure that the pipe is thoroughly dry, both inside and out, especially if it has been quenched after annealing. If a pipe is at all damp, steam will be generated and its pressure will drive out the molten lead with explosive force at grave risk to the fitter.

Pitch or resin loading

Pitch or resin or a mixture of both must be heated slowly until it is in the liquid state, taking care that it does not ignite. If this does occur, quickly smoother the mouth of the container with a wet sack which should always be kept to hand when using this technique.

Never try to anneal a pipe already loaded since there is a high risk of a dangerous explosion.

When the bend is completed the loading has to be melted out. Care should be taken to start heating at an end so that the pitch or resin can run out without building up any back-pressure. If heating is commenced away from an end, gases will form to drive out a semi-molten plug of resin or pitch with great force and risk to anyone in the near vicinity. At worst it is possible to cause an explosion and burst the pipe.

Sand loading

Before use, the sand should be thoroughly dried as moisture in the sand may generate steam if the pipe is heated and this may cause an explosion. Further, fine dry sand flows better when filling and emptying the pipe.

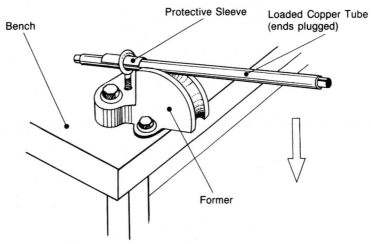

Bench

Protective Sleeve

Loaded Copper Tube
(ends plugged)

Former

A former with a curvature to correspond with the required
radius of the bend is bolted to the bench.
One end of the loaded pipe is passed through the screwed
eyelet provided, and the other end is forced downwards
around the grooved former.
Pipes up to 38 mm diameter may be formed in this manner.

Figure 8.18 Pipe bending – simple bench former

8.8.2 Pipe bending by hand

If standard bending formers or bending machines are not available, small-diameter
pipes can be bent by hand using a wood plank with a hole through it. A hole equal
to the outer diameter of the pipe is drilled through one end of a plank three to four
inches thick. The hole is gouged out to the bend radius. The pipe is threaded through
the hole to the required position and gradually bent by hand. This technique should
only be used for easy bends in loaded pipes. Preferably a bench former as shown in
Fig. 8.18 should be used for hand bending.

Thin-walled copper pipes are usually supplied cold drawn in the 'half-hard'
condition (temper). This ensures sufficient rigidity and straightness when installed
and minimizes damage in transit. In order to bend 'half-hard' pipe it is necessary to
anneal it in the bend zone. The cold-work done on the pipe during cold-bending has
the effect of restoring the half-hard temper to the bend zone.

Thick-walled (heavy-gauge) copper pipes can be forge-bent while red-hot, in a
similar manner to that employed when bending thick walled mild steel and wrought
iron pipes using a bending table as shown in Fig. 8.19(a). Forge-bending is a highly
skilled operation but, when correctly executed, produces a bend having a smooth
curve with a minimum of thinning and ovality. Pipes to be forge-bent are loaded
with compacted sand to prevent them from collapsing during the bending operation. When the pipe has been raised to the forging temperature at the bend zone,
it is transferred to a bending table where it is located by retaining pins (pegs) as
shown in Fig. 8.19(b). The pipe is pulled round by a steel sling or lever. Very-large-
diameter pipes are heated in a specially designed furnace and a winch is used when
forming the bend. The skill lies in local quenching of the red-hot pipe so that it only
'gives' and bends at the required point.

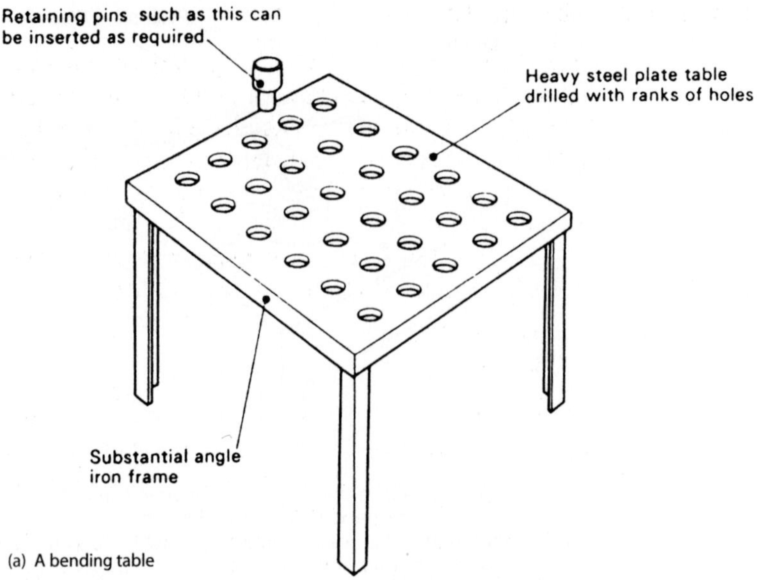

Retaining pins such as this can
be inserted as required

Heavy steel plate table
drilled with ranks of holes

Substantial angle
iron frame

(a) A bending table

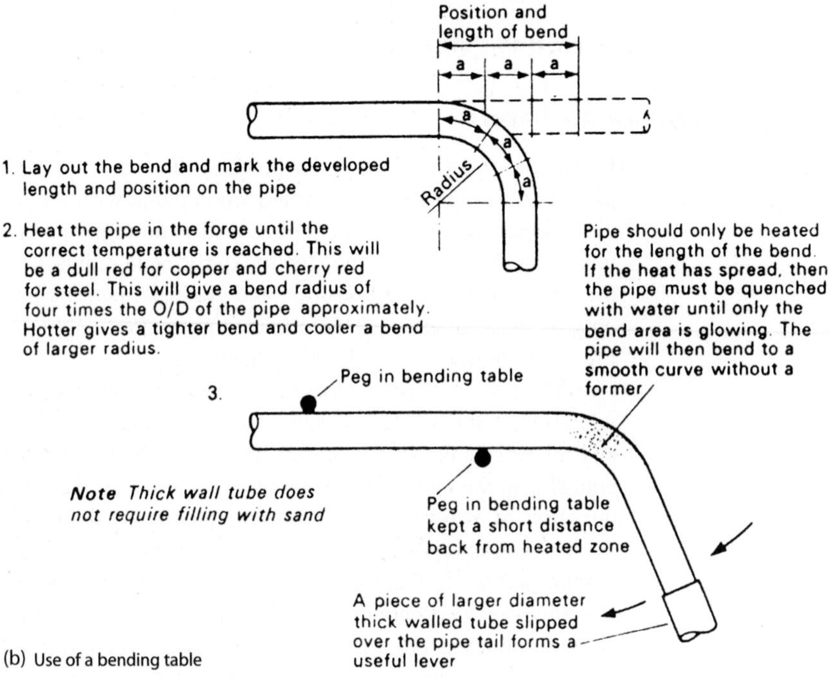

Position and
length of bend

a | a | a

1. Lay out the bend and mark the developed
   length and position on the pipe

Radius

2. Heat the pipe in the forge until the
   correct temperature is reached. This will
   be a dull red for copper and cherry red
   for steel. This will give a bend radius of
   four times the O/D of the pipe approximately.
   Hotter gives a tighter bend and cooler a bend
   of larger radius.

Pipe should only be heated
for the length of the bend.
If the heat has spread, then
the pipe must be quenched
with water until only the
bend area is glowing. The
pipe will then bend to a
smooth curve without a
former.

Peg in bending table

3.

Note *Thick wall tube does*
not require filling with sand

Peg in bending table
kept a short distance
back from heated zone

A piece of larger diameter
thick walled tube slipped
over the pipe tail forms a
useful lever

(b) Use of a bending table

Figure 8.19 Hot bending

8.8.3 Pipe bending by machine

Machine bending on site is usually performed in bending presses or in rotary bending machines. A typical hand-powered bending press is shown in Fig. 8.20. This type of bending machine is made in various sizes and is capable of bending pipe up to 50 mm diameter. Larger machines are similar in principle but employ a motor-driven hydraulic pump. These machines are capable of bending steel pipes in the largest sizes.

Rotary-type machines enable pipes to be bent without loading. They have two principle components as shown in Fig. 8.21(a), namely:

- A quadrant former.
- A back guide.

The former supports the throat of the pipe, whilst the back guide rotates round the bend as pressure is applied by the roller as shown in the sectional view in Fig. 8.21(b). This shows how these components support the pipe during the bending process. This enables the bend to be made without thickening the throat metal which would reduce the bore of the pipe. It is usual to anneal heavy-gauge pipe when using a hand-pull machine in order to make it easier for the operator and to reduce the load on the machine.

Some rotary-type machines are fitted with a pointer on the lever arm which, by proper adjustment, determines automatically the exact point at which the roller pressure is best applied to the pipe. Figure 8.21(c) shows the pointer correctly adjusted for bending. The pointer should be parallel with the pipe when the roller is making tight contact with the back of the guide. Figure 8.22 shows how incorrect positioning of the roller will result in defects in the bend.

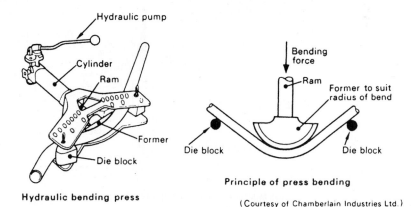

Hydraulic bending press

Principle of press bending

(Courtesy of Chamberlain Industries Ltd.)

A former to suit the size of the pipe to be bent is selected and secured to the end of the ram.
When the pump is operated, the ram moves forward pressing the pipe against the die blocks which are secured by pins in a steel frame.
A series of locating holes in the frame permits the die blocks to be positioned correctly to accommodate pipes of various sizes.

Figure 8.20 Bending press

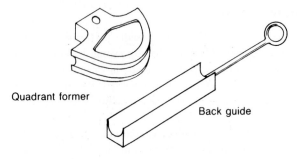

Quadrant former

Back guide

(*a*) **Basic components**

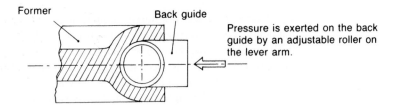

Former   Back guide

Pressure is exerted on the back
guide by an adjustable roller on
the lever arm.

(*b*) **Section showing hot pipe walls
are supported during bending**

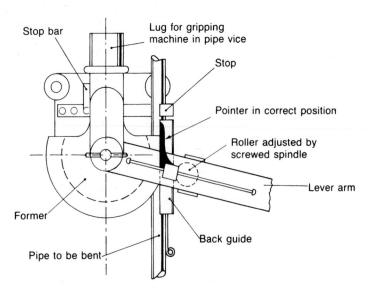

Stop bar   Lug for gripping
                  machine in pipe vice

                                    Stop

                        Pointer in correct position

                        Roller adjusted by
                        screwed spindle

                                    Lever arm

Former

Back guide

Pipe to be bent

(*c*) **Machine correctly adjusted for true bending**

Figure 8.21 Rotary bending machines

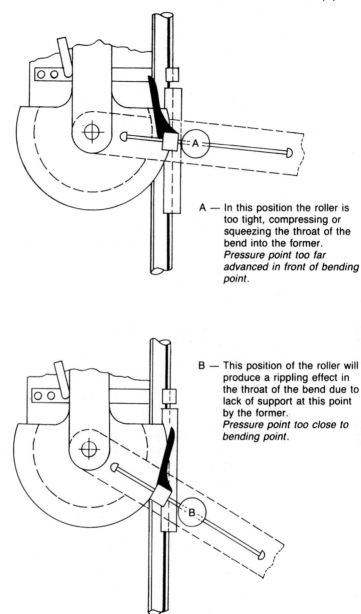

A — In this position the roller is too tight, compressing or squeezing the throat of the bend into the former. *Pressure point too far advanced in front of bending point.*

B — This position of the roller will produce a rippling effect in the throat of the bend due to lack of support at this point by the former. *Pressure point too close to bending point.*

Figure 8.22 Bending machines – incorrect adjustment

8.9 Pipe fitting

Light-gauge copper pipes with compression joint fittings as found in domestic water systems are satisfactory up to pressures of $1.0\,\text{MN/m}^2$ (10.0 bar) providing its temperature does not exceed that of boiling water. Central heating systems and natural gas distribution systems generally have soldered fittings. Industrial hydraulic systems, however, often have to withstand pressures up to $41.4\,\text{MN/m}^2$ (414 bar) and high-pressure superheated steam systems have to withstand temperatures many times that of boiling water. Every engineering installation has to be carefully designed to ensure that the pipework, fittings and installation are suitable for the service under consideration. The more important factors to be considered include:

- The working pressure.
- Whether the pressure is steady or pulsating.
- The working temperature.
- The rate of flow and therefore the bore of the pipework.
- The type of fluid being conducted (possibility of corrosion).
- The cost of the installation and subsequent maintenance.
- Safety – this includes the legal requirements of inspection, testing, maintenance and insurance.

The systems considered in this section will be limited to copper and steel piping with screwed or compression joints.

8.9.1 Safety

Because of the many fatal and serious accidents that have occurred through failures in high-pressure and high-temperature installations, stringent codes of practice have been drawn up which are enforceable by law. This ensures almost complete safety for those working in the vicinity of such installations and almost complete safety for the premises housing such installations. All installations operating at high pressures and, possibly, high temperatures should be:

- Designed by persons qualified to meet the requirements of the Health and Safety at Work Act and any associated legislation both national and international.
- Constructed only from equipment that has been tested and certified.
- Constructed from equipment that complies with the recommendations of the British Standards Institution.
- Inspected and tested by a specialist insurance surveyor upon completion.
- Colour-coded to indicate the content of the pipework in accordance with the appropriate British Standard recommendations.

8.9.2 Types of pipe (copper)

For pressures up to $0.7\,\text{MN/m}^2$ (7 bar), and temperatures up to 100°C, copper pipe is usually used. It will be solid drawn as shown in Fig. 8.23(a) and will not be seamed. Copper pipe has the advantage that it can be easily bent to shape and can be easily joined with compression-type fittings and also be soft- or hard-soldered. Further, it is resistant to corrosion either by the fluid flowing through it or the atmosphere surrounding it. Copper pipe has two main disadvantages:

1. It is much more expensive than steel pipe; however, its advantages can offset this disadvantage in the smaller sizes.
2. It is thin-walled and cannot, therefore, be threaded for use with screwed fittings without appreciably weakening the pipe.

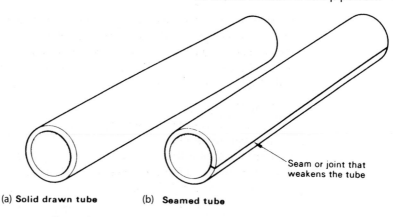

(a) **Solid drawn tube**   (b) **Seamed tube**

Figure 8.23 Effect of thread on wall strength

8.9.3 Types of pipe (steel)

Both plain carbon steel and alloy steel pipes are available for high-pressure installations or where copper would be too costly and the risk of corrosion is low. Note that the wall thickness for water pipes is increased to allow for corrosion. The types of pipe available are:

- Hot-finished seamless (HFS).
- Cold-finished seamless (CFS).
- Electric-resistance-welded (ERW).

Hot-finished seamless pipe can be used at the highest pressures and temperatures, depending upon its composition and specification. It has a rough finish and cannot be used with compression fittings. It is used with screwed and welded joints.

Cold-finished seamless pipe is used for the same purposes as hot-finished pipe but, as it is smooth, it can be used with compression fittings. It is more costly than hot-finished pipe. Its smooth surface and dimensional accuracy make it easier to thread.

Electric-resistance-welded pipe is rolled up from strip steel and electrically welded along the seam as shown in Fig. 8.23(b) in a totally automatic process. This makes it much cheaper than solid drawn pipe but it is not so strong and can only be used for low-pressure installations. It has a good finish and is accurate enough for use with compression fittings. ERW stainless steel pipe is often used for domestic central heating systems, whilst ERW low-carbon-steel pipe is used for electrical conduit. It is only available up to 50 mm outside diameter.

8.9.4 Threaded (screwed) joints

Both parallel and tapered pipe threads still have a conventional 55° Whitworth thread form although they are dimensioned in metric units.

Parallel threads

BS EN 228-1 is used for parallel pipe threads where pressure-tight joints are not made on the threads. Where pressure-tight joints are made on the threads, then BS EN 228-2 should be referred to. Pipe threads employ a constant pitch system so that all threads from 1 inch to 6 inches have a constant pitch of 2.309 mm. Therefore, the depth of the thread is also constant as is the wall thickness of the pipe, and the thread will not unduly weaken the pipe. The effect of threading a pipe on the strength of that pipe is shown in Fig. 8.24.

Tapered threads

Tapered threads are the subject of BS 2779. They are self-sealing when pulled up tight and are used with fittings that have an internal parallel thread. A tapered pipe thread is shown in Fig. 8.25. Tapered threads are also a constant pitch system in the

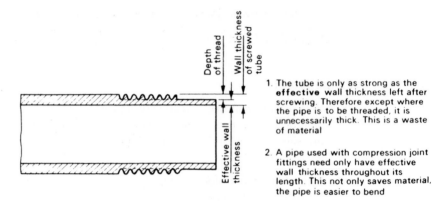

1. The tube is only as strong as the **effective** wall thickness left after screwing. Therefore except where the pipe is to be threaded, it is unnecessarily thick. This is a waste of material

2. A pipe used with compression joint fittings need only have effective wall thickness throughout its length. This not only saves material, the pipe is easier to bend

Figure 8.24 Solid drawn and seamed tube

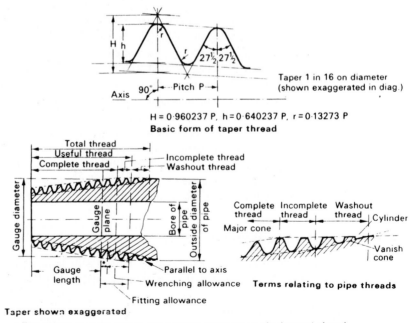

Figure 8.25 British Standard pipe threads

larger sizes for the same reason as explained above and also so that tapered and parallel threads can be used in conjunction with each other.

Figure 8.26 shows some typical screwed fittings. To ensure pressure tightness, the joint is usually sealed by brushing a 'jointing compound' onto the threads before screwing the joint together. Although the threaded joints shown are simple to make and neat in appearance, they have some fundamental limitations.

- Valves and other fittings cannot be readily broken out of the pipe run for maintenance or replacement.
- If the pipe is cranked, a large amount of room is required to swing it round as it is screwed home.

To overcome these limitations a number of unions, as shown in Fig. 8.27, have to be used. Unfortunately, these are a source of weakness and the pipe must be supported either side of a union.

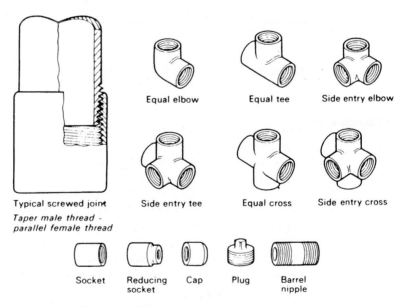

Figure 8.26 Typical screwed fittings

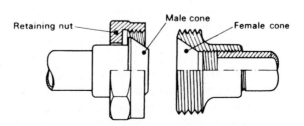

Figure 8.27 Union joint

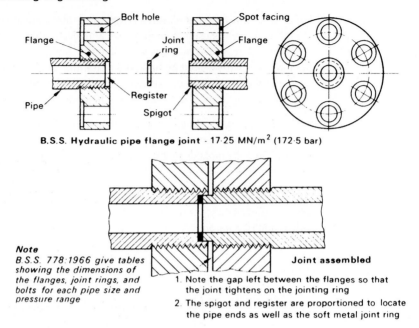

B.S.S. Hydraulic pipe flange joint - 17·25 MN/m² (172·5 bar)

Note
B.S.S. 778:1966 give tables
showing the dimensions of
the flanges, joint rings, and
bolts for each pipe size and
pressure range

Joint assembled

1. Note the gap left between the flanges so that the joint tightens on the jointing ring
2. The spigot and register are proportioned to locate the pipe ends as well as the soft metal joint ring

Figure 8.28 High-pressure flanged joint

8.9.5 Flanged joints

Flanged joints overcome the disadvantages of plain screwed joints, but are more costly to produce and more bulky. Figure 8.28 shows a standard high-pressure and high-temperature joint with an all-metal seal. The treaded ends of the pipes only locate the flanges and in no way contribute to the sealing of the joint. The flanges do not touch and so do not require machining on their faces. It is usual to spot-face the nut and bolt faces. This type of joint is not specified for pipework above six inches (152 mm) outside diameter. Above this size the flange is usually welded to the pipe. In some instances welded flanges may be specified for the smaller sizes. Wherever possible, bolt-holes should not lie on the vertical centreline. This is so that any seepage will not corrode the lower bolt. Figure 8.29 shows a simpler type of flanged joint for use at lower temperatures and pressures.

8.9.6 Compression joints

These are used in place of screwed joints for the smaller sizes of pipes, particularly copper pipes. Not only do they save time in installation but thinner-walled and less costly pipe can be used since it is not weakened by the screw thread as shown previously in Fig. 8.24. Traditionally, compression joints are made mainly from hot-pressed brass for use with copper pipework with pressures not exceeding 1 MN/m² (10 bar) and temperatures not exceeding 100°C. However, nowadays, compression joint fittings are being made from steel for pressures up to 41.4 MN/m² (414 bar) and temperatures up to and including superheated steam. Figure 8.30 shows a typical compression joint.

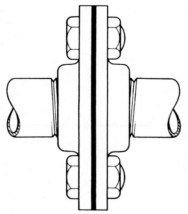

In this type of joint the flanges are pulled down tight onto the gasket or packing. The pipes are screwed into the flanges which are only located by the retaining bolts, there being no spigot and register

Figure 8.29 Low pressure flanged joint

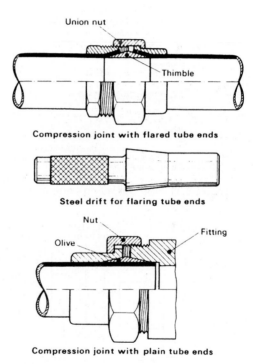

Compression joint with flared tube ends

Steel drift for flaring tube ends

Compression joint with plain tube ends

Figure 8.30 Compression joint

8.9.7 Threading equipment

Pipework may be threaded by hand on site, or machine-threaded in the workshop if the pipework is prefabricated. Solid 'button' dies, as used for electrical conduit installation, are largely unsuitable for pressure-tight joints since the thread being cut cannot be adjusted and the finish is too rough. Adjustable die-heads are used for pipework as shown in Fig. 8.31(a). The thread may be produced in a number of cuts.

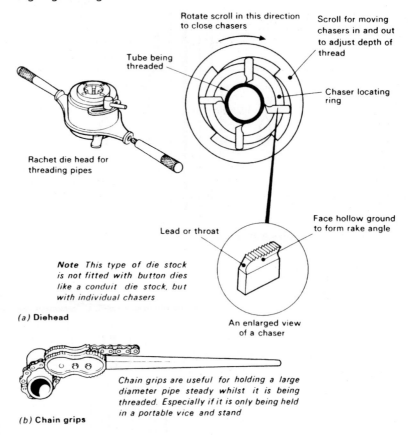

Rotate scroll in this direction to close chasers

Scroll for moving chasers in and out to adjust depth of thread

Tube being threaded

Chaser locating ring

Rachet die head for threading pipes

Face hollow ground to form rake angle

Lead or throat

Note *This type of die stock is not fitted with button dies like a conduit die stock, but with individual chasers*

(a) **Diehead**

An enlarged view of a chaser

Chain grips are useful for holding a large diameter pipe steady whilst it is being threaded. Especially if it is only being held in a portable vice and stand

(b) **Chain grips**

Figure 8.31 Pipe threading equipment

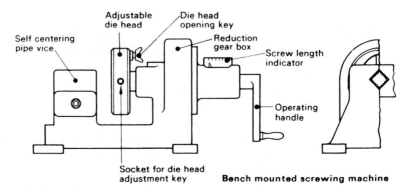

Adjustable die head

Die head opening key

Self centering pipe vice

Reduction gear box

Screw length indicator

Operating handle

Socket for die head adjustment key

Bench mounted screwing machine

Figure 8.32 Screwing machine

This prevents the metal tearing and allows the size of the thread to be adjusted in size so that the pipe will tighten up in the fitting to which it is assembled.

For this reason, a long 'tommy-bar' is fitted to the diehead so that the fitter can obtain sufficient leverage. For convenience, the tommy-bar is attached to the diehead by a ratchet so that the thread can be cut by rocking the bar back and forth through a small arc and eliminating the need to swing the bar round the tube.

Table 8.3 Colour code for general services

Pipe contents	Ground colour	BSS Colour No*	Colour band	BSS Colour No*
Water				
Cooling (primary)	Sea green	217	–	–
Boiler feed	Strong blue	107	–	–
Condensate	Sky bue	101	–	–
Drinking	Aircraft blue	108	–	–
Central heating (less than 40°C)	French blue	166	–	–
Central heating (0–100°C)	French blue	166	Post Office red	538
Central heating (over 100°C)	Crimson	540	French blue	166
Cold water from storage tank	Brilliant green	221	–	–
Domestic hot water	Eau-de-nil	216	–	–
Hydraulic power	Mid-Brunswick green	226	–	
Sea, river, untreated	Green grass	218	–	–
Air				
Compressed (under 1.38 MN/m^2 or 13.8 bar)	White	–	–	–
Compressed (over 1.38 MN/m^2 or 13.8 bar)	White	–	Post Office red	538
Vacuum	White	–	Black	–
Steam	Aluminium or crimson	540	–	–
Drainage	Black	–	–	–
Electrical services	Light orange	557	–	–
Town gas	Canary yellow	309	–	–
Oils				
Diesel fuel	Light brown	410	–	–
Furnace fuel	Dark brown	412	–	–
Lubricating	Salmon pink	447	–	–
Hydraulic power	Salmon pink	447	Sea green	217
Transformer	Salmon pink	447	Light orange	557
Fire installations	Signal red	537	–	–
Chemicals	Dark grey	632	See BSS 349:1319	–

*See BSS 381C – Ready mixed paints

When working with a portable pipe vice and stand, the torque of the die-head must be balanced. The fitter's mate will hold the pipe to prevent it turning with adjustable chain grips. An example of a set of chain grips is shown in Fig. 8.31(b). Hand- and power-operated threading machines are available both in the workshop and on site as portable units. They usually have self-opening dieheads so that they do not have to be reversed whilst the pipe is removed. A typical manually operated machine is shown in Fig. 8.32. In order to provide sufficient torque, the operating handle is geared to the diehead..

Chain grips can also be used for tightening threaded pipes into their fittings

8.9.8 Colour coding

For safety reasons all pipework installations should be colour-coded in accordance with the appropriate British Standard so as to identify the contents of the pipework system. The recommendations are given in Table 8.3.

Exercises

8.1 Structural steelwork
  a) Describe the basic differences between a British Standard Beam (BSB) and a rolled steel joist (RSJ).
  b) Figure 8.33 show a section through a steel-framed building. Copy out and complete the table naming the various features of the structure.
  c) List the more commonly used connections in structural steelwork and discuss the relative advantages and limitations of each of the systems you have listed.
  d) Sketch and describe two types of stanchion base, explain why such bases are required and state where each would be used.
  e) Sketch and describe:
     i) a stanchion splice
     ii) a beam-to-stanchion connection
     iii) a beam-to-beam connection.
  f) Describe under what circumstances web stiffeners are required and sketch a typical example.

  g) With the aid of sketches show the difference between N-type fabricated beams and Warren-type fabricated beams. List the advantages and limitations of such beams and describe the method of construction.
  h) Explain what is meant by a 'castellated' beam. What are the advantages and limitations of such a beam, and how can it be fabricated with minimum waste of metal?
     i) Because of the size of structural steelwork it cannot be marked out on a surface table in the usual workshop manner; therefore, describe how roof trusses can be marked out and pre-assembled.

8.2 Pipework
  a) Pipework and pipe bends are usually set out on drawings by centre lines only. What alternative methods can be used when specifying measurements for machine bending?

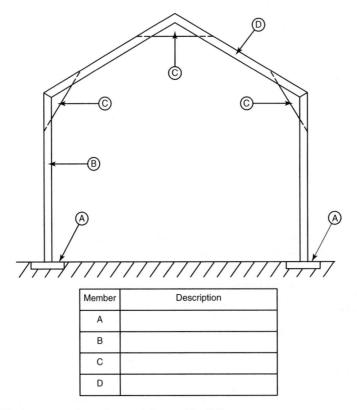

Member	Description
A	
B	
C	
D	

Figure 8.33 Simplified cross-section of a steel framed building

b) With the aid of sketches describe the deformation that can occur when a pipe is bent. Describe TWO common methods of preventing pipes from collapsing during bending. Describe the safety precautions that must be observed when loading and unloading pipes before and after bending.

c) Describe the difference between seamed and seamless pipe and list the advantages and limitations of both types.

d) Compare the advantages and limitations of copper pipe and steel pipe and give examples where it would be appropriate to use both types.

e) With the aid of sketches describe the process of hot-bending pipe on a bending table.

f) Compare the advantages and limitations of hand-bending and machine-bending pipework.

g) Explain why compression joints are preferable to screwed joints when using thin-walled tubing.

h) When assembling screwed joints, state which member of the joint has a parallel thread and which has a tapered thread, the pipe or the fitting.

i) Discuss the uses of unions and flanged joints in pipework.

j) Discuss the threading of pipework, illustrating your answer with sketches.

k) What colour coding should be used for the following pipework applications:
   i) electrical conduit
   ii) cold water
   iii) hot water
   iv) steam
   v) natural gas.

9

Joining processes (mechanical connections)

When you have read this chapter you should be able to:

- Describe the difference between temporary and permanent joints
- Describe the various types of screwed fastenings
- Describe what is meant by high-strength friction grip bolts and where they are used
- Describe what is meant by fitted bolts and where they are used
- Describe the various types of washer and where they are used
- Describe how bolt holes are aligned in structural steelwork on site
- Describe the use of captive nuts
- Describe the methods of cutting internal and external screw threads
- Describe how bolt holes are reamed for fitted bolts
- Describe the various types of riveted joints
- Calculate the hole and rivet dimensions for riveted joints
- Describe the need for and applications of POP® rivets
- Describe what is meant by self-secured joints
- Describe the techniques for making grooved seams, paned-down seams, knocked-up seams and the use of the grooving tool
- Calculate the allowances for self secured joints
- Describe the comparative uses of self-secured joints
- Describe the Pittsburgh lock joint

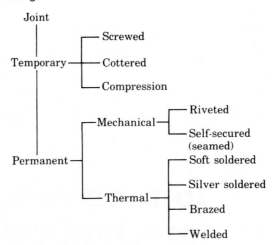

Figure 9.1 Types of joint

9.1 Mechanical connections (threaded)

Fabricated components can be joined together in a variety of ways. The type of joint may be temporary or permanent. Temporary joints may be assembled and dismantled as often as is required without damage to the members being joined or the joining device (e.g. a nut and bolt). Permanent joints cannot be dismantled after assembly without damage to, or the destruction of, the joint device (e.g. drilling out a rivet) or the members being joined (e.g. flame cutting a welded joint). Figure 9.1 shows the different ways in which fabrications may be joined.

9.1.1 Threaded fastenings

Various types of threaded fastenings are used where components must be assembled and dismantled regularly. They are also used for making connections in structural steelwork on site. The fastenings used in making connections in fabricated steelwork will all have a thread with a 'V' form. This has many advantages which is why it is so widely used:

- It is the easiest and cheapest form to manufacture.
- It is easily cut with taps and dies.
- It is the strongest thread form.
- It is self-locking and only works loose when subjected to extreme vibration.

Nowadays the ISO metric thread-form is normally used. This has a 60° included thread angle and is dimensioned in millimetres. For maintenance work on existing steelwork you will most likely meet with threaded fastenings having a Whitworth thread-form with a 55° included thread angle and is dimensioned in inches. Figure 9.2 shows a section through a male screw thread illustrating the basic elements essential to the understanding of screwed fastenings. Figure 9.3 shows a selection of screwed

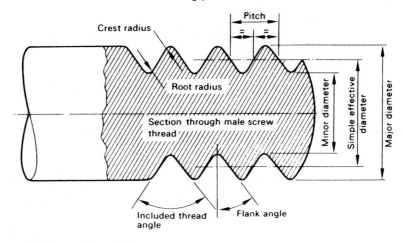

Figure 9.2 Screw thread elements

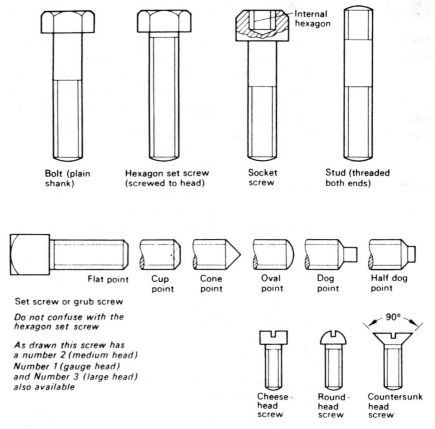

Figure 9.3 Types of screwed fastenings

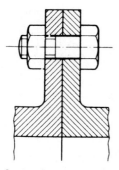

Section through a bolted joint

Plain shank extends beyond joint face

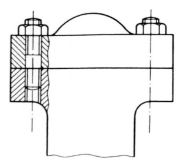

Stud and nut fixing for an inspection cover

This type of fixing is used where a joint is regularly dismantled. The bulk of the wear comes on the stud which can eventually be replaced cheaply. This prevents the wear falling on the expensive casting or forging

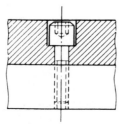

Cap head socket screw

Although much more expensive than the ordinary hexagon head bolt, the socket screw is made from high tensile alloy steel. They are heat treated to make them very strong, tough and wear resistant. They are widely used in the manufacture of machine tools. The above example shows how the head is sunk into a counterbore to provide a flush surface

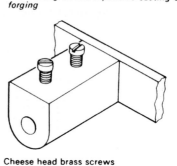

Cheese head brass screws

These are used in small electrical appliances for clamping cables into terminals

Figure 9.4 Use of screwed fastenings

fastenings and Fig. 9.4 shows some typical applications. Locking devices are used to prevent treaded fastenings working loose when subjected to vibration. A selection of locking devices is shown in Fig. 9.5.

Coarse thread systems

These are the strongest threads and unlikely to 'strip' when being pulled up tight. Unfortunately they tend to work loose when subjected to vibration. They are usually used for general purpose fabrication work and also in conjunction with soft and/or low strength materials.

Fine thread systems

These are not as strong as coarse thread screwed fastenings and care must be taken not to over-tighten them so as to avoid stripping the threads. They lock-up more

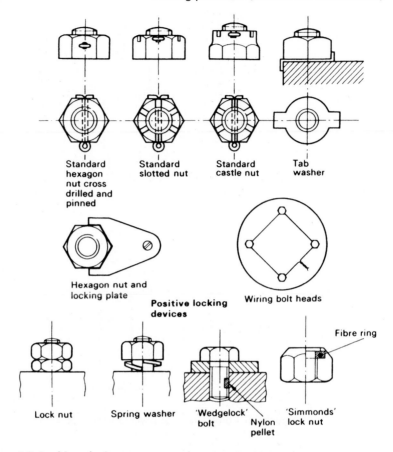

Standard
hexagon
nut cross
drilled and
pinned

Standard
slotted nut

Standard
castle nut

Tab
washer

Hexagon nut and
locking plate

**Positive locking
devices**

Wiring bolt heads

Fibre ring

Lock nut

Spring washer

'Wedgelock'
bolt

Nylon
pellet

'Simmonds'
lock nut

Figure 9.5 Locking devices

tightly and are less inclined to work loose. They are usually used in conjunction with tough and/or hard materials.

9.1.2 Threaded fastenings for structural steelwork (black and turned bolts)

Hexagon bolts and nuts are normally used for most screwed connections as they provide sufficient convenient spanner positions when tightening or loosening a connection in a restricted space. The cheapest type of bolt is the 'black bolt' – so-called because it is coated with a black scale from the forging process. These are made by the up-set forging process and the only machining is for the screw thread which may be cut or rolled. Rolled threads are the strongest.

Because of the manufacturing process, the shanks of black bolts lack dimensional accuracy and are always used in clearance holes that are about 1.6 mm larger in diameter than the bolt. They transmit their load by direct bearing (or shear) at relatively low levels of stress. Therefore, where high levels of stress have to be transmitted, black bolts cannot be used. For high levels of stress transmission 'turned' or 'fitted' bolts are used. These are more expensive since the forged shank is finished turned to a high level of dimensional accuracy so that they are a close fit in precision reamed holes.

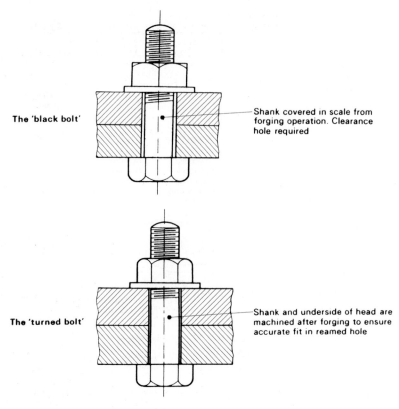

The 'black bolt'   Shank covered in scale from forging operation. Clearance hole required

The 'turned bolt'   Shank and underside of head are machined after forging to ensure accurate fit in reamed hole

Figure 9.6 Comparison of black and turned bolts

This makes them capable of carrying much higher loads. Figure 9.6 shows the difference between black and turned bolts.

9.1.3 High-strength friction grip bolts

When using turned or fitted bolts in connections, the major problem is the insertion of the bolts in very close fitting holes. One way to overcome this problem is to have the holes drilled slightly undersize in the workshop and make the connection with a limited number of temporary 'tacking' bolts. Finally, ream the holes to size on site and insert the fitted bolts one by one as the tacking bolts are removed. However, this is time-consuming and expensive since it is labour intensive.

Alternatively, *high-strength friction grip bolting* is now frequently used as it is less labour intensive and therefore cheaper. High-strength friction grip bolts are forged from high-tensile steel and the underside of the head and the bearing surface of the nut are semi-finished – special identification symbols are stamped on the head of the bolt. Such bolts can be easily inserted in normal clearance holes (usually 1.6 mm clearance) just as if the connection was going to be made with standard black bolts. Each friction grip bolt is provided with two special hardened washers which must be fitted under the head of the bolt and under the nut, as shown in Fig. 9.7. These washers spread the load, preventing the head and the nut becoming embedded in the relatively softer steel of the structural members.

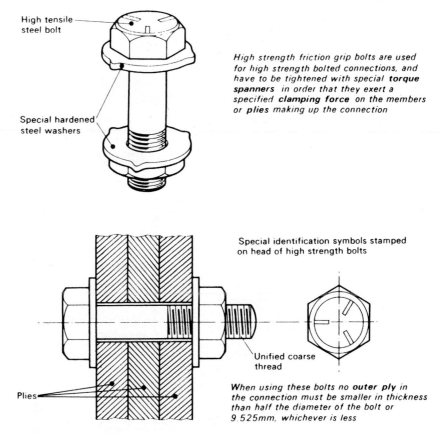

High tensile
steel bolt

*High strength friction grip bolts are used
for high strength bolted connections, and
have to be tightened with special **torque
spanners** in order that they exert a
specified **clamping force** on the members
or **plies** making up the connection*

Special hardened
steel washers

Special identification symbols stamped
on head of high strength bolts

Unified coarse
thread

*When using these bolts no **outer ply** in
the connection must be smaller in thickness
than half the diameter of the bolt or
9.525mm, whichever is less*

Plies

Figure 9.7 High-strength friction grip bolts

When high-strength bolting is employed, the bolts are tightened to such an extent that they carry the shear and bearing loads imposed on them by *friction* between the faces of the connecting parts. To ensure the correct degree of friction is exerted, the nut must be tightened using a torque spanner so that a consistent and specified degree of tightness can be applied. The differences between connections using black-bolts and high-strength friction grip bolts are shown in Fig. 9.8.

When steel structures are connected by bolting, it is common practice to descale and/or remove rust and then paint the contact surfaces with zinc oxide to prevent rusting. When using high-strength bolts *no paint should be applied to the contact surfaces* until after assembly to ensure a metal-to-metal contact. The bolts must be square to the bearing surfaces under the bolt head and nut, hardened 'taper washers' should be used where necessary. Figure 9.9(a) shows how the corners of the nut can bite into the structural member if a washer of the correct size is not fitted. Figure 9.9(b) shows the use of a bolt that is too short so that the full strength of the connection is not achieved. Figure 9.9(c) shows what happens if a taper washer is not inserted between the bolt head and the tapered flange of a joist or channel. The bending of the bolt severely weakens it. The taper of the washer and the taper of the member must match.

In a riveted or black bolt connection, the plies give until they bear on the side of the rivet or bolt.

The result is that in each case the rivet or bolt is subjected to 'shear' loads on the sections marked A-A and B-B when acted upon by a 'tensile' or pulling force, as shown in the diagrams.

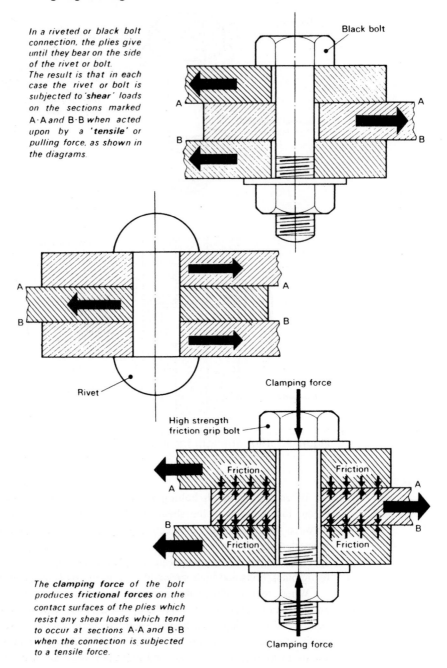

The **clamping force** of the bolt produces **frictional forces** on the contact surfaces of the plies which resist any shear loads which tend to occur at sections A-A and B-B when the connection is subjected to a tensile force.

Figure 9.8 The essential differences between riveted or black bolt connections and high-strength bolts

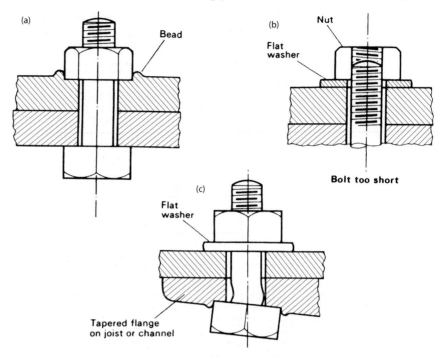

Figure 9.9 Defects in bolted connections

9.1.4 Washers in screwed connections

The use of a washer between the face of a nut and the part being fastened has the following functions:

- It increases the *frictional grip* thus reducing *slip* between the components being joined.
- It improves the locking action of the nut on the screw thread.
- The prevention of damage to the workpiece when the nut is being tightened.
- In the case of taper washers, it ensures the bolt head and/or nut can tighten down on a smooth surface that is at right-angles to the axis of the bolt.

In addition to the commonly used 'plain' or 'flat' washer, various other types are used; for example, the taper washer and the spring washer. Taper washers are used to ensure that the contact surface of the bolt head and nut are square with the axis of the hole so that the bolt is not bent and weakened. They are available with a 5° taper for use on the inside of joist or channel flanges, and with an 8° taper for use on the inside of 'T' section flanges.

Spring washers are made from hardened and tempered square or rectangular section spring steel and, when compressed by the nut, promotes 'tension' between the threads of the bolt and the nut so as to increase the locking friction to prevent the nut becoming loose when subjected to vibration. Further, the 'turned-out' ends of a

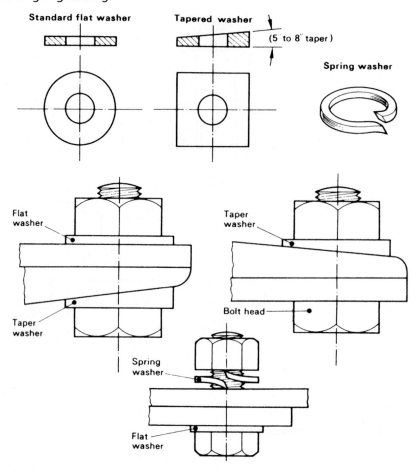

Figure 9.10 Flat, taper and spring type washers

spring washer tend to bite into the contact surfaces and this also provides additional resistance to any slackening off of the nut due to vibration. Figure 9.10 shows some washer types and their applications.

9.1.5 Alignment of holes

'Drifts' are used when assembling connections for riveting and bolting, especially large steel fabrications. They are made from hardened and tempered steel and are available to suit all sizes of hole. The three basic types of drift are described below and each has a specific function when used for aligning parts coming together so that the holes will be 'faired' to admit bolts or rivets as shown in Fig. 9.11(a).

Taper drift

Taper drifts are used for 'fairing' or aligning bolt holes. The holes in the members move into the correct alignment as the drift is hammered into the first hole.

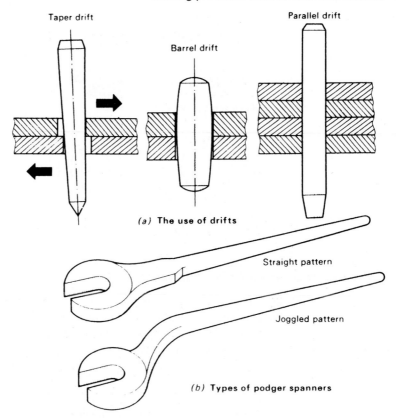

Figure 9.11 Fairing or aligning holes

Barrel drift

Barrel drifts are used when aligning holes in confined spaces the drift is hammered until it passes through the hole. Barrel drifts can only be used where the amount of movement required to bring the members into alignment is minimal.

Parallel drift

Parallel drifts are mainly used for maintaining the alignment of members during riveting or bolting operations, the members having been 'faired up' initially with the aid of a taper drift or a barrel drift. Parallel drifts are generally 0.08 mm smaller in diameter than the hole into which they are inserted.

Podger spanners

Podger spanners, as shown in Fig. 9.11(b), are used when making bolted connections. There are two patterns available – straight and joggled – and in a range of bolt and nut sizes. The joggled pattern has a cranked shank to allow clearance space between the fitter's hand and the workpiece. The shank is tapered so that it can be used as a drift when 'podgering' or fairing holes into alignment.

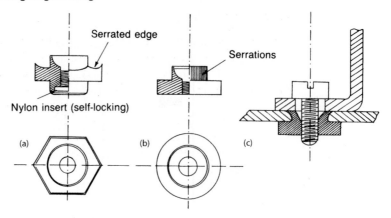

Figure 9.12 Hank bushes

9.1.6 Screwed fastenings (miscellaneous)

A hole punched in sheet material does not offer sufficient thickness (depth of metal) necessary for cutting an efficient and strong screw thread into it. Therefore, the sheet metal has to be reinforced at the point of connection. Let's now consider some of the alternative methods of doing this.

'Hank' bushes

'Hank' bushes are one method of producing effective screw threads in sheet metal. Some examples of hank bushes are shown in Fig. 9.12. The rivet shank, which is an integral part of the 'hank' bush, is inserted into a hole of suitable diameter previously punched or drilled in the sheet metal. With the bush supported on a suitable flat surface or anvil the protruding part of the bush is hammered down spread it and secured it in the sheet. When secured, the serrations on the under surface of the bush head (Fig. 9.12(a)) or on the rivet shank (Fig. 9.12(b)) bite into and grip the sheet metal and prevent the bush from turning. Figure 9.12(c) shows a countersunk type 'hank' bush which can be used where the two components have be in contact with each other.

POP® Nut threaded inserts

Nut threated inserts are a more sophisticated system of inserts used for connecting sheet metal components.

The following is an abstract from the wide range of sizes and types of these fastening devices manufactured by Tucker Fasteners Ltd., Emhart Technologies, reproduced here with permission.

Blind rivet nuts

Blind rivet nuts are especially designed to offer a means of providing a stronger threaded joint in sheet and other materials and, like blind rivets, only require access to one side of the workpiece for installation. They are not only a form of captive nut but, unlike some other types, they also allow components to be riveted together as well providing a screw thread anchorage. The principle of their use is shown below.

The POP® blind rivet nut is screwed onto the threaded mandrel of the setting tool and is then inserted into the drilled or punched hole.

The tool is operated, retracting the mandrel. The unthreaded part of the nut expands on the blind side of the workpiece to form a collar and applies a powerful clenching force that rivets the components firmly together.

With the POP® nut firmly in position the tool mandrel is simply unscrewed from the nut. The threaded portion is now ready to act as a secure anchorage point.

Blind rivet nuts are generally available in thread sizes M3 to M12 in numerous combinations of head style, body form and material. Three head styles are available, flat head, 90° countersunk and, for thin gauge materials, a small flange which provides a near flush appearance without the need for countersinking.

Standard bodies are round with open or closed ends. The closed end prevents the ingress of moisture or vapour through the bore of the nut. Where a higher torque resistance is required, body forms may be fully or partially hexagonal (set in a hexagonal hole), or splined (set in a round hole).

POP® Nut Threaded Inserts: application

POP® Nut Threaded Inserts provide a simple and effective way to join materials with the benefit of an internal thread, in a variety of applications.

POP® Nut Threaded Inserts are the perfect solution for providing high-quality, loadbearing threads in various materials where alternative methods cannot maintain torque and pull out loads. **POP**® Nut Threaded Inserts are suitable for single sheets down to 0.5 mm.

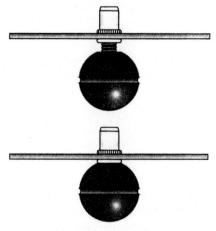

POP® Nut Threaded Inserts enable components, which are assembled later in the production cycle, to be adjusted.

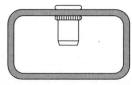

POP® Nut Threaded Inserts are ideally suited to applications where access is only available from one side of the workpiece.

POP® Nut Threaded Inserts: installation

POP® Nut Threaded Inserts are easily installed from one side of the workpiece without damaging surrounding surfaces of previously finished or delicate components and are suitable for use with all materials in today's manufacturing environment:

- Available in a variety of materials
- Wide range of styles
- Complete range of Hand and Power setting tools
- Bulk and small pack available.

Install/screw the **POP**® Nut Threaded Insert to the tool's threaded mandrel.

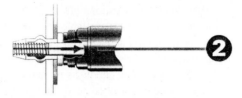

Operating the tool then retracts the mandrel. The unthreaded part of the nut then compresses to form a collar on the blind side of the workpiece, applying a powerful clenching force that firmly joins the components together.

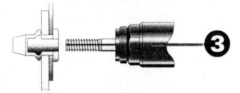

With the **POP**® Nut Threaded Insert firmly in position, the tool mandrel is simply demounted/unscrewed from the insert.

The threaded part of the insert then acts as a secure anchorage point for subsequent assembly work.

POP® Nut: Steel

Flat head open end (with knurls)

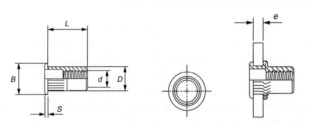

Thread d	Description	Length (mm) L	Grip (mm) e	Hole dia. (mm)	Barrel dia. (mm) D	Flange dia. (mm) B	Flange thickness (mm) S	Bulk box quantity	Small pack quantity
M4	PSFON430	10.0	0.3–3.0	6.0	5.9	9.0	1.0	10 000	500
	PSFON440	11.5	3.1–4.0					10 000	500
M5	PSFON530	12.0	0.3–3.0	7.0	6.9	10.0	1.0	8000	500
	PSFON540	15.0	3.1–4.0					5000	500
M6	PSFON630	14.5	0.5–3.0	9.0	8.9	12.0	1.5	4000	500
	PSFON645	16.0	3.1–4.5					4000	500
M8	PSFON830	16.0	0.5–3.0	11.0	10.9	15.0	1.5	2000	250
	PSFON855	18.5	3.1–5.5					2000	250
M10	PSFON1030	17.0	0.5–3.0	12.0	11.9	16.0	2.0	1500	200
	PSFON1060	22.0	3.0–6.0					1500	200
M12	PSFON1240	23.0	1.0–4.0	16.0	15.9	22.0	2.0	1500	200

Self tapping screws

Sheet metal fabrications are often held together by means of self-tapping screws. A range of sizes are available and are supplied with a variety of head shapes.

Self-tapping screws with Phillips or cross-head recessed type heads are usually preferred to screws with a straight screwdriver slot.

A self-tapping screw, as its name implies, cuts its own thread in a hole of suitable diameter punched or drilled in sheet metal. Figure 9.13 shows two types of self-tapping screw. Figure 9.13(a) shows a suitable type for fastening thin sheet to a thicker section. Figure 9.13(b) shows a suitable type for joining thin sheets of metal. In both cases the top sheet has a clearance hole. Figure 9.13(c) shows how a greater depth of thread can be achieved by dimpling the hole in the bottom sheet.

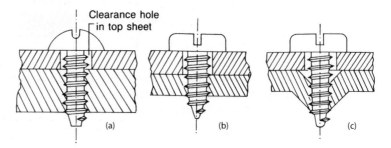

Figure 9.13 Self-tapping screws

9.1.7 Cutting internal screw threads (use of taps)

Figure 9.14(a) shows a section through a thread-cutting tap and how rake and clearance angles are applied to a thread-cutting tap. Since the 'teeth' are *form relieved*, the clearance face is curved and the *clearance angle* is formed by the tangent to the clearance face at the cutting edge. The rake angle is formed by the flute, so we still have our metal cutting wedge. Figure 9.14(b) shows a typical thread-cutting tap and names its more important features. Figure 9.14(c) shows a set of three taps:

1. The *taper* tap is tapered off for the first 8 to 10 threads and is used first. The taper helps to guide the tap into the previously drilled tapping size hole with its axis parallel to the axis of the hole. The taper also helps to increase the depth of cut gradually and helps to prevent overloading the teeth.
2. The *intermediate* or *second* tap has only 3 to 4 threads tapered to guide it into the threaded hole started by the taper tap. This tap can be used to finish threading a through hole. It also helps to cut full threads near to the bottom of a blind hole.
3. The *plug* tap does not have any tapered threads and is used for cutting a full thread to the bottom of a blind hole.

Thread-cutting taps are rotated by means of a tap-wrench. Various types of tap wrench are shown in Fig. 9.15(a) and (b) shows how a tap wrench should be used. The tap is rotated in a clockwise direction and it should be reversed every one or

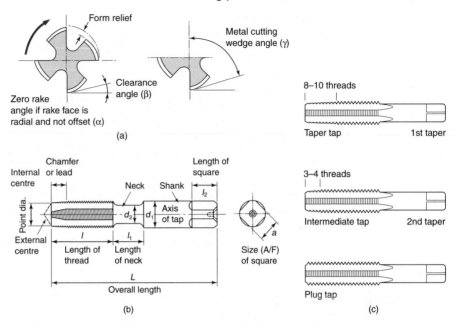

Figure 9.14 Screw thread taps: (a) cutting angles applied to a thread cutting tap; (b) nomenclature for taps; (c) set of thread cutting taps

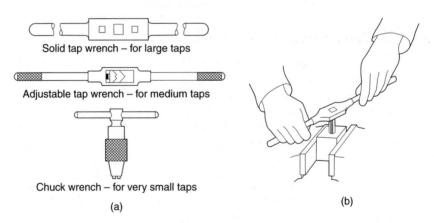

Figure 9.15 Tap wrenches: (a) types of tap wrench; (b) use of tap wrench

two revolutions to break up the swarf. It is essential to start and keep the axis of the tap parallel to the axis of the hole. Normally this means that the axis of the tap will be at right-angles to the work as shown. If the tap is started at an angle other than a right-angle, the tap will cut more heavily on one side of the hole than on the other. At best this will produce a drunken thread, at worst it will cause the tap to break off in the hole. It is usually impossible to remove a broken tap and the work is scrapped.

Before you can cut an internal screw thread, you have to decide on the size of the hole to be used. Theoretically this should be the same as the *minor diameter* of the thread to be cut. In practice, the hole is always somewhat larger in diameter than the minor diameter for the following reasons:

- A thread with 80% engagement is adequate for most general engineering purposes. This considerably eases the load on the tap which is a fragile cutting tool that is easily broken if overloaded.
- The nearest standard drill size available. A smaller one cannot be used or the tap will jam and break, so the nearest larger size has to be used.

Published sets of workshop tables provide information regarding tapping drill sizes. Table 9.1 shows part of such a screw thread table. To cut an M10 × 1.5 metric screwthread the table recommends the use of an 8.50 mm diameter drill to give the 80% engagement, or an 8.60 mm diameter drill if 70% engagement would be

Table 9.1 Screw thread data

	ISO metric tapping and clearance drills, coarse threat series				
	Tapping drill size (mm)		Clearance drill size (mm)		
Nominal size	Recommended 80% engagement	Alternative 70% engagement	Close fit	Medium fit	Free fit
M1.6	1.25	1.30	1.7	1.8	2.0
M2	1.60	1.65	2.2	2.4	2.6
M2.5	2.05	2.10	2.7	2.9	3.1
M3	2.50	2.55	3.2	3.4	3.6
M4	3.30	3.40	4.3	4.5	4.8
M5	4.20	4.30	5.3	5.5	5.8
M6	5.00	5.10	6.4	6.6	7.0
M8	6.80	6.90	8.4	9.0	10.0
M10	8.50	8.60	10.5	11.0	12.0
M12	10.20	10.40	13.0	14.0	15.0
M14	12.00	12.20	15.0	16.0	17.0
M16	14.00	14.25	17.0	18.0	19.0
M18	15.50	15.75	19.0	20.0	21.0
M20	17.50	17.75	21.0	22.0	24.0
M22	19.50	19.75	23.0	24.0	26.0
M24	21.00	21.25	25.0	26.0	28.0
M27	24.00	24.25	28.0	30.0	32.0
M30	26.50	26.75	31.0	33.0	35.0
M33	29.50	29.75	34.0	36.0	38.0
M36	32.00	–	37.0	39.0	42.0
M39	35.00	–	40.0	42.0	45.0
M42	37.50	–	43.0	45.0	48.0
M45	40.50	–	46.0	48.0	52.0
M48	43.00	–	50.0	52.0	56.0
M52	47.00	–	54.0	56.0	62.0

adequate. Compare these sizes with the minor diameter of this thread which is 8.376 mm (minimum).

9.1.8 Hints when tapping holes

- Make sure the taps are sharp and in good condition (no chipped or missing teeth) or it will jam and break off in the hole scrapping the job.
- Use a cutting compound that has been formulated for thread cutting. Lubricating oil is useless since it cannot withstand the cutting forces involved.
- Select the correct size of tap wrench to suit the size of the tap you are using. The wrong size will inevitably lead to a broken tap. A range of tap wrenches should be available.
- Make sure the tap is at right angles to the surface of the component, so that the axis of the tap is co-axial with the axis of the tapping size hole.

9.1.9 Cutting external screw threads (use of dies)

Figure 9.16(a) shows how the basic cutting angles are applied to a thread-cutting, button die. You can see that a die has rake, clearance and wedge angles like any other cutting tool. Figure 9.16(b) shows the main features of a button die, whilst Fig. 9.16(c) shows a typical die-stock that is used to rotate the die. Figure 9.16(d) shows how the die is positioned in the die-stock.

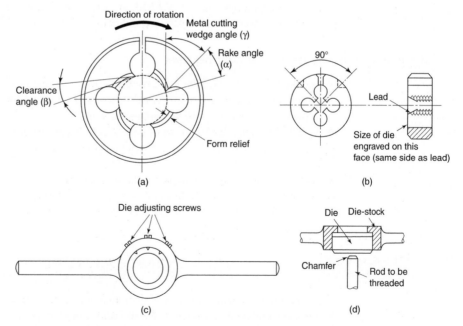

Figure 9.16 Split button dies: (a) cutting angles applied to a thread-cutting die; (b) split die; (c) die stock; (d) positioning of die in stock – the engraved face of the die is visible, ensuring that the lead of the die is in the correct position

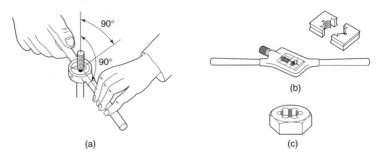

Figure 9.17 Screw thread dies: (a) die stock must be aligned with the axis of the work; (b) rectangular split dies; (c) solid die nut

Screw thread dies are used to cut external threads on components such as bolts. The split-button die, as shown in Fig. 9.16, is the type most widely used. The die-stock has three adjusting screws. The centre screw engages the slot in the die and spreads the die to reduce the depth of cut. The other two screws close the die and increase the depth of cut. Just as when using a tap, the die must be started square with the work axis as shown in Fig. 9.17(a). It is difficult to control a screw-cutting die and any attempt to cut a full thread in one pass will result in a 'drunken' thread. It is better to open up the die to its fullest extent for the first cut. (This will also produce a better finish on the thread.) Then close the die down in stages for the subsequent cuts until the thread is the required size. The thread size can be checked with a nut. The diestock is rotated in a clockwise direction and should be reversed after every one or two revolutions to break up the swarf. A thread cutting lubricant must be used.

Die nuts, as shown in Fig. 9.17(b), are used for cleaning up bruised threads on existing bolts and studs when carrying out maintenance work. They are not adjustable.

9.1.10 Hints when using screw-cutting dies

- The die has its size and the manufacturer's name on one face only. This is the face that should show when the die is in the stock. This will ensure that the taper lead of the die will engage with the end of the work.
- A chamfer on the end of the work will help to locate the die.
- Start with the die fully open and gradually close it down to the required size in successive cuts.
- Select the correct size of stock for the die. This is largely controlled by the diameter of the die. A range of diestocks should be available.
- Use a cutting compound that has been formulated for thread cutting.

9.2 Hand reamers and reaming

When producing a hole with a twist drill, that hole will invariably have a poor finish, be out-of-round, and be oversize. These faults can be overcome by drilling the hole very slightly undersize and correcting it for finish, roundness and size by *reaming*. This can be done by using a hand reamer rotated by a tap wrench. Reamed holes are usually used in structural steelwork for fitting bolts with turned shanks (fitted bolts).

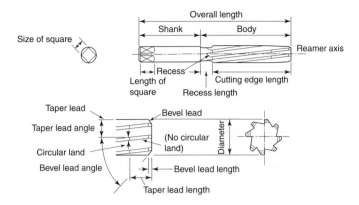

Figure 9.18 Hand reamer

Figure 9.18 shows a typical parallel hand reamer and names its main features. Hand reamers are rotated by means of a tap wrench and they are rotated in a clockwise direction. When withdrawing the reamer it must continue to be turned in the same direction. *It must not be reversed.* A reamer will always follow the original hole. It *cannot* be used to correct the *position* of a hole.

9.2.1 Hints when reaming

- Use a suitably formulated cutting compound. Lubricating oils are not suitable for reaming.
- Always turn the reamer clockwise. Never reverse it.
- Reamed dowel holes should always be 'through' holes so that the dowel can be knocked out from the reverse side when the assembly has to be dismantled.
- Leave the minimum of metal in the drilled hole to be removed by the reamer. A reamer is only a finishing tool.

9.3 Riveted joints

Permanent joints are those where one or more of the components and/or the fastening itself have to be destroyed to dismantle the assembly. For example, the only way to separate two components that have been riveted together is by chiselling the head off the rivet or by drilling it out. The rivet is destroyed.

Riveting is a method of making permanent joints. This process consists of drilling or punching holes in the components to be joined, inserting a rivet through the holes and, finally, closing the rivet by forming a second head. Some typical riveted joints are shown in Fig. 9.19. Unlike a nut and bolt where the connection is equally strong in both tension and shear, a rivet must only be subjected to shear forces. The rivet head is only designed to hold the rivet in place.

When choosing a rivet you must consider the following factors:

- The strength of the rivet (size and material).
- The material from which the rivet is made as it must not react with the components being joined causing corrosion.

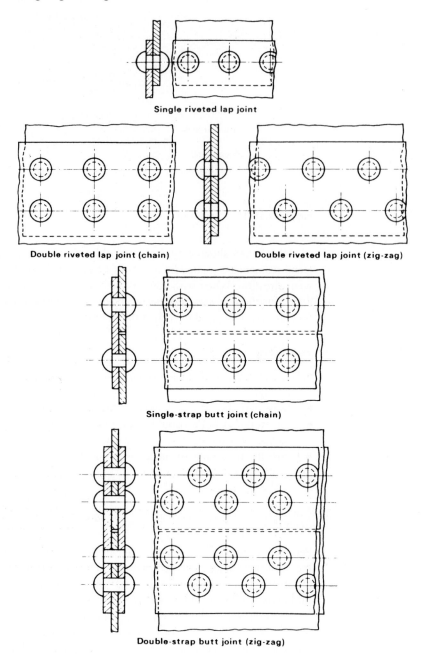

Single riveted lap joint

Double riveted lap joint (chain)   Double riveted lap joint (zig-zag)

Single-strap butt joint (chain)

Double-strap butt joint (zig-zag)

Figure 9.19 Types of riveted joint

- The ease with which the rivet can be closed.
- The appearance of the rivet when the joint is complete.

Some typical rivet heads are shown in Fig. 9.20. Where strength alone is required the snap head or the pan head is normally used.

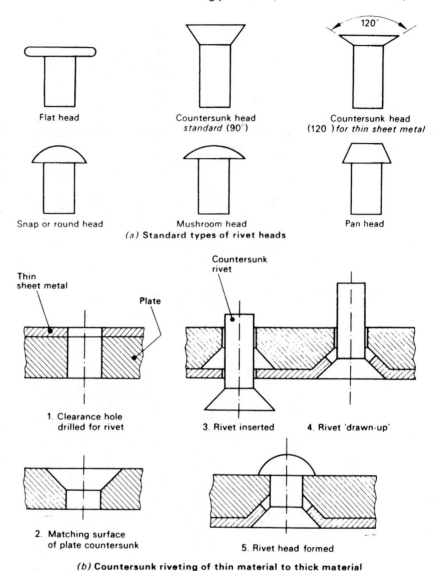

Flat head

Countersunk head
standard (90°)

Countersunk head
(120°) *for thin sheet metal*

Snap or round head

Mushroom head

Pan head

(a) **Standard types of rivet heads**

Thin
sheet metal

Plate

Countersunk
rivet

1. Clearance hole
drilled for rivet

3. Rivet inserted

4. Rivet 'drawn-up'

2. Matching surface
of plate countersunk

5. Rivet head formed

(b) **Countersunk riveting of thin material to thick material**

Figure 9.20 Rivet heads and applications

The success of a riveted joint depends upon the correct riveting procedure being used. The clearance of the hole will depend upon the diameter of the rivet specified. You should drill or punch the rivet hole so that its diameter is 1.0625 times the rivet shank diameter. If you are going to close the rivet with a snap head, the length of the rivet should be 1.5 times the diameter of the rivet plus the combined thicknesses of the components being joined. These proportions are shown in Fig. 9.21 together with those for a countersunk rivet.

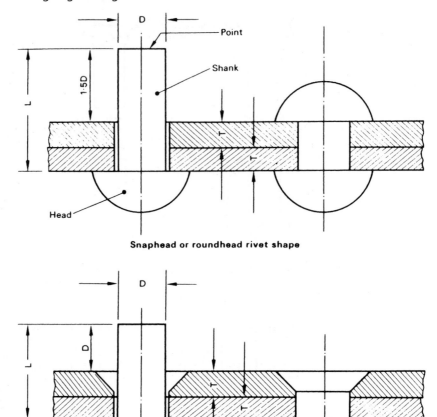

Snaphead or roundhead rivet shape

Countersunk rivet shape

Figure 9.21 Factors influencing choice of rivet lengths

The correct procedure for closing a rivet with a snap head is shown in Fig. 9.22, and the more common defects that occur in riveted joints together with their causes are shown in Fig. 9.23. Even if the rivet is correctly closed, the joint may still fail if it has not been correctly proportioned. *Urwin's formula* states that for metal plate work the diameter of the rivet can be related to the thickness of the plates being joined as follows:

$$D = 1.25\sqrt{T}$$

where: D = diameter of rivet in inches
T = plate thickness in inches

When working in metric units: 1 inch = 25.4 mm. Some common causes of failure are shown in Fig. 9.24.

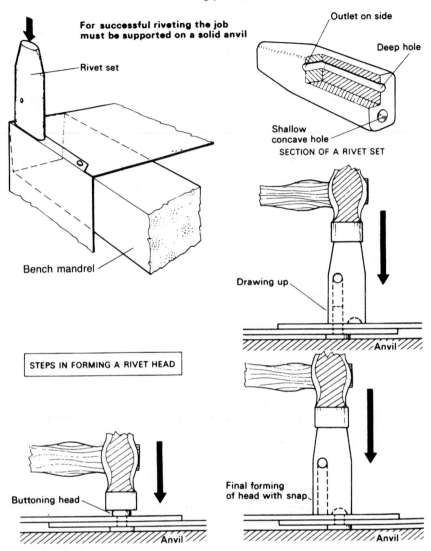

For successful riveting the job
must be supported on a solid anvil

Rivet set

Bench mandrel

Outlet on side

Deep hole

Shallow
concave hole

SECTION OF A RIVET SET

Drawing up

Anvil

Final forming
of head with snap

Anvil

STEPS IN FORMING A RIVET HEAD

Buttoning head

Anvil

Figure 9.22 Method of riveting

Hollow components present special problems when riveting because of the difficulty of getting a hold-up or 'dolly' behind the rivet being headed. Cylindrical components can be riveted using a bench mandrel Box shaped components, where a hold-up cannot be inserted, can be joined using POP® rivets.

CAUSE OF RIVETING DEFECT	RESULTANT EFFECT
Rivet too long	Too much shank protruding to form required head 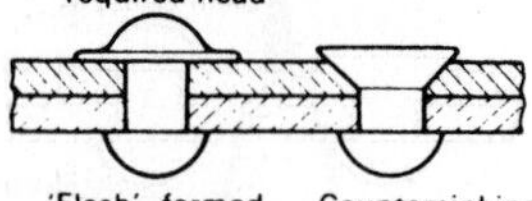 'Flash' formed around head (Jockey cap) — Countersinking over-filled
Rivet set or dolly not struck square	Badly shaped head - off centre 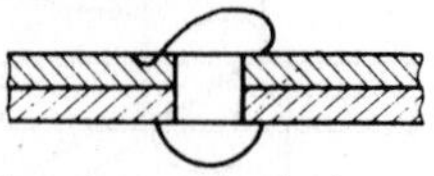 Sheet damaged by riveting tool
Drilling burrs not removed	Not enough shank protruding to form the correct size head  Plates or sheets not closed together. Unequal heads
Sheets not closed together — rivet not drawn up sufficiently	Weak joint. Rivet shank swells between the plates 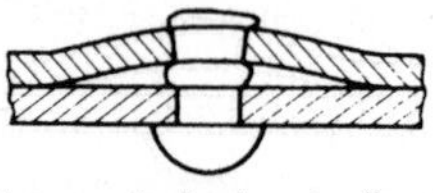 Not enough shank protruding to form correctly shaped head
Rivet holes not matched	Weak mis-shapened head 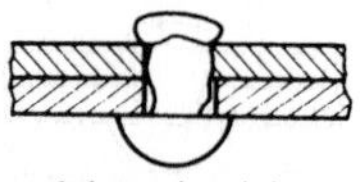 Rivet deformed and does not completely fill the hole
Insufficient hole clearance	Rivet not completely 'drawn through'. Not enough shank protruding to form head  Original head of rivet 'stands proud', the formed head is weak and mis-shapened
Hole too large for rivet	Hole not filled  Rivet tends to bend and deform. Head weak and poorly shaped
Rivet too short	Not enough shank protruding to produce a correctly shaped head 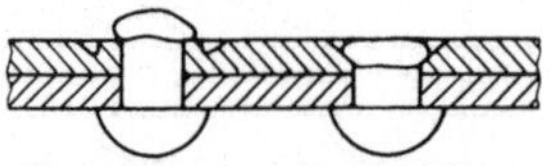 Plate surface damaged — Countersinking not completely filled

Figure 9.23 Common defects in riveting

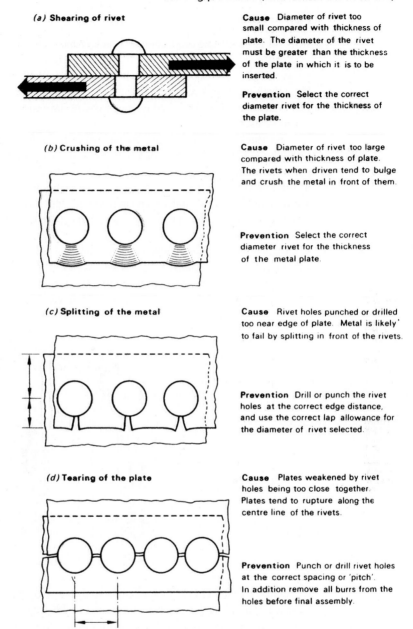

(a) Shearing of rivet

Cause Diameter of rivet too small compared with thickness of plate. The diameter of the rivet must be greater than the thickness of the plate in which it is to be inserted.

Prevention Select the correct diameter rivet for the thickness of the plate.

(b) Crushing of the metal

Cause Diameter of rivet too large compared with thickness of plate. The rivets when driven tend to bulge and crush the metal in front of them.

Prevention Select the correct diameter rivet for the thickness of the metal plate.

(c) Splitting of the metal

Cause Rivet holes punched or drilled too near edge of plate. Metal is likely to fail by splitting in front of the rivets.

Prevention Drill or punch the rivet holes at the correct edge distance, and use the correct lap allowance for the diameter of rivet selected.

(d) Tearing of the plate

Cause Plates weakened by rivet holes being too close together. Plates tend to rupture along the centre line of the rivets.

Prevention Punch or drill rivet holes at the correct spacing or 'pitch'. In addition remove all burrs from the holes before final assembly.

Figure 9.24 Common causes of failure in riveted joints

9.4 POP® Riveting

The following information is provided by Tucker Fasteners Ltd., Emhart Technologies, reproduced here with permission.

POP® rivets

POP® or blind riveting is a technique which enables a mechanical fastening to be made when access is limited to only one side of the parts to be assembled, although reliability, predictability, reduction of assembly costs and simplicity in operation, mean that blind rivets are also widely used where access is available to both sides of an assembly. POP® and other brands of blind riveting systems have two elements, the blind rivet and the chosen setting tool.

The blind rivet is a two-part mechanical fastener. It comprises of a headed tubular body mounted on a mandrel which has self-contained features that create (when pulled into the body during setting) both an upset of the blind end and an expansion of the tubular body, thus joining together the component parts of the assembly. The setting tool basically comprises an anvil which supports the head of the rivet and jaws which grip and pull the mandrel to cause it to set the rivet before the mandrel breaks at a predetermined load.

Many different styles of blind rivet are available but the most widely used is the Open End Rivet Body type defined in BS EN ISO 14588: 2000 as: 'A blind rivet having a body hollow throughout its length and able to use an standard mandrel'. The principle of blind riveting using open style rivets is shown in the following figure.

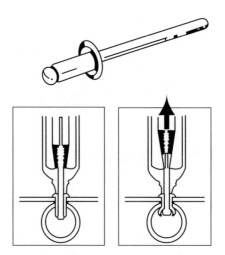

With the mandrel held in the setting tool, the rivet body is inserted into the prepunched or pre-drilled component. Operation of the setting tool pulls the mandrel head into the rivet body causing it to expand on the blind side of the assembly, whilst drawing the components together and eliminating any gaps between them as it does so. At a predetermined point, when the blind side head is fully formed, continued operation of the setting tool causes the mandrel to break, the spent portion of the mandrel is pulled clear and the installation of the rivet is complete.

The Closed End Rivet Body type is defined in BS EN ISO 14588: 2000 as: 'A blind rivet body which is closed and remains closed after setting'. This type is also commonly known as 'sealed'. The closed end rivet prevents ingress of vapour or moisture through the bore of the installed rivet and also ensures mandrel head retention, particularly important in electrical equipment, for example.

POP® blind rivets are available in a variety of materials, body styles and head forms to provide fastening options for a broad spectrum of assembly and environmental requirements from brittle and fragile materials such as acrylic plastics through to stainless steel. A summary is shown in Section 4.2.9. This figure and the following tables are taken from the publications of Emhart Teknologies (Tucker Fasteners Ltd.) from whom further information can be obtained. This company's address is listed in Appendix 3.

POP® range guide

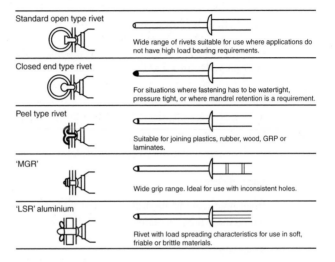

Standard open type rivet	Wide range of rivets suitable for use where applications do not have high load bearing requirements.
Closed end type rivet	For situations where fastening has to be watertight, pressure tight, or where mandrel retention is a requirement.
Peel type rivet	Suitable for joining plastics, rubber, wood, GRP or laminates.
'MGR'	Wide grip range. Ideal for use with inconsistent holes.
'LSR' aluminium	Rivet with load spreading characteristics for use in soft, friable or brittle materials.

Head style

The low-profile domed head is suitable for most applications but, where soft or brittle materials are fastened to a rigid backing, the large flat head variety should be considered.

The 120° countersunk head style should be used wherever a flush surface is required.

Mandrel types

POP® open type rivets are normally supplied with 'Break stem' mandrels (code BS) designed to retain the mandrel head when the rivet is set. 'Break head' (code BH) mandrels, designed to eject the mandrel head from the rivet body, can be supplied for most open rivets and are particularly useful in the pre-assembly of electrical circuit boards.

Finish

- *Rivet body standard finishes*: Steel and nickel-copper rivet bodies are normally supplied zinc plated.
- *Paint and other finishes*: Rivets with differing surface finishes and paint colours can be provided on request. Aluminium alloy rivets are available anodized and dyed, matt or gloss for aesthetic and environmental reasons.

Good fastening practice

Blind riveting is a highly reliable and proven method of fixing material together permanently. To achieve a superior fastening, the following principles should be considered.

Workpiece materials

When materials of different thickness or strengths are being joined, the stronger material – if possible – should be on the blind side. For example, if plastic and metal are to be joined, the plastic sheet should be beneath the rivet head and the metal component should be on the blind side.

Hole size and preparation

Achieving a good joint depends on good hole preparation, preferably punched and, if necessary, de-burred to the sizes recommended in the POP® blind rivet data tables.

Rivet diameter

As a guide for load-bearing joints, the rivet diameter should be at least equal to the thickness of the thickest sheet and not more than three times the thickness of the sheet immediately under the rivet head. Refer to data tables for rivet strength characteristics.

Edge distance

Rivet holes should be drilled or punched at least two diameters away from an edge – but no more than 24 diameters from that edge.

Rivet pitch

As a guide to the distance between the rivets in load-carrying joint situations, this distance should never exceed three rivet diameters. In butt construction it is advisable to include a reinforcing cover strip, fastening it to the underlying sheet by staggered rivets.

Rivet material

Choosing rivets of the correct material normally depends on the strength needed in the riveted joint. When this leads to rivets of material different to the sheets being joined it is important to be aware that electrolytic action may cause corrosion. (See Section 4.2.12.)

Setting and safety

The type of setting tool is usually selected to suit the production environment. The tool must be cleared of spent materials before setting the next rivet and, in the case of power operated tools, **must not** be operated without the mandrel deflector or mandrel collection system being in position. Safety glasses or goggles should always be worn.

Selection of POP® (or blind) rivets

Joint strength

First assess the tensile strength and the shear-load strength required by the joint, both of which can be achieved by the correct number and spacing of fastenings, and by choosing a rivet with a body of the correct material and diameter. The strength columns on the data pages enable a rivet of the correct strength to be chosen.

Joint thickness

The next stage is to work out the combined thickness of the materials to be joined, remembering to allow for any air gaps or intermediate layers such as sealants. Then identify the selected rivet in the size with the necessary grip by consulting the data page. It is important to do this because a rivet with the incorrect grip range cannot satisfactorily grip the back of the workpiece or assembly.

Corrosion acceleration (nature of materials)

Finally, follow the general rule that the rivet chosen should have the same physical and mechanical properties as the workpiece. A marked difference in properties may cause joint failure through metal fatigue or galvanic corrosion. Corrosion is accelerated by certain combinations of materials and environments. Generally, avoid contact between dissimilar metals. The significance of the letters A, B, C and D in the following chart is as follows:

A The corrosion of the metal considered is not accelerated by the contact metal.

B The corrosion of the metal being considered may be slightly accelerated by the contact metal.

C The corrosion of the metal considered may be markedly accelerated by the contact metal.

D When moisture is present, this combination of metal considered and contact metals is inadvisable, even under mild conditions, without adequate protective measures.

Where two symbols are given (for instance **B** or **C**) the acceleration is likely to change with changes in the environmental conditions or the condition of the metal.

Rivet material	Contact metal					
	Nickel Copper Alloys	Stainless Steel	Copper	Steel	Aluminium and Alloy	Zinc
Nickel Copper Alloy	–	A	A	A	A	A
Stainless Steel	A	–	A	A	A	A
Copper	B or C	B or C	–	A	A	A
Steel	C	C	C	–	B	A
Aluminium and Alloys	C	B or C	D	B or C	–	A
Zinc	C	C	C	C	C	–

Design guidelines

Soft materials to hard

A large flange rivet can be used with the flange on the side of the soft material. Alternatively, POP® LSR type rivets spreads the clamping loads over a wide area so as to avoid damage to soft materials.

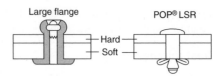

Plastics and brittle materials

For fragile plastics and brittle materials, POP® riveting offers a variety of application solutions. Soft-set/All Aluminium rivets offer low setting loads, whereas both the 'Peel' and 'LSR' ranges afford enhanced support to the materials being joined. For stronger plastics, standard POP® rivets – open or closed end products – may be used.

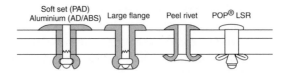

Channel section material

An extended nosepiece can be used to reach to the bottom of a narrow channel section (A) (see figure below). A longer mandrel rivet should be used and the maximum nosepiece diameter should be equal to that of the rivet flange. Alternatively, the rivet can be set from the other side (B) or the channel widened to accept a standard rivet and setting tool (C).

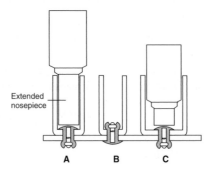

Thick/thin sheet

When materials of different thickness are to be fastened, it is best to locate the thicker plate at the fastened side (A) (see figure below). When the hole diameter in the thinner plate is large, a large flange rivet should be used (B). When the thinner plate is located at the fastened side, either use a backing washer (C) or ensure that the diameter of the hole in the thicker plate is smaller than the one in the thinner plate (D).

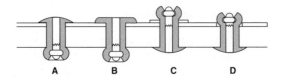

Blind holes and slots

The setting of a POP® or blind rivet against the side of a blind hole, or into and against a milled slot, intersecting hole or internal cavity, is possible because of the expansion of the rivet body on installation.

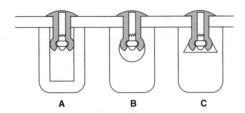

Pivot fastening

Use of a special nosepiece will provide a small gap between the rivet flange and the assembly, so providing for pivot action.

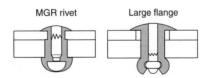

Hole diameters

Whilst standard hole diameters are the rivet body diameter plus 0.1 mm, component hole sizes may not always be this accurate, for example in pre-punched components. In cases when the hole on the fastened (blind) side is larger, POP® MGR rivets should be selected because of its superior hole filling characteristics. POP® LSR and POP® Peel rivets are also possible alternative solutions in these circumstances, especially when working with friable or fragile materials. Alternatively, when then larger hole is on the flange side, a large flange rivet should be chosen.

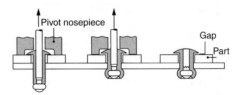

When, however, good hole filling and retained mandrels are used to give high shear and vibration resistance POP®'F' series rivets should be specified. (See Section 4.2.13.)

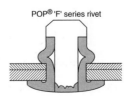

Despite the fact that welding is a more skilful process and despite the fact that it requires much more costly equipment, welding has largely superseded riveted joints for many applications when making permanent connections (see Chapter 10). This is because:

- It is difficult to make riveted joints fluid tight.
- Welded joints are lighter than riveted joints.
- The joint stresses are distributed uniformally in a welded joint, whereas they are concentrated around each rivet in a riveted joint.
- Riveting is slower than welding.
- The components don't have to have rivet holes drilled or punched in them. This is not only time-consuming but the reduction in cross-sectional area at each hole weakens the components being joined.

9.5 Self secured joints

These joints, as the name implies, are formed by folding and interlocking thin sheet metal edges together in such a manner that they are made secure without the aid of any additional jointing process. Their use is confined to fabricated articles made from thin sheet metal less than 1.6 mm thick. A selection of these joints is shown in Fig. 9.25. Of these, the following are the most widely used:

- The grooved seam.
- The paned-down joint.
- The knocked-up joint.

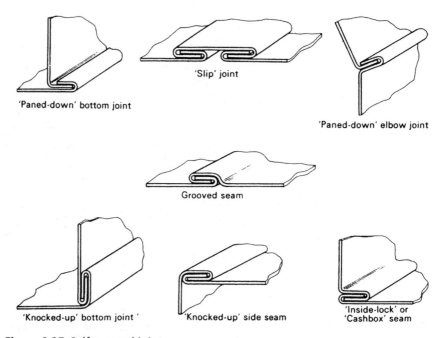

Figure 9.25 Self-secured joints

9.5.1 The grooved seam

This seam consists of two folded edges called 'locks' as shown in Fig. 9.26(a). The two locks (one up and one down) are hooked together as shown in Fig. 9.26(b). They are finally locked together by use of a *hand grooving tool*, as shown in Fig. 9.26(c), or by means of a grooving machine.

The internal grooved seam

In reality this is a *countersunk grooved seam* and it is made in a similar way to the normal grooved seam. The only difference is that where a grooving tool cannot be used inside the seam, the interlocking edges are placed over a special *grooving bar* and the groove is sunk with the aid of a mallet or hammer as shown in Fig. 9.27(a).

The double grooved seam

The double grooved seam makes use of a 'locking strip' to hold the folded edges together. This strip or cap combines strength with a pleasing appearance. The seam

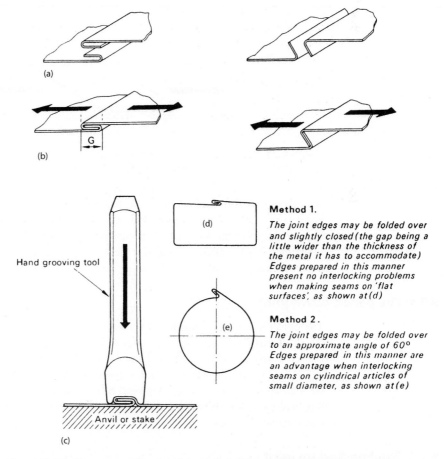

Method 1.

The joint edges may be folded over and slightly closed (the gap being a little wider than the thickness of the metal it has to accommodate) Edges prepared in this manner present no interlocking problems when making seams on 'flat surfaces', as shown at (d)

Method 2.

The joint edges may be folded over to an approximate angle of 60° Edges prepared in this manner are an advantage when interlocking seams on cylindrical articles of small diameter, as shown at (e)

Hand grooving tool

Anvil or stake

Figure 9.26 The grooved seam: (a) edges of joint folded to form 'Locks'; (b) folded edges inter locked (G represents the width of Groove); (c) joint finally locked

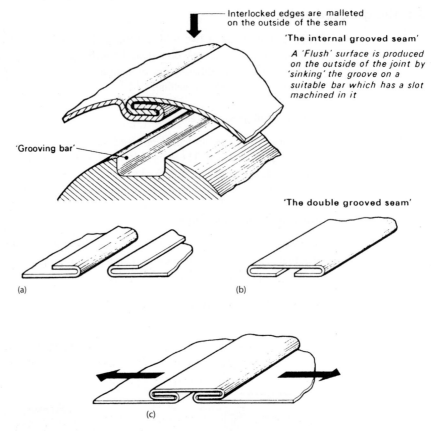

Figure 9.27 Variations of the grooved seam: (a) joint edges folded to form 'locks'; (b) locking strip with edges folded; (c) the assembled joint

is made as usual by folding 'locks' to the desired size, as shown in Fig. 9.27(b). A metal strip of the correct width, with its edges folded as shown in Fig. 9.27(c), is then slipped over the locks of the two pieces to be joined as shown in Fig. 9.27(d). In practice it is usually necessary for the locking strip to be driven onto the joint by lightly tapping one end with a mallet or hammer.

9.5.4 The paned-down joint

The construction of a paned-down joint is shown in Fig. 9.28. The two edges of the sheet metal components to be joined are each folded at right-angles and to the desired width as in Fig. 9.28(a). These 'flanges' are placed in position as shown in Fig. 9.28(b) and one flange is *paned down* (hammered down) to close it over the other as shown in Fig. 9.28(c). A special hammer called a *paning-hammer* is used for this final operation. Where a large number of these seams are to be made a *paning-down machine* is used.

9.5.5 The knocked-up joint

This is a useful joint that is much stronger than a simple paned-down joint. In principle, it is a paned-down joint that has been 'knocked-up' as shown in Fig. 9.28(d).

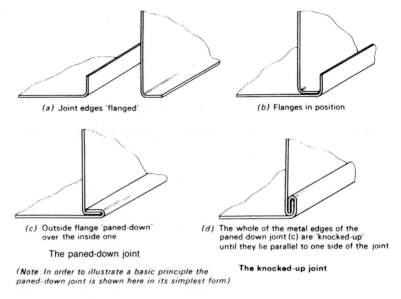

(a) Joint edges 'flanged'

(b) Flanges in position

(c) Outside flange 'paned-down'
over the inside one

The paned-down joint

(d) The whole of the metal edges of the
paned-down joint (c) are 'knocked-up'
until they lie parallel to one side of the joint

The knocked-up joint

(Note: In order to illustrate a basic principle the
paned-down joint is shown here in its simplest form)

Figure 9.28 The paned-down and knocked-up joint

Whilst the paned-down joint can be slipped apart unless it forms the base of a cylindrical container or similar, the assembled knocked-up joint is secured such a way that it cannot be easily pulled apart.

9.6 Folding and jointing allowances

When making self-secured joints or seams, it is necessary to make allowance for the amount of metal that is to be added for the fold or 'locks'. This allowance depends largely upon two basic factors:

- The width of the folded edge.
- The thickness of the metal.

Figure 9.29 shows enlarged cross-sections through commonly used self-secured joints which will now be considered in detail.

9.6.1 Allowance for a grooved joint

If the edges of the metal is folded over to the width W, and the joint is formed as shown in Fig. 9.29(a), the final completed width of the joint G will be seen to be greater than W. In fact, the final width of the grooved seam will have a minimum value of $W + 3T$, where T represents the metal thickness.

The normal allowance for a grooved seam is three times the specified width of the seam (or the width of the grooving tool). This allowance may be used in one of two ways:

1. Half this allowance is made on each side of the pattern, or
2. One third of the allowance is made on one side of the pattern and the remaining two thirds is made on the other side.

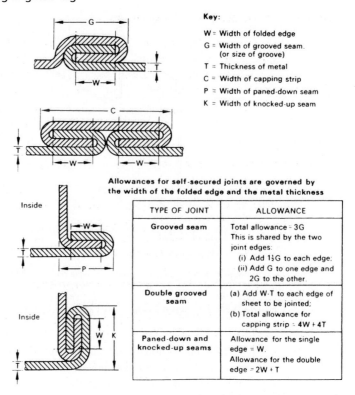

Key:

W = Width of folded edge
G = Width of grooved seam.
    (or size of groove)
T = Thickness of metal
C = Width of capping strip
P = Width of paned-down seam
K = Width of knocked-up seam

Allowances for self-secured joints are governed by the width of the folded edge and the metal thickness

TYPE OF JOINT	ALLOWANCE
Grooved seam	Total allowance = 3G This is shared by the two joint edges: (i) Add 1½G to each edge; (ii) Add G to one edge and 2G to the other.
Double grooved seam	(a) Add W-T to each edge of sheet to be jointed; (b) Total allowance for capping strip = 4W + 4T
Paned-down and knocked-up seams	Allowance for the single edge = W. Allowance for the double edge = 2W + T

Figure 9.29 Allowances for self-secured joints

The thickness of the metal has to be taken into account when selecting a grooving tool for closing the seam. For example, if the joint is to be made on 0.6 mm thick metal with a finished groove width of 6.35 mm, it follows that the width of the folded edge (the distance in from the edge to the 'fold line') must be 6.35 mm minus 1.5 times the thickness of the metal.

Allowance for folding = width of groove minus 1.5 time thickness of metal
= 6.35 mm − (1.5 × 0.6 mm)
= 6.35 mm − 0.9 mm
Width of folded edge = 5.45 mm
= **5.4 mm** (rounded off downwards to ensure clearance is maintained)

If the metal had been folded over 6.35 mm from the edge, then reference to Fig. 9.29(a) shows that the finished width of the grooved seam will be somewhat greater ($G > W$).

Width of groove = width of folded metal plus three thicknesses of metal
= 6.35 mm + (3 × 0.6 mm)
= 6.35 mm + 1.8 mm
Size of grooving tool = 8.15 mm (minimum)
= **8.2 mm** (rounded off upwards to ensure clearance is maintained)

9.6.2 Allowance for double grooved seam

The allowance for the edges of the sheet is the width of the folded edge minus one thickness of the metal. If this allowance was not made for the metal thickness, the dimension C across the seam would be increased by two metal thicknesses. Figure 9.29(b) shows that the width of the capping strip equals twice the width of the folded edges plus four thicknesses of metal, i.e. $C = 2W + 4T$. Thus the *total allowance for the capping strip will be 4 times the width of the folded edge plus four thicknesses of metal.*

9.6.3 Allowance for paned-down and knocked-up joints

As stated previously, the knocked-up joint is an extension of the paned-down joint. Therefore, the allowances will be identical for both these types of joint. As shown in Fig. 9.29(c) and (d), P represents the size of the paned-down joint and K represents the size of the knocked up joint.

The size of paned-down joints and knocked-up joints are determined by the width of a single folded edge. Examination of the sectional views shows that the size of the paned down joint P is equivalent to $2W + 2T$, whilst K which represents the size of the knocked-up joint is equivalent to $2W + 3T$.

The allowances for making both these joints is the width of the folded edge for the single fold and twice the width of the folded edge plus one thickness of metal for the double edge.

9.6.4 The hand-grooving tool

This simple tool, which is used for closing grooved seams by hand, is made from hardened and tempered medium carbon steel. A rectangular slot with radiused corners is machined into its working face.

The width of this groove denotes the size of the tool. Hand grooving tools are available in sizes varying from 3.2 mm to 19 mm in steps of about 1.6 mm. Reference back to Fig. 9.26(c) shows that a completed grooved seam will have a width equal to the size of a single folded edge plus three times the metal thickness and a total thickness of four times the metal thickness. Thus, when making a grooved seam there are two possible approaches:

1. To determine the correct width of the single folded edge, i.e. the folding allowance, for a specific size of groove by deducting one an a half tines the metal thickness.
2. To make the interlock with a specific width of a single folded edge, and the select a suitable size of grooving tool to close the seam.

In practice, the second approach of making the seam to fit a standard grooving tool is the one usually adopted and reference to Fig. 9.29(a) shows why. The sectional view of the seam shows that the edges of the metal (which are square) do not fit accurately into the radiused corners of the locks. This obviously affects the tolerance between the groove allowance and the size of the grooving tool. Hence, in the initial stages of closing the seam, a further clearance is essential between the metal and the tool. It is normal practice to select a tool about 1.6 mm larger than the total width of the grooved seam. As a general rule the thickness of the metal determines the size of the seam. Therefore, it follows that the thinner the metal the smaller the folded edge as shown in Table 9.2.

Table 9.2 Relationship between metal thickness and width of seam

Metal thickness		Minimum width for single folded edge (mm)
mm	SWG	
0.315	30	3
0.400	28	3.5
0.500	26	3.5
0.600	24	4.0
0.800	22	4.0
1.000	20	5.0
1.250	18	6.0
1.600	16	6.0

9.6.5 Comparative uses of self-secured joints

Self-secured joints are employed for *straight seams*, *corner seams* and *bottom joints*. Let's now examine these applications in greater detail.

Straight seams

These are longitudinal joints on cylindrical and conical articles and those between flat surfaces. The most universal type of joint for this purpose is the *grooved seam*. Because of the manner in which the grooved seam is made, it has a very special feature. The tensile load applied to the seam is evenly distributed along its entire length. The *internal grooved seam* is used where it is necessary to avoid having any projections on the outer surface of the article. It is ideal for telescopic articles. The *double grooved seam with cap strip* may be used for either temporary or permanent joints, depending upon which is required. This variation of the grooved seam is an excellent joint for firmly holding together the edges of metal that is too strong and thick to groove in the usual way. It is an ideal joint for assembling and connecting lengths of ducting together, especially on site. Thermal insulation covers, which have to be periodically dismantled, use this type of joint. The capping strip is easily removed and restored as and when the need arises. It is also known as a 'slip joint'.

Corner seams

These seams are either straight or curved and are usually employed on the corners of trunking, square and rectangular duct elbows, boxes and tanks with flat surfaces. When used for joining the corners of square duct elbows, *the single folded edge is always made on the flat sides or cheeks.* The double edge is formed on the 'throat' and back which are the curved sides. The types of joints used for securing corners are 'paned-down' and 'knocked-up' joints. The knocked-up joint is the strongest of all self-secured joints.

Bottom seams

These can be made on articles of any cross-section providing it is possible to fold up an edge. The type of self-secured joint employed for making strong bottom joints

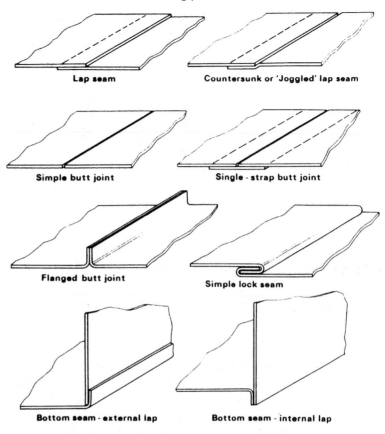

Figure 9.30 Types of joint requiring additional processes, i.e., not self-secured

are the paned-down and the knocked-joint joints. Figure 9.30 shows some practical examples of the use of self-secured joints.

Sealing

Self-secured joints are frequently employed in the manufacture of containers that have to hold foodstuffs and liquids. For such applications it is necessary to seal the joints and seams to make them *fluid-tight*. For 'potable' applications where food stuffs and drinks are involved the sealant is either a latex rubber insert or a *lead-free* solder. For 'non-potable' applications the seams and joints can still be sealed with traditional tin/ lead soft-solder (see Fig. 10.12).

9.7 The Pittsburgh lock

The 'Pittsburgh lock' or 'lock seam' is used extensively in ductwork shops. These seams are normally produced on special *lock forming machines*. However, they can be successfully produced (but more slowly) on a standard folding machine providing the correct sequence of operations is performed. The need for adopting a planned

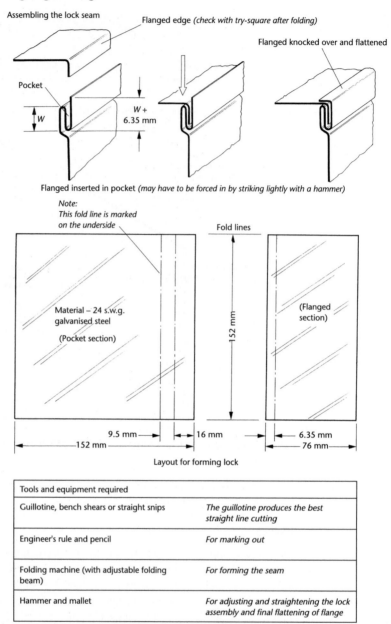

Figure 9.31 The Pittsburgh lock

sequence of operations and the necessary skill required to make a Pittsburgh lock can be appreciated by practising with scrap metal off-cuts. Figure 9.31 shows the lock and the layout for forming the lock.

A pencil should be used for marking the fold lines because a scriber would cut through any anti-corrosion coating that the metal sheet may have. Figure 9.32 shows a typical sequence of operations for making a lock using two off-cuts of 0.56 mm galvanized sheet steel (that is, steel protected with a coating of zinc). The first operation is to cut two pieces of sheet metal to the required sizes. The fold lines are then marked in

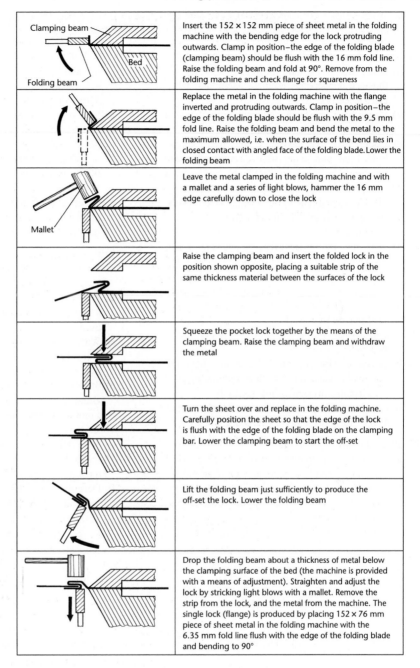

	Insert the 152 × 152 mm piece of sheet metal in the folding machine with the bending edge for the lock protruding outwards. Clamp in position–the edge of the folding blade (clamping beam) should be flush with the 16 mm fold line. Raise the folding beam and fold at 90°. Remove from the folding machine and check flange for squareness
	Replace the metal in the folding machine with the flange inverted and protruding outwards. Clamp in position–the edge of the folding blade should be flush with the 9.5 mm fold line. Raise the folding beam and bend the metal to the maximum allowed, i.e. when the surface of the bend lies in closed contact with angled face of the folding blade. Lower the folding beam
	Leave the metal clamped in the folding machine and with a mallet and a series of light blows, hammer the 16 mm edge carefully down to close the lock
	Raise the clamping beam and insert the folded lock in the position shown opposite, placing a suitable strip of the same thickness material between the surfaces of the lock
	Squeeze the pocket lock together by the means of the clamping beam. Raise the clamping beam and withdraw the metal
	Turn the sheet over and replace in the folding machine. Carefully position the sheet so that the edge of the lock is flush with the edge of the folding blade on the clamping bar. Lower the clamping beam to start the off-set
	Lift the folding beam just sufficiently to produce the off-set the lock. Lower the folding beam
	Drop the folding beam about a thickness of metal below the clamping surface of the bed (the machine is provided with a means of adjustment). Straighten and adjust the lock by stricking light blows with a mallet. Remove the strip from the lock, and the metal from the machine. The single lock (flange) is produced by placing 152 × 76 mm piece of sheet metal in the folding machine with the 6.35 mm fold line flush with the edge of the folding blade and bending to 90°

Figure 9.32 Sequence of folding operations (Pittsburgh lock)

accordance with the dimensions provided on the drawing shown in Fig. 9.31. The width of the flanged edge is normally made *slightly less than the depth of the pocket*. The allowance for the pocket is equal to twice the width of the pocket plus an allowance for 'knocking over'. In this example this allowance is $W + W + 6.35$ mm. Lock seams are used as longitudinal corner seams for ducts of various shapes and the necessary allowances have to be added to the pattern or layout template.

19.1 Threaded connections

a) State why the V-form thread is used for the threaded connections used in assembling fabricated steelwork.

b) Show FIVE methods of preventing threaded connections working loose.

c) Under what circumstances are coarse thread bolts used, and under what circumstances are fine thread bolts and nuts used.

d) Compare the advantages and limitations of using:
   i) Hot forged (black) bolts with rolled threads
   ii) High-strength friction grip bolts
   iii) Fitted bolts.

e) With the aid of sketches explain why washers have to be used under the heads of bolts and between the nut and the workpiece. Under what circumstances do taper washers have to be used?

f) Describe THREE methods of aligning bolt holes in structural steel work.

g) Under what circumstances do bolt holes have to be reamed.

h) Describe with the aid of sketches how the threads of nuts and bolts can be reconditioned if they become bruised during assembly.

i) Under what circumstances are 'Hank' bushes POP® Nut-threaded inserts used?

19.2 Riveted joints

a) List the factors that have to be considered when selecting a rivet for a particular application.

b) State Urwin's formula and use it to calculate the shank diameter of a rivet for connecting 20 mm thick steel plates.

c) Calculate the shank length of a snap-head rivet for making a lap joint in 20 mm thick plates. What diameter should the rivet holes be drilled to receive the rivet in this example?

d) Sketch the following riveted joints:
   i) A single row lap-joint
   ii) A double row butt-joint with a single butt-strap, zig-zag riveted
   iii) A double row butt-joint with two butt-straps, chain riveted.

e) Describe with the aid of sketches what is meant by POP® Riveting and state under what circumstances it would be used.

19.3 Self-secured joints

a) With the aid of sketches describe how to make a grooved seam and state where it would be used.

b) With the aid of sketches describe how to make a paned-down joint and state where it would be used.

c) With the aid of sketches describe how to make a knocked-up joint and state where it would be used.

d) State the formula and describe how the folding allowance is calculated for a grooved seam for 1 mm thick sheet metal if the width of the seam is 6 mm.

e) State why self secured joints must be sealed to make them fluid tight and list the factors that must be considered when sealing such joints.

f) Describe the Pittsburgh lock, how it is made, and under what circumstances it is used.

10 Joining processes (soldering, brazing and braze-welding)

When you have read this chapter you should be able to:

- Distinguish between soft soldering, hard soldering (brazing) and braze-welding
- Understand the principles of soldering as a means of joining metal components and the purpose of and need for a suitable flux
- Relate various soft-soldering alloys to specific applications
- Describe the various soft-soldering processes and their heat sources
- Describe the need for and use of lead-free soft solders
- Describe the silver-soldering process, the fluxes use and the heat sources required
- Describe the various brazing processes, the filler alloys (spelters) used and the fluxes required
- Describe the heat sources appropriate for brazing processes
- Describe the aluminium brazing process
- Explain what is meant by braze-welding
- Describe the joint types and preparation for braze-welding processes
- Identify the materials that can be joined by hard-soldering processes

10.1 Soft soldering

Traditionally, soft soldering always used tin/lead alloys as the joining medium. The composition of the solder varied according to the application of the joint; some solder alloys and their uses are listed in Table 10.1.

More recently tin/lead alloys have fallen out of favour because of the lead content which causes *environmental problems* when disposing of articles containing soft soldered joints at the end of their useful life. The safe disposal of such articles is now the responsibility of the manufacturer. Similarly tin/lead alloys cannot be used for *potable* applications since it introduces *health problems*. For example, tin/lead alloy solders can be used for making joints in domestic central heating systems where the joints will remain undisturbed for very long periods of time and the water in the system will not come into contact with foodstuffs nor will it be used for drinking. This is *not a potable* application.

However, tin/lead solders *cannot be used* for plumbing for drinking water or for utensils used in cooking because these *are potable* applications and the lead content can cause health problems. For potable applications alternative solders have to be used that are safe from a health perspective (see Section 10.2). Although not potable applications, the printed circuit boards used in the control systems of such goods as washing machines, television sets and computers are now made using *lead-free* solders. This is because of the environmental issues that occur when such appliances wear out and have to be discarded. It is the responsibility of the manufacturer to ensure that such goods do not cause environmental harm. The simplest way to ensure that no harm occurs when such appliances are discarded is to *only use lead-free solders* in their manufacture.

10.1.1 Soft soldering (tin/lead solders)

Soft soldering is a low-temperature thermal process in which the metals being joined are not themselves melted. The process involves the use of one of the solder

Table 10.1 Types of softer solder

BS Solder	Composition (%)			Melting range (°C)	Remarks
	Tin	Lead	Antimony		
A	65	34.4	0.6	183–185	Free-running solder ideal for soldering electronic and instrument assemblies. Commonly referred to as *electrician's solder*
K	60	39.5	0.5	183–188	Used for high-class tin-smith's work, and is known as *tinman's* solder
F	50	49.5	0.5	183–212	Used for general soldering work in coppersmithing and sheet-metal work
G	40	59.6	0.4	183–234	*Blow-pipe solder.* This is supplied in strip form with a D cross-section 0.3 mm wide
J	30	69.7	0.3	183–255	*Plumber's solder.* Because of its wide melting range this solder becomes 'pasty' and can be moulded and wiped

alloys listed in Table 10.1 to form a bond between the metals (parent metals) being joined. For a successful joint, the molten solder must adhere to the surfaces of the parent metals being joined. It must do this by reacting with the surface layers of the parent metals to form an *intermetallic compound* that cannot be physically removed. The formation of this intermetallic compound is referred to as 'tinning' the parent metal surfaces. In a correctly soldered joint, microscopical examination shows that when metals such as brass, copper and steel are tinned the tin content of the solder reacts chemically with the parent metal to form an intermetallic compound which becomes part of the parent metal. This is shown in Fig. 10.1.

The stages in making a soft-soldered joint using a tin/lead solder are:

1. Tinning the joint surfaces of the parent metal.
2. Additional molten solder flowing between the heated parent metal surfaces which remain unmelted.
3. The molten solder completely filling the space between the surfaces and fusing with the tinned surfaces of the parent metal.
4. Solidification of the solder.

These stages in making a soft-soldered joint using a tin/lead solder are shown in Fig. 10.2. The ease with which the solder penetrates the joint increases with

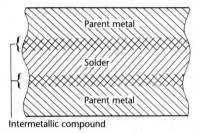

The intermetallic compound layer will continue to grow in thickness the longer the joint is kept at soldering temperature.
Solder cannot be completely wiped or drained off when in the molten state. The metal surface, therefore, remains permanently wetted or 'tinned' by a film of solder.
A film of solder cannot be mechanically prised off, leaving the surface of the parent metal bare in its original state, unlike lead which has been cast upon a metal surface.

Figure 10.1 A section through a soft-soldered joint

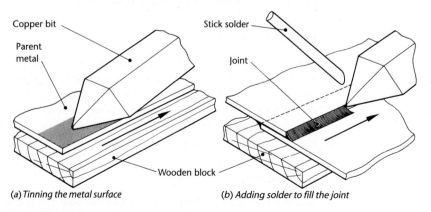

The metal to be soldered is supported on a wooden block (heat insulator) to prevent unnecessary heat loss by conduction

(a) Tinning the metal surface   (b) Adding solder to fill the joint

Figure 10.2 Basic stages in soft soldering

corresponding increases in the tin content. All tin/lead solders contain traces of antimony to add strength and stiffness to the joint.

10.1.2 Preparing the joint

It is essential that the joint surfaces of the parent metals being joined are both chemically and physically clean otherwise it is not possible to tin them. That is, not only must the surfaces be free from dirt, grease and oil, they must be free from oxide and other chemical films. Any of these substances would form a barrier between the solder and the parent metal and would, therefore, prevent the essential tinning processes from taking place. Table 10.2 lists some common methods of cleaning metal surfaces prior to soldering.

Most metals, when exposed to air at room temperature, will acquire a thin oxide coating within a few minutes of being cleaned

Raising the temperature of the metal accelerates the formation of surface oxide films and this must be prevented during the soldering process otherwise a sound joint will not be achieved. Prevention of this oxide film forming is achieved by the application of a suitable soldering flux immediately after preparatory cleaning.

10.1.3 Soldering fluxes

The essential requirements of a soldering flux are:

- It must remain liquid at soldering temperatures.
- In its liquid state it must cover the joint area and prevent oxidation by excluding the air.
- It must dissolve any residual oxide film on the surfaces being joined (this only applies to *active fluxes*).
- It must be readily displaced by the molten solder.
- It should not leave behind a corrosive residue.

Figure 10.3 shows the essential functions of a soldering flux, how it removes any residual oxides from the joint surfaces, how it protects the heated joint surfaces from atmospheric attack, and how it promotes the 'wetting' of the joint surfaces by the molten solder. 'Wetting' is the ability of the molten solder to flow evenly over the joint surface and not to collect in globules.

Passive fluxes

Passive (non-corrosive fluxes) such as resin are used for those applications where it is not possible to remove any corrosive residue by washing after the soldering operation is complete, for example electrical circuit boards. Unfortunately, passive fluxes do not remove any residual oxide film to any appreciable extent – they only prevent such films from reforming during the soldering process. Therefore, the initial chemical cleaning and mechanical scouring of the joint faces must be very thorough.

Active fluxes

Active fluxes such as 'Baker's fluid' (acidified zinc chloride solution) quickly dissolve any oxide film and prevent it from re-forming. They also etch the joint surfaces of

Table 10.2 Common methods of cleaning metal surfaces

Mechanical (abrasive) cleaning

Methods	Remarks
1. Rubbing stained or corroded surfaces with *emery cloth* 2. Use of files 3. Use of *scrapers* 4. Scouring with the aid of *wire wool* 5. *Sand-blasting*	Particular attention should be paid to any pitting or small depressions which may be in the surfaces of the metal to be soldered. These depressions tend to harbour dirt and oxides which, if not effectively removed, will render any application of soldering flux ineffective in such local areas. Sand-blasting can reach into fine recesses not normally accessible by other means

Metal surfaces should look bright and be free from stains, but need not be highly polished. In practice it has been found that very small scratches on a metal surface actually promote the spreading of the molten solder

Chemical cleaning

Methods	Remarks
1. Application of a suitable *solvent*. For 'one-off' or small batches, swabbing with either cotton waste or rag dipped in *white spirit* or *carbon tetrachloride* is usually adequate. For 'production lines', a suitable *trichloroethylene* degreasing apparatus is essential 2. Use of *hot alkaline solutions* 3. Use of *alkali detergents* of the soapless type 4. *Pickling* in acids	Coated metals such as *tinplate, terneplate* and *galvanized iron* or *steel* do not require mechanical cleaning, but it is essential that any *oil* or *grease* on their surfaces should be removed Trichloroethylene is effective for removing most oils or greases in common use Two chambers are used in the degreasing apparatus, the article first being dipped in liquid solvent, followed by immersion in a condensing vapour of the boiling solvent Alkaline degreasing is effective in removing the last traces of oil or grease, whether animal or vegetable in origin. A cold solution containing **equal parts** of *hydrochloric acid* and *water* is used for the effective removal of *rust* from degreased *mild steel*

After chemical cleaning by means of alkali or acid solutions, articles should be rinsed thoroughly in soft or distilled water and adequately dried

the parent metals thus ensuring good 'wetting' and bonding. Unfortunately, all active fluxes leave a corrosive residue behind that has to be thoroughly washed off immediately after soldering and the joint should be treated with a rust inhibitor.

General uses for soldering fluxes

Soldering fluxes are of two basic types:

1. Substances which yield an acid only when heated – these are classed as 'non-corrosive' because the flux residue is non-active and even protective.
2. Acids – these are classed as 'corrosive'.

Both the corrosive and non-corrosive types of soldering fluxes are obtainable in paste form. There are a number of paste-forming ingredients which may be added to fluxes, both of the active and non-active types. These include *petroleum jelly* (Vaseline), *tallow, lanolin* and *glycerine*.

A molten solder is said to 'wet' when it leaves a continuous permanent film on the surface of the parent metal instead of rolling over it.

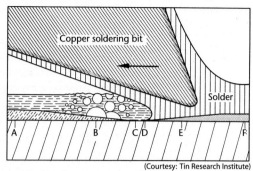

(Courtesy: Tin Research Institute)

Diagrammatic representation of the displacement of flux by molten solder.
A Flux solution lying above oxidised metal surface.
B Boiling flux solution removing the film of oxide (e.g. as chloride).
C Bare metal in contact with fused flux.
D Liquid solder displacing fused flux.
E Tin reacting with the basis metal to form compound.
F Solder solidifying.

Figure 10.3 The essential functions of a soldering flux

- **Non-corrosive type**. *Methylated spirits* or *industrial spirits* are used to digest crushed *resin* in the proportion of 1 part resin with 2 to 4 parts 'spirits' by weight. Resin paste is used as the flux in single and multi-cored solders.
- **Corrosive type**. *Petroleum jelly* is the main paste-forming ingredient and is mixed with a solution of *zinc chloride, ammonium chloride* and *water* in approximately the following proportions:

Petroleum jelly	65%
Zinc chloride	25%
Ammonium chloride	3.5%
Water	6.5%

The main advantage of paste fluxes is that, unlike liquid fluxes, they do not tend to drain off the surface and run over other parts of the work where flux would be harmful.

Typical soldering fluxes and their applications are listed in Table 10.3.

10.1.4 Heat sources for soft soldering

Many different heat sources are used in soldering processes, many of which are beyond the scope of this book. Some typical heat sources are:

1. Soldering with a copper bit (soldering iron).
2. Soldering using a blowpipe flame.
3. Hot-plate soldering, where cleaned and fluxed components are either assembled with solder paste applied to their adjoining surfaces, or with pre-placed solder foil or fine solder wire placed between the joint surfaces. The assemblies are placed on a heated plate or tray which brings them up to the soldering temperature.

Table 10.3 Soldering fluxes in general use

Flux	Remarks
Non-corrosive types	
Resin or rosin	In its natural form is the gum extracted from the bark of pine trees. It is an amber-coloured substance which is solid at room temperature and does not cause corrosion, but it reacts mildly at soldering temperatures. It is used mainly for electrical work, but because it is non-toxic, it is safe to use on food containers.
Tallow	This is a product of animal fat. It is virtually inactive at room temperature and, like resin, is only slightly active at soldering temperatures. The flux is used extensively with 'plumber's solder' for jointing lead sheets and pipes, and with 'body solder' on previously tinned steel for motor-vehicle repair work.
Olive oil	This is a natural vegetable oil. It forms a weak vegetable acid at soldering temperatures. A useful flux when soldering pewter
Corrosive types	
Zinc chloride	Commonly called 'Killed Spirits of Salts' and forms the base for more commercially produced fluxes. A good genera flux suitable for *mild steel, brass, copper, ternplate* and *tinplate*
Ammonium chloride	Commonly called 'Sal Ammoniac'. This substance in a paste form is used as an *electrolyte* in the cells of dry batteries. As a soldering flux, it is generally used in liquid form when tinning *cast iron, brass* or *copper*. It can be obtained in the form of a solid block (this is useful for the tinning of a soldering bit) or in the form of crystals.
Hydrochloric acid	This is known as 'Raw Spirits of Salts' and is extremely corrosive. It is used in dilute form when soldering *zinc* and *galvanized iron* or *steel*. **Zinc chloride (killed spirits of salts) is produced by the chemical action of hydrochloric acid on zinc**
Phosphoric acid	This flux is very effective when soldering *steel, copper* and *brass*. It tends to leave a 'glassy' residue

A mixture of equal parts of zinc chloride and hydrochloric acid provides a very active flux for soldering *stainless steels*. **This flux is dangerous because it gives off irritant fumes**.
A solution of zinc chloride and 10% ammonium chloride has a lower operating temperature than *zinc chloride* (262°C) and is more suitable for use with low-melting-point solders.

4. Induction heating, where the cleaned and fluxed components are placed within a coil of copper tube which is connected to a source of high-frequency alternating current. Water is passed through the coil to prevent it overheating. Eddy currents are induced in the components being joined, raising them to the correct soldering temperature within a few seconds.
5. Resistance soldering, where a low-voltage electric current is passed through the cleaned and fluxed components being joined. Due to the relatively high resistance that is present at the joint faces and the large current being passed through them, the joint quickly reaches the correct soldering temperature.
6. Dip soldering. This is particularly suitable for sheet metalwork where the joint clearances have consistent close tolerances. Previously cleaned and fluxed components are immersed in molten solder contained in a heated bath. The solder is drawn into the joints by capillary action. The components being dip-soldered should be preheated to ensure they are thoroughly dry to prevent an explosion. Copper should not be dip-soldered since the tin in the solder will dissolve the

copper. Hot dip galvanizing, which uses zinc instead of solder, is used to seal such articles as galvanized buckets and animal feed troughs.

Let's now consider the heat sources named in items 1 and 2 above in greater detail.

Soldering with a copper bit (soldering iron)

The traditional *soldering iron* is still widely used for light sheet metalwork in tin plate and thin brass and copper. It consists of a copper 'bit', a steel shank and a wooden handle. The 'bit' is made of copper for the following reasons:

- *Heat capacity*. The soldering 'bit' must be capable of storing and carrying sufficient heat energy and transferring that heat energy to the joint being soldered. Although other metals have a greater heat capacity they are unsuitable for other reasons.
- *Thermal conductivity*. As already stated the soldering 'bit' must be capable of transferring the heat energy stored in the 'bit' to the components being joined as quickly as possible. Copper has the highest thermal conductivity of all the common metals available in industry.
- *Reservoir of molten solder*. Copper can be readily 'wetted' or 'tinned' with molten solder. When manipulating the molten solder the facets of the 'bit' must be completely 'wet' with molten solder. The larger the area of the facets at the tip of the 'bit', the more solder it will hold.

Various types of soldering irons in general use are shown in Fig. 10.4.

The most usual methods of heating the 'bit' of the soldering iron are the *gas-fired soldering stove* and the *electrically heated* soldering iron. Soldering stoves are usually made from cast iron lined with fire brick. They can usually heat several irons at the same time so that soldering can be performed continuously. The temperature of the stove can be controlled by adjustment of the gas supply. Typical soldering stoves are shown in Fig. 10.5(a).

Electrically heated soldering irons are heated by a resistance element as shown in Fig. 10.5(b). Electrically heated soldering irons are used for small and fine work where little heat is lost from the bit and they are, therefore, limited in their use for sheet metalwork. They are mainly used for electrical and electronic assembly operations. For safety reasons electric soldering irons used for industrial applications should be used from a low-voltage supply, being connected to the mains via a low-voltage transformer.

Summing up the main functions of the soldering iron as a tool for making soft-soldered joints:

- Storing and carrying heat energy from the heat source to the work. This is the most important function because lack of heat energy is the most usual cause of poor soldering and joint failure. Therefore, the 'bit' must be made from copper and must be large enough to provide a suitable reservoir of heat energy. The larger the 'bit', the more heart it will store.
- Storing and carrying the molten solder.
- The size of the bit is determined by the *rate* at which the heat energy must be supplied to the work which, in turn, is governed by the *conductivity* of the parent metal to be joined by soldering and the *mass* of the parent metal in the soldering zone. Heat flow is shown in Fig. 10.6.

Standard workshop pattern

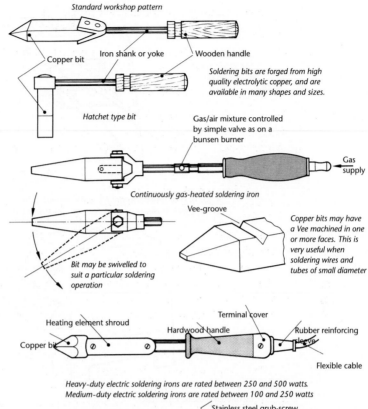

Copper bit   Iron shank or yoke   Wooden handle

Hatchet type bit

Soldering bits are forged from high
quality electrolytic copper, and are
available in many shapes and sizes.

Gas/air mixture controlled
by simple valve as on a
bunsen burner

Gas
supply

Continuously gas-heated soldering iron

Vee-groove

Copper bits may have
a Vee machined in one
or more faces. This is
very useful when
soldering wires and
tubes of small diameter

Bit may be swivelled to
suit a particular soldering
operation

Heating element shroud   Terminal cover

Hardwood handle   Rubber reinforcing
sleeve

Copper bit

Flexible cable

Heavy-duty electric soldering irons are rated between 250 and 500 watts.
Medium-duty electric soldering irons are rated between 100 and 250 watts

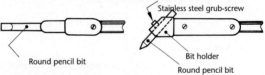

Round pencil bit

Stainless steel grub-screw

Bit holder
Round pencil bit

The 'pencil bit' electric soldering tools are suitable for very intricate or light
soldering work – their power rating starts at about 50 watts.

Irons with a 50 watt rating have bits not exceeding 225 gramme mass.

Irons with a 500 watt rating have bits not exceeding 2.3 kilogramme mass.

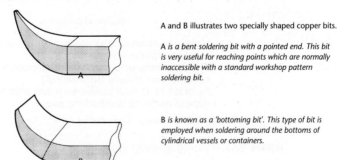

A and B illustrates two specially shaped copper bits.

A is a bent soldering bit with a pointed end. This bit
is very useful for reaching points which are normally
inaccessible with a standard workshop pattern
soldering bit.

B is known as a 'bottoming bit'. This type of bit is
employed when soldering around the bottoms of
cylindrical vessels or containers.

Figure 10.4 Typical soldering irons in common use

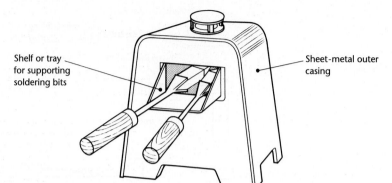

Shelf or tray
for supporting
soldering bits

Sheet-metal outer
casing

*A gas ring or row of burner jets is incorporated below the chamber. The flames
do not come in direct contact with the soldering bits in the oven or chamber.*

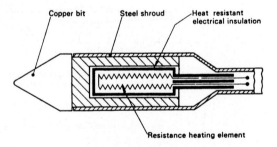

Cast-iron outer
case

(a) Gas–fired soldering stoves

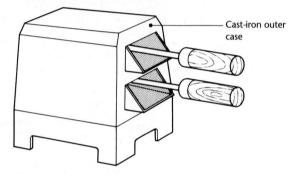

Copper bit   Steel shroud   Heat resistant
electrical insulation

Resistance heating element

Electric soldering iron

Heat is generated by an electric current overcoming the
resistance of the wire element.
The wire from which the heating element is wound, is an alloy
of nickel and chrome (NICHROME).
Because of its high resistance it can withstand high temperatures
without oxidising and burning away.

(b) Electric resistance heating – soldering iron

Figure 10.5 Heating a soldering iron

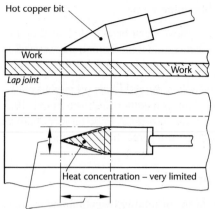

Hot copper bit

Work

Work

Lap joint

Heat concentration – very limited

Small area of contact with soldering bit

A small soldering bit is not suitable for soldering any heavy or comparatively large-size sheet-metal products.

Reason: When the copper bit is too small for the job, it cannot maintain an adequate temperature to allow the solder to flow freely and to penetrate into the seam or joint.

When an inadequate size of copper bit is heated to the correct temperature and applied to thick gauge sheet metal, the small amount of heat stored in it is rapidly conducted away by contact with the relatively large mass of cold metal.

Result: The small copper bit cools very quickly causing the molten solder to chill and render the joint ineffective.

When using a small copper bit for the soldering of small parts, it is very important not to rest the parts on a cold mass of metal. Where possible small parts to be soldered should be placed on some sort of insulating material such as a block of wood. Similarly when parts to be soldered are to be held in a vice, wood of fibre faced clamps should be used.

(a) Lack of heat when using too small & soldering iron

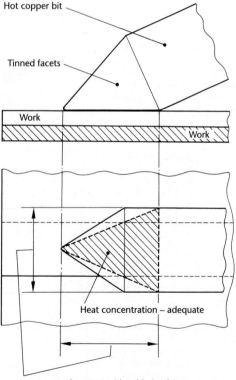

Hot copper bit

Tinned facets

Work

Work

Heat concentration – adequate

Large area of contact with soldering bit

Preheating parts prior to soldering will reduce heat loss by conduction

(b) Adequate heat concentration is provided by the use of a large enough soldering iron

The use of a large size copper bit provides a much greater heat source.

A large reservoir of heat will compensate any heat loss by conduction when the bit makes contact with a relatively cold mass of metal.

The tinned facets of the copper bit actually making contact with the surface of the metal to be soldered should be of sufficiently large area.

Reason: A large area, tinned face on the copper bit will provide a much grater heat concentration. A large concentration of heat will allow time for it to flow into the work while the bit is slowly drawn along the surface of the metal

Figure 10.6 Heat transfer

Soldering using a blowpipe flame

A greater rate of heat transfer can be accomplished by the direct use of a gas torch for heating the joint being soldered. Bottled gas is used on site work whilst natural gas from the mains supply is used in the workshop. Typical applications are:

- Filling in dents or sunken welded joints on motor vehicle bodies prior to painting.
- Sweating pre-tinned joints together as shown in Fig. 10.7. For example, pipe joints or long seams where the heat tends to be rapidly conducted away from the joint.
- The tinning of castings and some metal bearings.

10.1.5 Types of soft-soldered joint

Soft soldering as a means of joining two components together relies almost entirely on adhesion for its strength. Such joints are strongest at room temperature but their mechanical strength falls off rapidly as the temperature increases. The strength of a soft-soldered joint is determined by the following factors:

- The strength of the solder itself. This is governed by the tin/lead ratio of its composition. Surplus solder does not add strength to the joint.

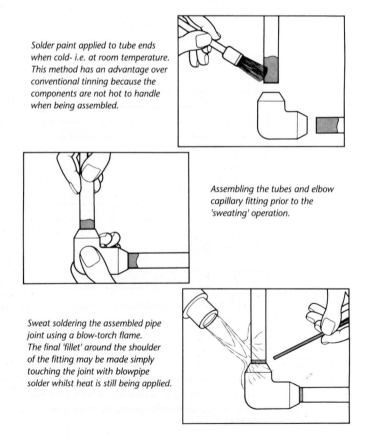

Solder paint applied to tube ends when cold- i.e. at room temperature. This method has an advantage over conventional tinning because the components are not hot to handle when being assembled.

Assembling the tubes and elbow capillary fitting prior to the 'sweating' operation.

Sweat soldering the assembled pipe joint using a blow-torch flame. The final 'fillet' around the shoulder of the fitting may be made simply touching the joint with blowpipe solder whilst heat is still being applied.

Figure 10.7 Sweating a pipe joint using a blow torch

- The strength of the bond between the solder and the tinned surfaces of the parent metal. Weakness can be caused by too thin a solder film and equally by too thick a solder film. The optimum film thickness of the solder film is 0.075 to 0.080 mm.
- The design of the joint. Where a strong joint is required the joint edges should be interlocked before soldering. Factors influencing typical joint designs for soft

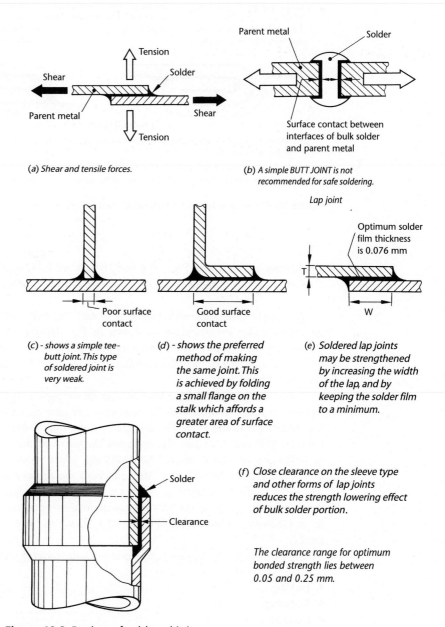

Figure 10.8 Design of soldered joints

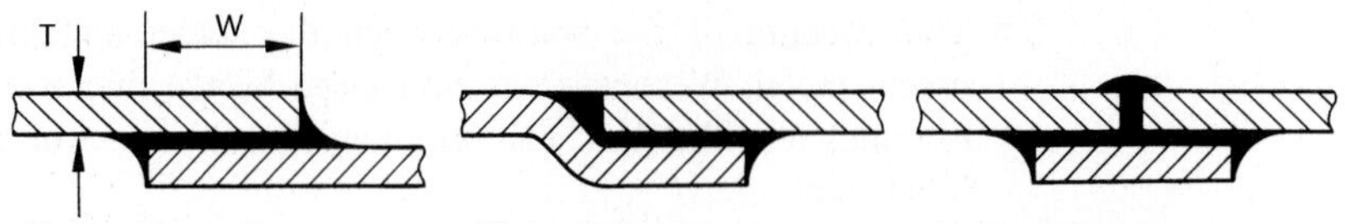

(a) *Simple lap joint.*

(b) *Joggled lap joint.*

(c) *Simple butt joint with cover strap.*

The strength of soldered joints is dependent upon the area of surface contact. With a simple soldered lap joint, the width of the lap should not be less than three times the metal thickness.

The width of the lap is usually determined by the thickness of the metal being joined, and, in general it ranges between 3 mm and 12 mm.

Provided the joint space is full and that there is sufficient solder to blend out sharp corners, an excess of solder does not add to the strength of the joint.

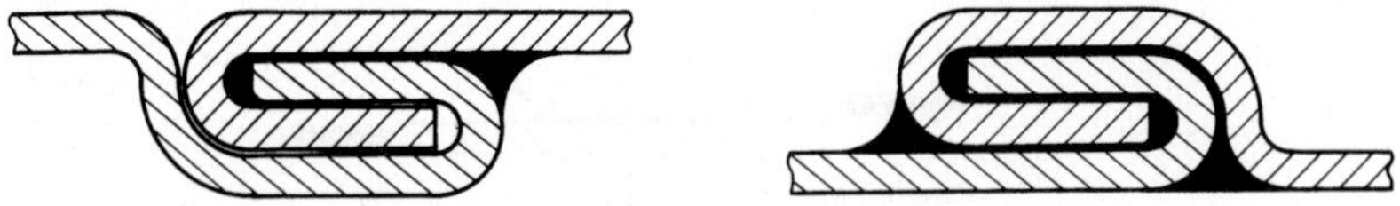

Interlocking joints.

(d) *shows a grooved seam soldered on the outside only, whilst* (e) *is soldered both inside and outside.*

Most soldered joints when subjected to stress have to rely on the strength of the bulk solder. Where really strong soldered joints are required the surfaces are interlocked before soldering.

Figure 10.9 Various types of soft-soldered joints

soldering are shown in Figs 10.8 and 10.9. A soldered joint is strongest in *shear* (42 to 49 MN/m^2) and weakest in *tension* (0.28 to 0.35 MN/m^2).

A straight butt joint under tension places the bulk solder between the joint interfaces under tensile stress as shown in Fig. 10.8(b). It can be seen that the stress-carrying capacity of this type of joint is limited to the tensile strength and cross-sectional area of the solder itself and, as has already been stated, soft solder is very weak in tension.

Joints designed for 'shear loading' can be strengthened by increasing the overlap area (surface contact area) as required, and also by reducing the thickness of the solder film in the joint. The thickness of the bulk solder in the joint should lie between 0.05 and 0.25 mm, the optimum thickness of the solder film being 0.075 to 0.080 mm as stated above. Soldered lap joints made with close clearances derive the greater part of their strength from the *bonded zones* at the interfaces where the solder is in intimate contact with the joint surfaces of the parent metal. Any tendency for the joint to yield under load is opposed by the force required to displace the solder bond. Figure 10.9 shows various types of soft-soldered joints.

Figures 10.10–10.12 show three common soft-soldering operations, namely:

- Tacking.
- Sweating.
- Floating.

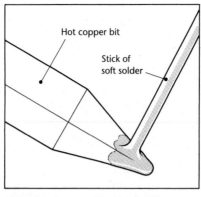
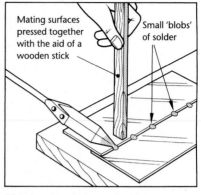

(a) Loading the copper bit with soft solder

(b) Tacking a seam prior to soft soldering

A little solder is picked up on the point of the hot soldering bit by touching it with a stick of solder. Any excess solder may be removed by simply shaking or flicking the bit.

Figure 10.10 Soft-soldering processes – tacking

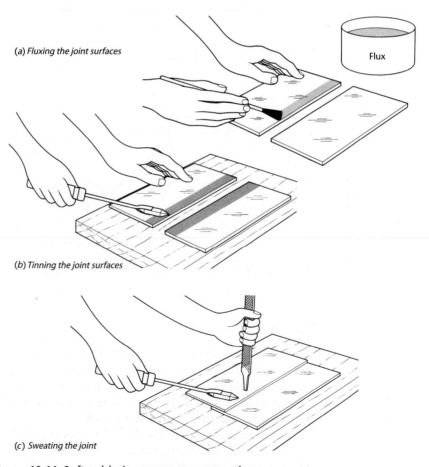

(a) Fluxing the joint surfaces

(b) Tinning the joint surfaces

(c) Sweating the joint

Figure 10.11 Soft-soldering processes – sweating

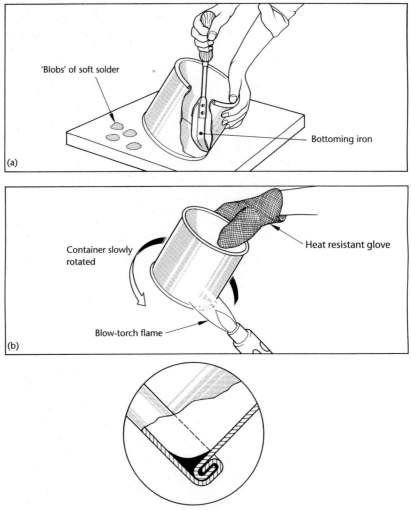

Section through a knocked-up bottom

Figure 10.12 Soft-soldering processes – floating

Tacking

This technique is shown in Fig. 10.10. When making long soldered seams it is difficult to stop the metal from distorting due to thermal expansion. Therefore, it is normal practice to make the first 'tack' at the centre of the seam and then 'tack' the seam at intervals from the centre to the ends of the seam. Alternate 'tacks' should be applied either side of the initial centre 'tack'.

The process of tacking (Fig. 10.10(b)) is carried out by placing small 'blobs' of molten solder at regular intervals along the seam in order to hold it in position prior to the main soldering operation. The tinned tip of the copper bit is used and the small amount of solder that it carries and its local heating effect causes the solder to penetrate the joint at the point of contact. During 'tacking' the mating surfaces of the seam are pressed together by use of a wooden stick or the 'tang' of a file so that the minimum heat energy is conducted away from the joint.

Sweating

The stages of this process are shown in Fig. 10.11.

1. Firstly, a thin film of soldering flux is applied to the previously cleaned joint surfaces as shown in Fig. 10.11(a). The seam to be soldered should be supported on a block of wood to prevent heat loss, never on metal.
2. The joint surfaces are 'tinned' with a film of soft solder. Solder flows more readily between surfaces that have been pre-tinned. Further, a pre-tinned surface does not require the use of a corrosive, active flux.
3. The tinned joint surfaces are placed together and held down by a wooden stick or the 'tang' of an old file. The heated soldering iron is placed on one end of the seam, ensuring that maximum contact is made with a facet at the end of the copper bit. As the solder between the two surfaces begins to melt and flow out from under the edges of the joint, the bit is drawn slowly along the seam followed by the hold-down stick pressing down on the metal. Success of this process depends upon having a constant supply of heat so the soldering 'bit' must be of ample size and capable of transmitting sufficient heat through the joint.

Floating

This process is shown in Fig. 10.12 and is used to seal self-secured joints to make containers fluid-tight. The method used in Fig. 10.12(a) shows how the joint can be sealed using a specially shaped 'bottoming iron'. The joint is fluxed both inside and outside and the container is supported at an angle of 45° on a wooden block to prevent heat loss and to cause the molten solder to flow into the joint. 'Blobs' of solder are prepared and dropped into the container where they are melted with the heated 'bottoming iron'. The molten solder is then carefully flooded along the inside of the joint as the container is slowly rotated. The quantity of solder is uniformly controlled by steady manipulation of the soldering iron, adding additional blobs of solder as and where required.

An alternative technique is to use a flame as a heat source as shown in Fig. 10.12(b). A heat-resistant glove should be worn as a protection from the flame and the radiated heat. The 'blobs' of solder are dropped into the container and the flame is applied to the outside of the joint. The container is carefully and slowly rotated to spread the molten solder evenly around the joint. With either process the solder not only penetrates the interlocking surfaces of the seam but presents a smooth and nicely rounded corner on the inside of the joint. If the container is to be used for foodstuffs (potable usage) then a *lead-free solder* must be used as required under current health and hygiene regulations.

10.2 Soft-soldered joints using lead-free solders

These are tin-based solders developed in the light of national and international legislation restricting the use of lead and its alloys in the workplace and in the home (safe drinking water systems) for health reasons, and restricting the disposal of goods containing such materials for environmental reasons. This includes the disposal of electronic printed circuit boards as found in television sets, computers and washing machines that have reached the end of their useful lives. The use of lead training and monitoring programs, special safe handling and hazardous waste disposal costs and safe drinking water requirements all add to the costs of using tin/lead solders in manufacturing. This makes the use of lead-free solders more attractive despite their higher initial costs and changes to processing equipment to provide the higher temperatures involved in their use.

The proprietary brands of lead-free solders are based on tin/copper/silver alloys and are designed to provide a safe and cost-effective alternative to the traditional tin/lead soft solders. Unfortunately, they are rather more difficult to use since they melt at higher temperatures and do not flow as freely as the traditional soft solders. Many other factors also have to be considered such as the type of flux required, type of joint, service temperature and corrosion prevention.

Most manufacturers of lead-free solders provide suitable fluxes for use with their products.

10.2.1 Properties of lead-free soldered joints

Joint strength

1. For successful joining of materials using lead-free solders there must be *no contamination of the joint zone by lead*. This not only weakens the joint but contravenes the drinking water legislation. Previously soft-soldered joints using traditional tin/lead solders cannot be successfully repaired using lead-free solders unless all traces of lead can be removed from the joint zone.

2. Further, it must be remembered that unlike the tin/lead soft solders that do not show a ductile-to-brittle transition at low temperatures, lead-free solders like most metals and filler alloys do show a pronounced ductile-to-brittle transition, at which point they lose their pliability and become very crack-sensitive and easily shatter at low temperatures.

3. When using tin/lead solders corrosion is never a problem since the joint is protected by the stable lead-oxide film that forms over the joint in air. Lead-free solders based upon tin form a tin oxide film that is easily eroded and/or mechanically damaged. The tin content is also subject to electrolytic corrosion in the presence of specific chemical solutions that form an electrolyte.

10.2.2 Lead-free solder alloys

The two most popular and readily available lead-free solders in common use are:

- A *tin/copper alloy* melting at about 227 to 230°C. Essentially this is a bronze but because of its relatively high tin content its melting point is sufficiently low for it to be classed as a soft solder. Although the cheapest and most widely used of the lead-free solders, there is some difficulty in getting it to 'wet' and flow as freely as a conventional tin/lead solder. However, by using an appropriate heat source and a suitably active flux provided by the manufacturers of the solder it does show useful gap-filling properties.

- A *tin/silver alloy* melting at about 210 to 220°C. The melting temperature is dependent upon the percentage of silver in the alloy. Because of the silver content this alloy has superior wetting and free-flowing properties. Although more expensive than the tin/copper solders it is more easily used with ordinary fluxes and heat sources and, where only small quantities are required and cost is not of prime importance, it is a suitable substitute for tin/lead solders for most purposes.

10.2.3 Lead-free soldering – heat sources

Care needs to be taken in order to prevent the higher process temperatures involved from causing distortion of the parent metals in the joint zone. This is not so important when flame-sweating pipe fittings and similar processes where a gas/air torch is employed since the flame temperature is well above the melting point for all soft solders, both tin/lead and lead-free. However, care must be taken when using a soldering iron as there is a tendency to 'dwell' on the joint too long and this can cause distortion. To avoid this happening, a soldering iron with the largest possible copper bit commensurate with the size of the job should be used. This will enable rapid heat transfer to the joint to take place so that the solder is melted before distortion occurs. Needless to say, the soldering bit must be heated to a high enough temperature for the solder alloy being used. This may be a problem with electrically heated soldering irons and these should be chosen with care and be at least two sizes more powerful than would normally be used for tin/lead solders.

10.3 Hard soldering (brazing)

Hard soldering is a general term for *silver soldering* and *brazing*. These are very similar thermal joining processes to soft soldering in as much that the parent metal does not become fused or molten and that the filler alloy has to have a lower melting temperature range than the metals being joined. However, the filler materials for hard soldering have a much higher tensile strength than soft solder and the process temperatures are much higher.

The term 'brazing' is derived from the fact that the filler material (spelter) is a brass alloy of copper and zinc and a typical spelter containing 40% copper and 60% zinc melts at about 850°C to produce a strong and malleable joint. This is very much higher than the temperatures required when soft-soldering.

Brazing is defined as a process of joining metals in which the molten filler material is drawn into the gap between the closely adjacent surfaces of the metals being joined by capillary attraction.

10.3.1 Basic principles of brazing

The success of all hard-soldering operations (brazing and silver soldering) is dependent upon:

- The selection of a suitable filler alloy which has a melting range that is appreciably lower that that of the parent metals being joined – but substantially higher than that used for soft soldering.

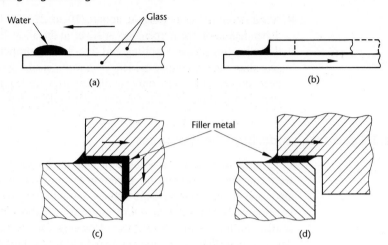

Figure 10.13 Capillary flow

- The thorough cleanliness of the surfaces being joined. Note that previously soft-soldered joints cannot be remade by hard soldering.
- The complete removal of the oxide film from the surfaces of the parent metal and the filler material by a suitable flux.
- The complete 'wetting' of the joint surfaces by the molten brazing alloy.

It is an essential feature of hard soldering that the filler material is drawn into the joint area by capillary attraction so that the joint is completely filled. The principle of capillary flow is shown in Fig. 10.13. Consider a clean piece of glass with a droplet of water lying on it: the water does not 'wet' the surface of the glass but remains in the same place as shown in Fig. 10.13(a). If a second piece of glass is laid on the first piece of glass and is made to slide until its edge touches the water droplet, some of the water will be drawn into the very narrow gap between the mating surfaces for quite an appreciable distance by capillary penetration as shown in Fig. 10.13(b).

The same effect occurs when a joint is hard soldered, except that instead of glass plates the mating surfaces are of metal – the joint surfaces of the parent metals being joined – and, in place of water, it will be the filler material that is drawn into the joint. The success of joints formed by capillary penetration is governed by the maintenance of an appropriate joint clearance to promote the necessary capillary attraction. The mating surfaces must be parallel and there should be no break in the uniformity of the clearance, as shown in Fig. 10.13(c). If a break occurs due to widening or closing of the clearance gap then the capillary flow will stop in that vicinity and the filler material will not penetrate further. Capillary attraction will draw the filler metal around the corner when the conditions are correct but the flow will cease at the chamfered corner, as shown in Fig. 10.13(d), resulting in a weak joint.

10.3.2 Metals suitable for hard-soldering

The following common metals and their alloys can be joined by hard soldering processes:

- Copper and copper-based alloys.
- Low, medium, and high carbon steels.

- Alloy steels including stainless steels.
- Malleable cast iron and wrought iron.
- Nickel-based alloys.
- Aluminium and certain aluminium alloys.

Metals and alloys of dissimilar compositions can be hard-soldered (brazed or silver-soldered) together, for example: copper to brass; copper to steel; brass to steel; cast iron to mild steel; and mild steel to stainless steel.

10.3.3 Filler materials

Hard soldering filler materials may be basically *copper/zinc* alloys for *brazing* and *copper/silver/zinc* alloys for *silver soldering*. These filler alloys are essentially dissimilar in composition to the parent metal being joined. Copper is frequently used for flux-free, furnace-brazing mild steel components together in furnaces having a reducing atmosphere. Brazing filler alloys may be classified into three main types:

- Silver solders.
- Brazing alloys containing phosphorus.
- Brazing brasses or 'spelters'.

Silver solders

These are substantially more expensive than normal brazing alloys since they contain a high percentage of the precious metal *silver*. They produce strong and ductile joints at much lower temperatures than brazing filler alloys with a tensile strength in the region of $500 \, MN/m^2$. Silver solders are free-flowing at their processing temperature which, being relatively low compared with brass spelters, have little thermal effect upon the properties of the parent metals.

Silver solders make it possible to speed up and increase the productivity of the hard-soldering process. Further, post-soldering finishing processes can be eliminated as silver-soldered joints are very neat. Because of the high cost of silver solders they are mainly used for small, delicate work.

Brazing fluxes based upon borax or upon boric acid are unsuitable for use with silver solders since these fluxes are not sufficiently fluid at the relatively low temperatures required for silver soldering.

Below 760°C a flux containing potassium fluoborate should be used as it is very active and completely molten at 580°C.

The flux should become completely molten at a temperature at least 50°C below the melting point of the filler material and retain its activity at a temperature at least 50°C above the melting temperature of the filler material. Most manufacturers of proprietary brands of silver solders also provide suitable fluxes for use with their solders and which will give the optimum results. Table 10.4 lists the compositions and melting ranges of some common silver solders and suggests some typical applications.

Brazing alloys containing phosphorus

Filler materials that contain phosphorus are usually referred to as self-fluxing brazing alloys. They contain silver, phosphorus and copper, or just copper and phosphorus, the

Table 10.4 Composition of silver solders

British Standard 1843 Type	Composition percentage				Approximate melting range (°C)
	Silver	Copper	Zinc	Cadmium	
	min — max	min — max	min — max	min — max	
3	49 to 51	14 to 16	15 to 17	18 to 20	620–640
4	60 to 62	27.5 to 28.5	9 to 11	–	690–735
5	42 to 44	36 to 38	18.5 to 20.5	–	700–775

Type 4 possesses a high conductivity and is, therefore, very suitable for making electrical joints. It is the most expensive because of its high silver contents.
Type 3 is extremely fluid at brazing temperatures which makes it ideal when brazing dissimilar metals. *A low melting point alloy.*
Type 5 is a general purpose solver solder which can be employed at much higher brazing temperatures.

Table 10.5 Composition of brazing alloys containing phosphorus

British Standard 1845 Type	Composition percentage			Approximate melting range (°C)
	Silver	Phosphorus	Copper	
	min — max	min — max		
6	13 to 15	4 to 6	Balance	625–780
7	–	7 to 7.5	Balance	705–800

former possessing a lower melting range of temperatures. The main feature of these alloys is their ability to braze copper components together in air without the need for a separate flux. When melted in air the products of oxidation react with the phosphorus content of the filler material to form an efficient flux. However, when brazing copper-based alloys a suitable flux is required. Brazing alloys containing phosphorus are only effective when used in an oxidizing atmosphere and should only be used for brazing articles made from copper and all copper-based alloys providing the alloy contains less than 10% nickel. Nickel, nickel-based alloys and ferrous metals should not be brazed (they can be silver-soldered or brazed with a copper/zinc spelter) with brazing alloys containing phosphorus. Although they will 'wet' the parent metal and flow into the joint interface, they will form brittle compounds that will weaken the joint.

Table 10.5 lists two common brazing alloys containing phosphorus. Type 6 filler material produces a joint with a tensile strength of about $450\,MN/m^2$ whilst the higher melting point type 7 filler material produces a joint with a lower tensile strength of $350\,MN/m^2$. These brazing alloys find their greatest use in resistance brazing operations, in refrigerator manufacture, electrical assemblies (armature windings to commutator segments), and for brazing the seams and fittings in domestic hot-water cylinders.

Brazing brasses (spelters)

Traditional brazing techniques involve the use of a brass alloy filler material called a 'brazing spelter' and borax as a flux. Such alloys melt at much higher temperatures

Table 10.6 Composition of brazing spelters

British Standard 1845 Type	Composition percentage		Approximate melting range (°C)
	Copper	Zinc	
	min max		
8	49 to 51	Balance	860–870
9	53 to 55	Balance	870–880
10	59 to 61	Balance	885–890

This group of copper alloys tends to lose zinc by vaporization and oxidation when the parent metal is heated above 400°C. This loss of zinc produces relatively higher tensile strength. The brazing alloys containing a high percentage of zinc, therefore, produce joints of the lowest strength.

Type 8 is used for medium strength joints, whilst the strongest joints can be produced by using *Type 10*.

than the silver solders and the copper filler materials containing phosphorus but produce sound joints having tensile strengths between $400\,MN/m^2$ and $480\,MN/m^2$ and are suitable for making high-strength joints between ferrous components. They were traditionally used in the manufacture of bicycle frames for brazing the steel tubes into the malleable cast iron fittings. The composition and melting ranges of three common brazing 'spelters' are listed in Table 10.6, where it can be seen that increasing the zinc content decreases the melting temperature range.

The zinc content is more volatile than the copper in the brazing alloy and evaporates to some extent during the brazing operation so that the joint finishes with a higher-strength bond than would appear possible for the high zinc content spelter used. For this reason, where brass components are to be joined by brazing, the parent metal must always have a higher copper content that the filler material. Thus a joint can be successfully made between two brass components manufactured from 60/40 or 70/30 brass alloys using a 50/50 brazing spelter.

10.3.4 Heat sources for brazing

The choice of heat source when brazing depends upon the workshop facilities available and the volume of work to be manufactured. These heat sources are:

- Flame brazing (torch).
- Furnace brazing.
- Dip brazing.
- Electric induction brazing.
- Electric resistance brazing.

Flame brazing (torch)

This is the most versatile process used to fabricate any assembly large or small and is widely used in jobbing and batch production shops. The brazing spelter can be applied to the joint in two ways:

1. The granulated spelter is 'calcined' by boiling in a borax solution. These calcined spelter granules are mixed with borax paste and spread on the joint. As heat is

applied and the joint zone is brought up to the brazing temperature the spelter melts. Additional borax is sprinkled on the joint which is tapped and scratched with a steel spatula to assist the flow of solder into and throughout the joint.

2. The tip of the spelter in wire or strip form is heated and dipped into the borax flux which adheres to it. When the joint, which is also fluxed, is brought up to the brazing temperature, the fluxed end of the spelter wire or strip is applied to the assembly whereupon it melts and flows by capillary attraction into the joint.

The heating flame is produced by a suitably designed torch with an oxygen–fuel gas mixture or an air–fuel gas mixture depending on the melting range of the filler material being used. The flame must *not be used to melt the filler material* as this will result in an unsound joint. The flame is used to bring the parent metals being joined up to the brazing temperature so the filler material is melted by contact with the parent metal.

A wide variety of gas mixtures are used for torch brazing. In decreasing order of flame temperature these are:

- Oxy-acetylene.
- Oxy-propane.
- Compressed air–natural gas.
- Naturally aspirated air with natural gas (methane) or bottled gas (propane or butane).

Because of the high cost of the gases used, oxy-acetylene is mainly used for small, fine work, aluminium brazing (see Section 10.4) and full fusion welding (see Chapter 11). Examples of brazing torches are shown in Fig. 10.14 and an example of a torch-brazing operation is shown in Fig. 10.15.

Furnace brazing

Furnace brazing is mainly used when the components to be joined are engineered so that they can be pre-assembled or jigged to hold them in position during the brazing processes. In addition the volume of work must justify the capital outlay on the plant required, and the filler material and flux (if required) can be pre-placed. Furnace brazing is also used when a controlled atmosphere is required. The heat source may be gas or oil in a muffle-type furnace so that the flame and products of combustion do not impinge on the work being brazed and also so that the atmosphere can be controlled. The majority of controlled atmosphere furnaces are electrically heated for cleanliness and accuracy of control, both of the temperature and the atmosphere.

The two types of furnace used are:

- Batch-type furnaces with either air or a controlled atmosphere.
- Conveyor-type furnaces with a controlled atmosphere.

Air-atmosphere furnaces are used for brazing assemblies that can be protected by the use of a suitable flux or for joints to be made with brazing alloys containing phosphorus. Furnace brazing with an air atmosphere and brazing alloys containing phosphorus are not suitable for joining pure copper components.

Controlled-atmosphere furnaces are usually electric-resistance-heated. The conveyor-type furnace has a heat-resistant wire mesh belt to carry the work through the furnace on metal trays. Alternatively a roller bed may be used so that each new

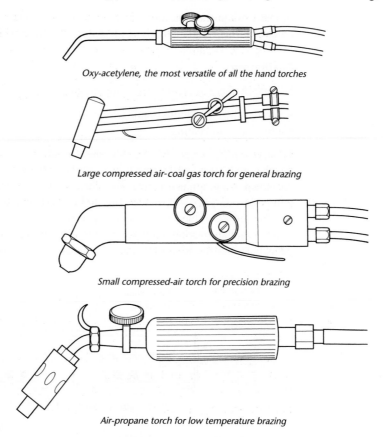

Oxy-acetylene, the most versatile of all the hand torches

Large compressed air-coal gas torch for general brazing

Small compressed-air torch for precision brazing

Air-propane torch for low temperature brazing

Figure 10.14 Typical hand torches used for brazing

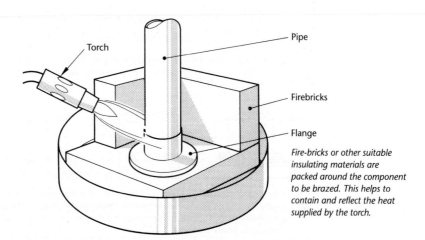

Torch

Pipe

Firebricks

Flange

Fire-bricks or other suitable insulating materials are packed around the component to be brazed. This helps to contain and reflect the heat supplied by the torch.

Figure 10.15 Hand torch in use with brazing hearth

tray of work pushes the previous tray through the furnace. Figure 10.16 shows both types of furnace and methods of pre-assembly of the parts to be joined.

Dip brazing

Essentially there are two types of dip brazing:

- Molten-filler-metal dip brazing.
- Flux-bath dip heating.

In *molten-filler-metal dip brazing* the components to be joined are assembled together and submerged into a bath of molten filler metal. The molten filler metal is contained in a graphite crucible that is heated externally to maintain the brazing alloy in liquid form. Flux is floated on the surface of the molten filler metal to exclude the air. The pre-assembled parts should be raised and lowered through the molten flux into the molten brazing alloy several times to ensure that the flux and brazing alloy penetrates into the joint clearance. Large assemblies are usually pre-heated before

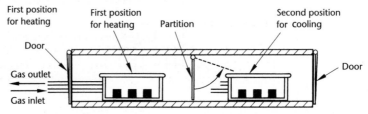

Schematic layout of batch brazing in a sealed container

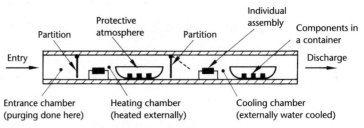

Schematic layout of continuous brazing furnace

Individual assemblies or components for batch brazing (in trays) are passed through the furnace on a conveyor system. Inert or reducing atmospheres can be used to protect the work from oxidation.

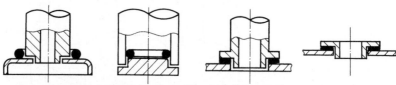

Filler metal for furnace brazing is used in the form of pre-placements. The components to be brazed are assembled with the filler alloy in the required position and fluxed if necessary. Pre-placed brazing alloy inserts are available in a variety of forms, such as wire rings, bent wire shapes, washers and foils.

Figure 10.16 Furnace brazing

dip brazing so as not to cool down the bath of molten brazing alloy and thus producing a faulty joint. This process is mainly used for brazing mild steel and malleable cast iron components using a traditional copper/zinc brazing spelter.

Flux-bath dip heating, also known as *salt-bath dip heating*, is a process that has been developed mainly for the brazing of aluminium components, although it is also used for joining mild steel components using copper/zinc brazing alloys. It is the molten salts contained in the salt bath and maintained in a molten condition either by gas or oil burners or electrically that supply the brazing heat. When heated electrically, a low-voltage alternating current is passed through the salts, which are electrically conductive, by means of carbon electrodes immersed in the salts themselves. This enables the temperature of the salt bath to be accurately controlled.

> The salt bath is a metal or ceramic container

Salt-bath brazing enables the parts being joined to be heated more rapidly to the brazing temperature than in a controlled-atmosphere furnace because of the greater heat conductivity of the liquid medium in the bath. The salts used are carefully chosen for their melting characteristics and their ability to have a fluxing action on the parent metal of the components being joined and on the filler material. Figure 10.17 shows the two dip-brazing techniques discussed above.

Electric induction brazing

Induction heating is used to braze parts that are self-jigging, where very rapid heating is required and where the volume of production warrants the capital outlay on the equipment required. The assembly to be brazed is placed in the magnetic field

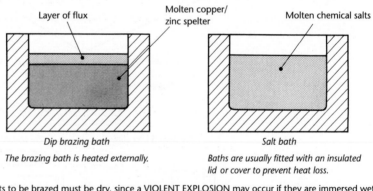

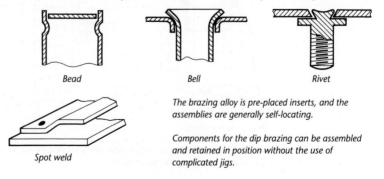

Figure 10.17 Two methods of dip brazing

of a solenoid carrying a high-frequency alternating electric current. If there is no alternative to the use of a jig to hold the parts, the jig must not interfere with the effective heating of the parts being joined.

In this process, which is normally operated continuously, the heat required for brazing is generated within the joint material itself. The solenoid coil is made from copper tube and cooling water is circulated through the copper tube solenoid to keep it cool. The heating effect of an alternating current is a surface effect, the higher the frequency of the current the shallower the heating effect. Most of the intense heat generated by this technique is near the surface of the joint materials. The heat so generated travels throughout the joint materials by conduction.

The principle of induction heating is shown in Fig. 10.18 together with internal and external solenoid coils. The filler material and flux is pre-placed in this process. Induction heating is widely used for silver soldering.

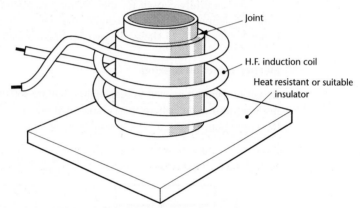

Solid copper coils are also sometimes employed. The coils are normally insulated with glass fibre. In practice the work to be brazed is brought to the coils.

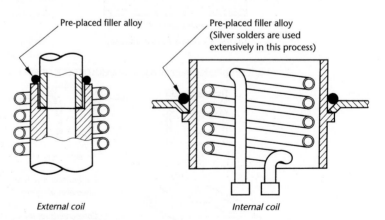

It is usual for induction coils to be designed to surround the joint, but internal coils can be used for certain applications.

Figure 10.18 Electric induction brazing

Electric resistance brazing

In this process the heat required for brazing is generated by the passage of a high-value electric current through the joint material at a low voltage. The resistance to the current flow by the joint materials and the carbon electrodes is responsible for the heat generated. The brazing heat can be precisely localized and this ensures no general loss of mechanical properties throughout the parent metal. The main advantage of resistance brazing is that, normally, no expensive jigs are required, since the carbon electrodes themselves hold the components being joined in the correct position.

In *direct heating* as shown in Fig. 10.19(a) the assembly is placed between the electrodes so that the current passes from one electrode through the parent metals and pre-placed brazing insert to the other electrode.

In *indirect heating* as shown in Fig. 10.19(b) the electrodes are placed so that the current does not pass through the joint, but passes through only one component of the assembly. The heat generated travels through the joint by conduction. This technique finds its greatest use where a small component has to be brazed onto a much larger component.

10.4 Aluminium brazing

Aluminium brazing uses a filler material that is itself an aluminium alloy having a lower melting temperature range than the aluminium alloy parent metals that are being joined. The various grades of commercially pure aluminium alloys containing

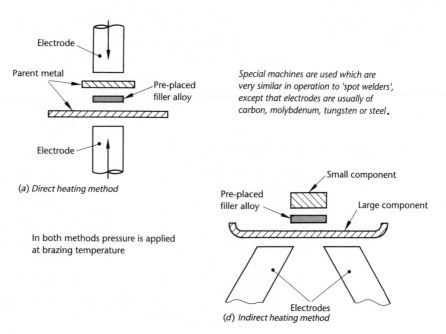

Electrode

Parent metal

Pre-placed filler alloy

Electrode

(a) Direct heating method

In both methods pressure is applied at brazing temperature

Special machines are used which are very similar in operation to 'spot welders', except that electrodes are usually of carbon, molybdenum, tungsten or steel.

Small component

Pre-placed filler alloy

Large component

Electrodes
(d) Indirect heating method

Figure 10.19 Electric resistance brazing

Table 10.7 Composition of filler alloys for aluminium brazing

British Standard 1843 Type	Composition percentage					Approximate melting range (°C)	Brazing range (°C)
	Aluminium	Copper		Silicon			
		min	max	min	max		
1	Balance	2 to 5		10 to 13		550–570	570–640
2	Balance	–		10 to 13		565–575	585–640
3	Balance	–		7 to 8		565–600	605–615
4	Balance	–		4.5 to 6.5		565–625	620–640

It is impossible to braze certain aluminium alloys whose melting points are below those of the available brazing alloys.

1.25% manganese, and certain aluminium/magnesium/silicon alloys, can be successfully joined by brazing. Alloys containing more that 2% magnesium are difficult to braze as their oxide films are very tenacious and difficult to dissolve with ordinary brazing fluxes. Borax-based fluxes are unsuitable for brazing aluminium and its alloys because of its high melting temperature. However, proprietary makes of flux are available and these are basically mixtures of alkali-metal chlorides (such as sodium, potassium and lithium chlorides) and fluorides.

These fluxes are *hygroscopic* in that they absorb water from the atmosphere so that it is important that all lids or covers are immediately and firmly replaced on flux containers when not in use to prevent deterioration of the contents. Care should be taken when handling fluxes containing fluorides as contact with them may cause skin complaints.

> Inhalation of the fumes given off when brazing with fluoride fluxes can have toxic effects

The melting range of the parent metals that can be aluminium-brazed range between 590°C and 600°C. Table 10.7 shows that the temperature margin between the melting ranges of the parent metal and the melting ranges of the aluminium brazing alloys used are very close. Great care must be taken when aluminium flame brazing because of this very small temperature margin if the joint is to be made without melting the parent metal during the operation. Techniques other than flame brazing are critically controlled to provide a uniform brazing temperature.

10.4.1 Aluminium flame brazing – procedure

Aluminium flame brazing with a gas torch is relatively simple but depends on the skill of the operator to maintain the correct brazing temperature without overheating the parent metal, and providing the following procedure is adopted:

- Assuming that an oxy-acetylene torch is used, the flame is adjusted to a slightly *carburizing condition*, that is, burning the acetylene with a reduced oxygen supply. This enables the joint zone to be kept in a *reducing atmosphere* to overcome aluminium's great affinity for oxygen. The oxy-acetylene flame is lit and the oxygen valve is slowly opened until the inner 'cone' is clearly defined. The oxygen valve is then slightly closed until there is a slight haze or 'feather' around its point as shown in Fig. 10.20(a).
- The joint assembly is preheated using the envelope of the flame.

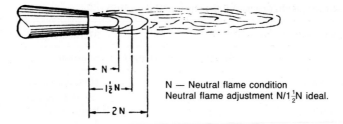

N — Neutral flame condition
Neutral flame adjustment N/1½N ideal.

(a) Essential flame condition

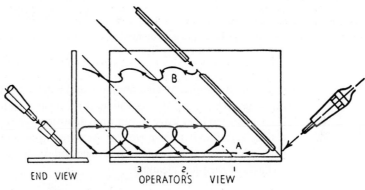

END VIEW   OPERATOR'S   VIEW

A — Movement of Torch.
B — Movement of Filler Rod.
1, 2 and 3 — Points of application of Filler Rod.

(b) Filler rod and torch manipulation

Figure 10.20 Aluminium flame brazing

- The filler rod is preheated at one end in the flame and dipped into the flux so that a 'tuft' of flux adheres to the filler rod.
- The joint is then heated with the inner flame of the torch held about 50 mm away until the flux starts to flow freely along the joint, the flux being applied from the tuft on the end of the filler rod.
- As soon as the flux begins to flow rapidly and smoothly along the metal, due to the capillary action of the joint clearance and the driving power of the flame, a small quantity of filler material (filler rod) is added and the rod is withdrawn. As in all brazing processes, it is good practice for the heat of the joint assembly to melt the filler material and the flame of the torch.
- As the filler material begins to flow, the torch is moved forward and away in a sweeping action. More filler material is added progressively along the joint as the movement of the torch is maintained in order to drive the filler metal along where the joint has been 'wetted' by the flux. This technique is shown in Fig. 10.20(b).

10.4.2 Aluminium flame brazing – flux removal

Flux removal is an essential part of the brazing process and must be done immediately after brazing is complete to prevent corrosion. Flux removal is achieved by

Table 10.8 Methods of flux removal by chemical cleaning

Stage	Method A			Method B		
	Solution	Temp	Time (min)	Solution	Temp	Time (min)
1	Concentrated nitric acid	Room	5–15	10% nitric acid, 0.23% hydrofluoric acid	Room	5–10
2	Water rinse	Room	–	Water rinse	Room	–
3	10% nitric acid, 5–10% sodium dichromate	Room	5–10	–	–	–
4	Water rinse	Hot	–	Water rinse	Hot	–
5	Dry using warm air	–	–	Dry using warm air	–	–

Method A is the most common; and Method B gives a good uniform appearance to the joint surfaces, as it counteracts the darkish appearance in the joint caused by the silicon content of the filler metal.

scrubbing the joint and surrounding metal with a stiff bristle or fine wire brush and hot soapy soft water. This is followed by rinsing with cold soft water and drying. Hard water tends to leave a thin lime deposit on the metal surface. Where the flux residue is difficult to remove, for example from crevices, chemical cleaning my have to be resorted to in addition to soap and water cleaning. Table 10.8 lists various methods of chemical cleaning.

10.5 Types of brazed joints

There are two basic types of brazed joint, namely the *lap joint* and the *butt joint* as shown in Fig. 10.21. However, there are many combinations and variations of these joints designed to meet specific requirements. The factors affecting the strength of hard soldered joints are similar to those that apply to soft soldering but, since the mechanical strength of the filler materials (brazing alloys) is very much greater than that for soft solders, the overall joint strength is very much higher.

10.6 Braze-welding

As for the hard soldering (brazing) processes discussed already, braze-welding can be used to join similar or dissimilar metals by using a filler material with a lower melting temperature range than the parent metals being joined. When *brazing*, the filler material is drawn into a close-fitting joint by *capillary attraction*. When *braze-welding* the joint gap is not so closely fitting, there is *no capillary penetration* and a relatively large quantity of filler material is used.

Traditionally, braze-welding was known as bronze welding because the process resembled welding and, originally, the filler material was a *copper–tin–phosphorus alloy*, that is, a *bronze*. However, the name (although often still used) is a misnomer because the filler material used nowadays is essentially a *copper–zinc alloy*, that is, a *brass*. This alloy may also contain traces of silicon, manganese and/or nickel depending upon the job in hand. Some commonly used filler materials are listed in Table 10.9.

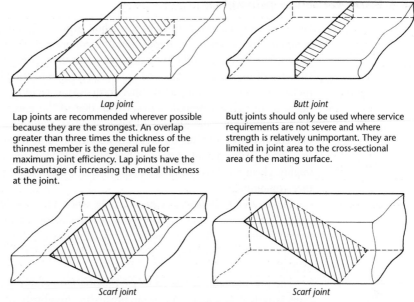

Lap joint

Lap joints are recommended wherever possible because they are the strongest. An overlap greater than three times the thickness of the thinnest member is the general rule for maximum joint efficiency. Lap joints have the disadvantage of increasing the metal thickness at the joint.

Butt joint

Butt joints should only be used where service requirements are not severe and where strength is relatively unimportant. They are limited in joint area to the cross-sectional area of the mating surface.

Scarf joint

Scarf joint

Scarf joints are a variation of the butt joint and are used to increase the cross-sectional area and provide a larger surface for the brazing filler metal. Scarf joints are more costly to produce, more difficult to hold in alignment, but have the advantage of no increased thickness at the joint. For maximum joint efficiency the joint area obtained should be over three times the normal cross-sectional area.

Figure 10.21 The two basic types of joint used in brazing

Table 10.9 Composition of copper alloy filler rods for braze welding

British Standard 1453 Type	Composition percentage											Melting point (°C)
	Copper		Zinc		Silicon		Tin	Manganese		Nickel		
	min	max	min	max	min	max	max	min	max	min	max	
C2	57 to 63		36 to 42		0.2 to 0.5		0.5	–		–		875
C4	57 to 63		36 to 42		0.15 to 0.3		0.5	0.05 to 0.25		–		895
C5	45 to 53		34 to 42		0.15 to 0.5		0.5	0.5		8 to 11		910
C6	41 to 45		37 to 41		0.2 to 0.5		1.0	0.2		14 to 16		

Type C2 is termed a *silicon-bronze* filler rod and is specially recommended for the braze welding of *brass* and *copper* sheet and tubes as are used for sanitary and hot water installations. It is also suitable for the braze welding of *mild steel* and *galvanized* steel.

Type C4 is termed a *manganese-bronze* filler rod. It has a higher melting point and is especially suitable for braze welding *cast* or *malleable iron*, and also for building up worn parts such as gear teeth.

Types C5 and *C6* are *nickel-bronze* rods which are recommended for braze welding *steel* or *malleable iron* where the **highest mechanical strength** is required. These are the high melting point welding rods and have a valued application in the reclaiming and building up of wearing surfaces.

10.6.1 Braze-welding – principles

Braze-welding can be defined as a method of joining materials with a gas-welding technique by means of the deposition of a copper-rich filler metal on the parts to be joined. The bonding of the joint depends upon the wetting (tinning) of the unmelted surfaces of the parent metals and the interdiffusion of the filler metal and the parent metals. In braze-welding as distinct from fusion welding the melting temperature range of the filler material is lower than that of the parent materials being joined. As a result of this, the process of braze-welding requires a lower temperature than that required for full fusion welding which, in turn, results in less distortion and oxidation of the parent metals. When braze-welding, the workpiece must be thoroughly clean and free from scale, oil/grease, paint, metal plating or any other surface contaminant.

Fluxes of the borax type are suitable when braze-welding. A paste or a powdered flux worked into a paste with water can be more thoroughly and evenly spread than a dry flux in powder form. If a powder flux is used the tip of filler rod should be heated and dipped into the flux powder in order to pick up a 'tuft' of flux.

The parent metal is heated to the brazing temperature taking care not to melt it, especially if it is copper. As in the processes previously described, the filler material should be melted by the heat energy stored in the job and not by the flame. The purpose of the flame is to maintain the job at the brazing temperature. A globule of filler metal is melted off and deposited in the joint. Circular motion of the brazing torch will cause the deposited filler metal to spread and 'tin' the surface of the parent metals in the joint zone. When local tinning of the joint has taken place, the filler material should be reheated to a fluid condition and additional filler rod applied to the weld deposit already in the joint *but not in front of it*. This fresh globule of molten filler material is manipulated with the flame so that it is pushed forward to create an advanced 'wet edge' to the joint. This operation is repeated until the joint is complete.

The strength of the deposited filler metal will be found to be equal to and, in some cases, greater than the strength of the parent metal. The overall strength of a braze-welded joint is, therefore, entirely dependent on the soundness of the bond between the filler material and the parent metal. Therefore, the operator should ensure, by careful observation, that the filler metal is actually 'tinning' the surface of the joint immediately in front of the braze-welding flame. This will indicate that the joint is at the correct temperature and that it has been properly prepared. Figure 10.22 shows the formation of a correctly 'wetted' and 'tinned' braze-weld.

Braze-welding utilizes the concentrated heat of the oxy-acetylene flame. By careful control of the torch this heat can be localized to bring any area of the joint up to the braze-welding temperature. This enables perfect control of the filler material to be achieved without capillary attraction. The *leftward* method of welding is employed, with a torch nozzle about two sizes smaller than would be used for fusion welding than for butt joints in metal of the same thickness. The flame condition should be slightly oxidizing, that is, the torch is set to a normal neutral flame with a well-defined cone and the oxygen feed is slightly increased by careful adjustment of the oxygen valve on the torch to give a slightly noisier flame and a slightly

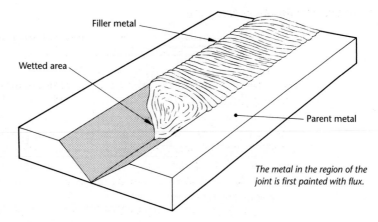

Figure 10.22 The wetting action in braze welding

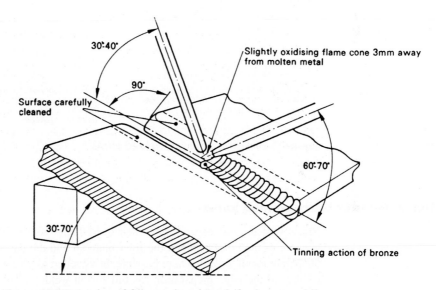

Figure 10.23 Angles of filler rod and torch for braze welding

more pointed cone. Figure 10.23 shows the angles of the filler rod and welding torch for the braze-welding of ferrous metals. The following metals can be joined by braze-welding:

- Cast iron.
- Copper.
- Galvanized mild steel.
- Malleable cast iron.
- Mild steel.
- A combination of any two of the above metals.

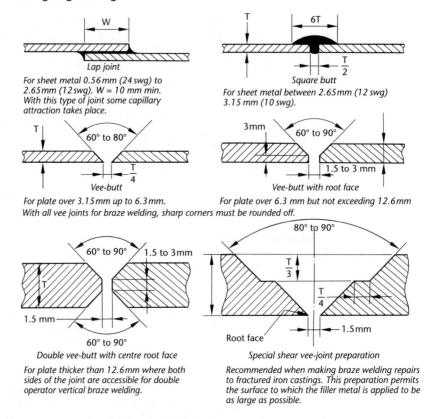

Figure 10.24 Forms of braze welded joints

10.6.2 Types of braze-welded joints

Braze-welded joints depend upon their overall strength on the relatively high *shear strength* of the bond between the filler material and the parent metal. The edges of the parent metal should, where possible, be bevelled to give a 'vee' of 90° in a similar way to the edge preparation for full fusion welding. The upper and lower surfaces of the parent metal should be cleaned by wire brushing for a short distance either side of the joint edges. Where bevelling is not possible, the width of the filler metal deposit at the top of the joint should be at least twice the thickness of the parent metal, and the deposit should penetrate just over the edge of the parent metal on the underside. Typical joints for braze-welding are shown in Figs 10.24 and 10.25.

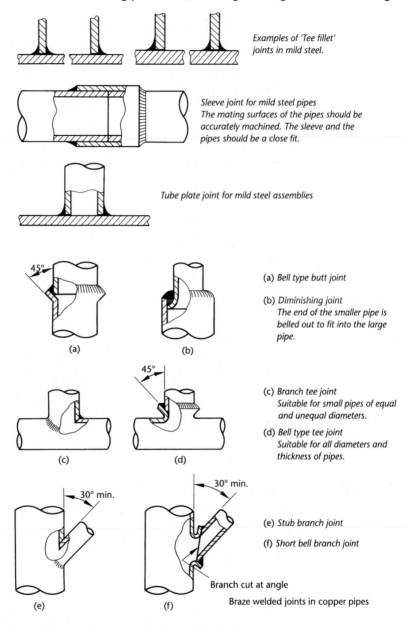

Examples of 'Tee fillet' joints in mild steel.

Sleeve joint for mild steel pipes
The mating surfaces of the pipes should be accurately machined. The sleeve and the pipes should be a close fit.

Tube plate joint for mild steel assemblies

(a) Bell type butt joint

(b) Diminishing joint
The end of the smaller pipe is belled out to fit into the large pipe.

(c) Branch tee joint
Suitable for small pipes of equal and unequal diameters.

(d) Bell type tee joint
Suitable for all diameters and thickness of pipes.

(e) Stub branch joint

(f) Short bell branch joint

Branch cut at angle

Braze welded joints in copper pipes

Figure 10.25 Applications of braze welded joints

Exercises

10.1 Soft soldering

a) Under what circumstances may traditional tin/lead solder still be used and under what circumstances must lead-free solders be used?

b) Describe how the surfaces of the metals being joined are prepared ready for soldering in order to ensure a sound joint.

c) Describe the purpose of a soldering flux and state the difference between an active and a non-active flux when using a traditional tin/lead solder.

d) With the aid of sketches, describe the technique of soft-soldering with a copper bit soldering iron.

e) Explain the procedure for sweating a pipe fitting onto a copper pipe using a blow-pipe flame.

f) With the aid of sketches explain how self-secured joints may be sealed by 'floating'.

g) List the problems associated with using lead-free solders and explain how they can be overcome.

10.2 Hard soldering (brazing)

a) Describe the fundamental differences between 'hard' and 'soft' soldering.

b) Describe the basic principles of brazing.

c) Describe what is meant by capillary penetration of a brazed joint and explain what precautions have to be taken to ensure this is achieved.

d) List the advantages and limitations of hard soldering.

e) Briefly describe the essential differences between silver solders, brazing alloys containing phosphorus and brazing 'spelters' and give typical applications of each of these filler materials.

f) List the materials that may be joined by the brazing process. Can the brazing process be used to join different metals?

g) With the aid of sketches, describe THREE different heat sources for brazing and compare their advantages and limitations.

h) Briefly describe the process of aluminium brazing, paying particular attention to the problems associated with this process.

10.3 Braze-welding

a) Describe the fundamental differences between brazing and braze-welding.

b) Describe, in detail, how two mild-steel components may be joined by braze-welding.

c) With the aid of sketches show the different types of joint that may be braze-welded.

d) State the fundamental principles of braze-welding.

e) List the advantages and limitations of braze-welding compared with the hard-soldering processes.

11

Joining processes (welding)

When you have read this chapter you should be able to:

- Explain what is meant by fusion welding
- Describe the process of oxy-acetylene welding of low carbon steels
- Differentiate between, identify and describe oxygen and acetylene equipment
- Discuss the safe handling of oxy-acetylene equipment and the appropriate safety procedures in its use
- Describe the types and applications of oxy-acetylene flames
- Describe the structure of the weld zone
- Describe the various oxy-acetylene welding techniques
- Explain the importance of edge preparation prior to welding
- Describe the effect of oxide formation and the means of its avoidance
- Describe the process of manual metallic-arc (stick) welding
- Discuss the types and uses of electrode coatings
- Differentiate between welding with a direct-current source and an alternating-current source
- Describe manual metallic arc-welding techniques
- Describe the types of arc-welded joints and the edge preparation of the parent metal
- Discuss the cause and effect of weld defects
- Discuss the workshop testing of welded joints
- Describe the principles of submerged arc welding
- Describe the principles of gas shielded arc welding (TIG and MIG)
- Describe the principles of automated welding
- Describe the principles of resistance welding (spot, seam, projection and butt welding)

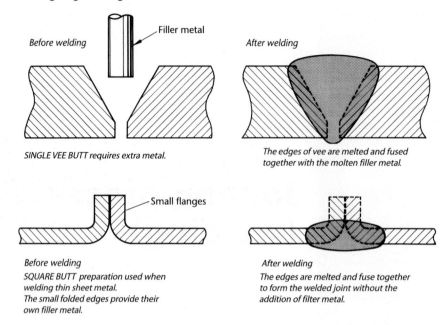

Before welding

Filler metal

After welding

SINGLE VEE BUTT requires extra metal.

The edges of vee are melted and fused together with the molten filler metal.

Small flanges

Before welding
SQUARE BUTT preparation used when welding thin sheet metal. The small folded edges provide their own filler metal.

After welding
The edges are melted and fuse together to form the welded joint without the addition of filter metal.

Figure 11.1 Fusion welding with and without the addition of filler metal

11.1 Fusion welding

Full fusion welding, as the name implies, is a thermal joining process in which the edges of the parent metal to be joined are melted and caused to fuse together with or without the use of a filler material. The basic principle of fusion welding is shown in Fig. 11.1, which outlines two typical joint configurations: one using a separate filler rod and the other self-supplying and not needing a filler rod. The filler rod is a thin steel rod or wire of similar composition to the parent metals being joined. Filler rods are usually copper plated so that they do not rust.

The two basic welding processes are shown in Fig. 11.2 and they are commonly referred to as:

- Gas-welding (oxy-acetylene welding).
- Metal-arc welding (electric), also commonly known as 'stick' welding.

So that the welding operative can produce the correct type of weld in the correct place, it is necessary to provide the necessary information on the fabrication assembly drawing. Standard welding symbols based on BSEN 22553: 1995 were discussed in Section 4.14.

11.2 Oxy-acetylene welding

An oxy-fuel gas flame provides the heat required at a high enough temperature to melt most engineering materials in common use. The fuel-gas most widely used is

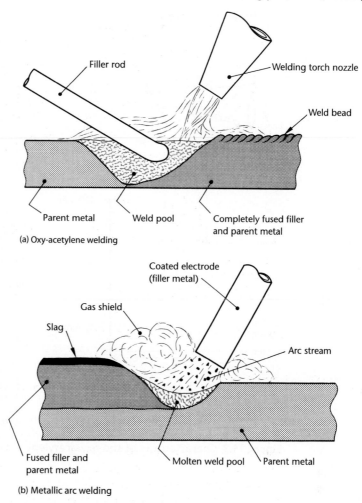

Figure 11.2 Comparison of oxy-acetylene and metallic arc welding

acetylene and the process is then referred to as oxy-acetylene welding. Acetylene is the most economical gas to use in conjunction with commercially pure oxygen supplied from high-pressure cylinders to give a flame with a maximum temperature of 3200°C. Figure 11.3 shows the main characteristics of the oxy-acetylene flame and, by comparison, lists the flame temperatures provided by other oxy-fuel gas mixtures.

This chapter only refers to the equipment required for *high-pressure* oxy-acetylene welding equipment as shown in Fig. 11.4 since this is the most widely used. The safe use of welding equipment was discussed in Chapter 1. The essential differences between oxygen and acetylene cylinders are shown in Fig. 11.5.

Oxygen

This is supplied in a thick-walled solid drawn steel cylinder painted black. The gas is stored in the cylinder under very high pressure, usually 136.6 bar (13660kN/m^2) in

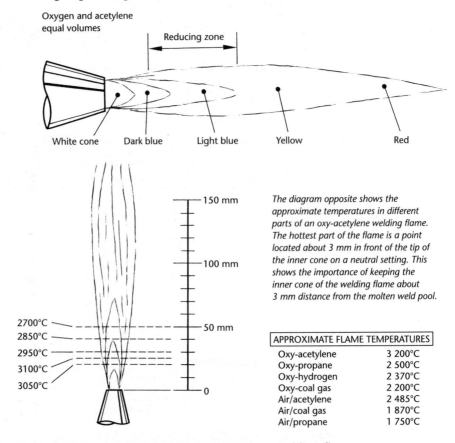

Figure 11.3 Characteristics of the oxy-acetylene welding flame

Oxygen and acetylene equal volumes

Reducing zone

White cone   Dark blue   Light blue   Yellow   Red

150 mm

100 mm

50 mm

0

2700°C
2850°C
2950°C
3100°C
3050°C

The diagram opposite shows the approximate temperatures in different parts of an oxy-acetylene welding flame. The hottest part of the flame is a point located about 3 mm in front of the tip of the inner cone on a neutral setting. This shows the importance of keeping the inner cone of the welding flame about 3 mm distance from the molten weld pool.

APPROXIMATE FLAME TEMPERATURES	
Oxy-acetylene	3 200°C
Oxy-propane	2 500°C
Oxy-hydrogen	2 370°C
Oxy-coal gas	2 200°C
Air/acetylene	2 485°C
Air/coal gas	1 870°C
Air/propane	1 750°C

mild steel cylinders and 172 bar (17240 kN/m^2) in alloy steel cylinders. The capacity of oxygen cylinders is normally 3.4 m^3, 5 m^3 and 6.8 m^3. The oxygen volume remaining in a cylinder is directly proportional to its pressure in the cylinder. Therefore, if the pressure of the oxygen in the cylinder falls by 10% during a particular welding operation, then 1/10 of the cylinder content has been used. Note that oxygen cylinders have an outlet socket with a *right-hand* thread.

Acetylene

This is also supplied in a thick-walled solid drawn steel cylinder painted maroon. As can be seen in Fig. 11.5 these are shorter and squatter than the oxygen cylinders to assist in identification. High-pressure acetylene is dangerously unstable and, for this reason, it is dissolved in acetone which is capable of absorbing a large volume of gas and releasing it as the pressure in the cylinder falls. One volume of acetone can dissolve 25 volumes of acetylene at 15°C. The acetylene is stored in the cylinder at a pressure of 15.5 bar (1152 kN/m^2). Compressed acetylene is susceptible to dangerous explosions and for this reason the cylinder is filled with an inert porous

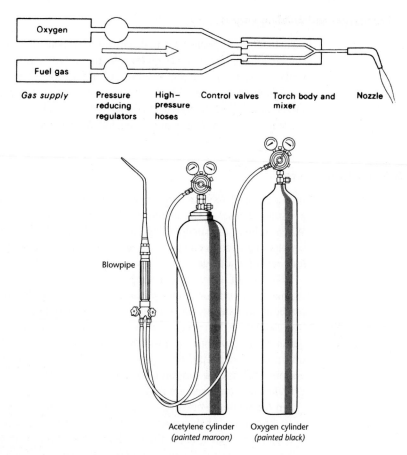

| Gas supply | Pressure reducing regulators | High-pressure hoses | Control valves | Torch body and mixer | Nozzle |

Oxygen

Fuel gas

Blowpipe

Acetylene cylinder
(painted maroon)

Oxygen cylinder
(painted black)

Figure 11.4 Basic welding equipment

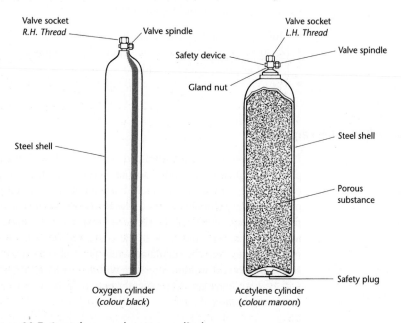

Valve socket
R.H. Thread

Valve spindle

Valve socket
L.H. Thread

Safety device

Valve spindle

Gland nut

Steel shell

Steel shell

Porous substance

Safety plug

Oxygen cylinder
(colour black)

Acetylene cylinder
(colour maroon)

Figure 11.5 Acetylene and oxygen cylinders

substance that can absorb the dissolved acetylene. This is shown in Fig. 11.5. The storage of the dissolved acetylene in the small pores in the cylinder filling prevents the sudden and catastrophic decomposition of the acetylene throughout the mass should it be started by local heating or some other cause. Note that as a further safety precaution, acetylene cylinder outlet nozzles have a *left-hand* thread. This prevents the equipment being connected up incorrectly when changing cylinders.

11.2.1 Discharge rate

In the event of oxygen being drawn off from its cylinder at too great a rate there is a distinct danger of the cylinder valve freezing due to the sudden drop in pressure. When welding or flame-cutting heavy sections, sufficient oxygen cylinders should be coupled together to complete the work without any one cylinder suffering an unduly high rate of discharge.

The rate of acetylene consumption must also be kept below the safe limit; that is, the maximum rate of discharge should never exceed 20% of the total cylinder volume per hour. In other words, the rate of discharge should not empty the cylinder in less than five hours. The reason for this limit on the rate of discharge of an acetylene cylinder is not to prevent freezing as in the case of oxygen, but to prevent acetone being drawn off mixed with the acetylene. Acetylene mixed with even small quantities of acetone will have a substantially lower flame temperature. The volume of acetylene cylinders in normal use are $2.8\,\mathrm{m}^3$ and $5.7\,\mathrm{m}^3$.

In fabrication workshops where welding gases are needed in several places and/ or at high rates of consumption, it is normal to use a manifold system as shown in Fig. 11.6. This enables the cylinders to be housed in a custom-built store with adequate ventilation for safety together with easy access and handling facilities. The gases are then distributed to the various work stations by means of a pipe line. The advantages of a manifold system are:

- More space and less clutter in the workplace.
- No spare cylinders have to be kept in the workplace.
- Less transportation of cylinders.
- The cylinders are easily reached and removed in case of fire.

11.2.2 Pressure regulators

In order to reduce the pressure and control the flow of the gases to the welding torch, pressure regulators are fitted to the oxygen and the acetylene cylinders. Examples of these are shown in Fig. 11.7. Each regulator is fitted with two pressure gauges: one pressure gauge indicates the pressure of the gas in the cylinder and the other pressure gauge indicates the reduced pressure at the outlet being fed to the welding torch. The pressure regulator screw is slacked off when not in use. When the equipment is to be used the regulator screw is tightened until the welding pressure gauge shows the correct welding pressure with the welding torch valves open. This pressure will be kept constant automatically providing it is less than the pressure of the gas remaining in the cylinder.

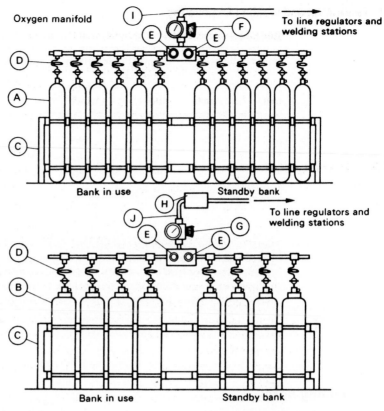

Oxygen manifold

To line regulators and welding stations

Bank in use   Standby bank

To line regulators and welding stations

Bank in use   Standby bank

A Oxygen cylinders
B Acetylene cylinders
C Storage rack
D High-pressure coupling pipe ('pig-tail')
E Separate valve for each bank

F Oxygen output regulator and pressure gauge (Line pressure 4.15 bar)
G Acetylene output regulator and (Line pressure 620 millibar)
H Anti-flashback device (acetylene)
I Oxygen supply line (copper pipe)
J Acetylene supply line (steel pipe)

Figure 11.6 The manifold system for gas welding

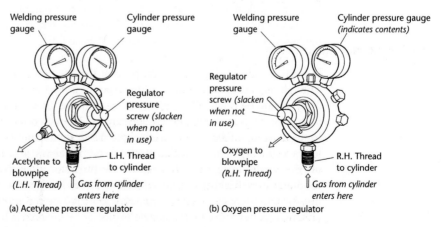

Welding pressure gauge   Cylinder pressure gauge

Regulator pressure screw (slacken when not in use)

Acetylene to blowpipe (L.H. Thread)   L.H. Thread to cylinder   Gas from cylinder enters here

(a) Acetylene pressure regulator

Welding pressure gauge   Cylinder pressure gauge (indicates contents)

Regulator pressure screw (slacken when not in use)

Oxygen to blowpipe (R.H. Thread)   R.H. Thread to cylinder   Gas from cylinder enters here

(b) Oxygen pressure regulator

Figure 11.7 Pressure regulators

Table 11.1 Sizes of welding hose in general use

Internal diameter		Approximate outside diameter		Used for oxygen and acetylene supplies with:
mm	inch	mm	inch	
4.8	3/16	13.5	17/32	Light-duty high-pressure torches
6.3	1/4	15.1	19/32	Light-duty high-pressure torches
8.0	5/16	16.7	21/32	Heavy-duty high-pressure torches
9.5	3/8	18.3	23/32	Large cutting torches and where *low-pressure generated acetylene* is used

Note: to prevent dirt entering the regulator and preventing it from working properly the gas cylinder valves should be 'cracked' before fitting a regulator to a cylinder. That is, the cylinder valve is opened momentarily by a small amount to blow any dust and dirt from the valve socket before installing the regulator (see Section 1.17.4, Fig. 1.41). Remember that all *oxygen equipment* has *right-hand threads* and the all *acetylene equipment* has *left-hand threads* to prevent incorrect connection which could lead to dangerous accidents.

11.2.3 Welding hoses

The welding gases are fed from the regulator to the welding torch by means of flexible hoses. These are specially manufactured for this purpose and no other sort of hose should be used. Welding hose has a seamless rubber or rubber compound lining that is chemically resistant to the normal welding gases. It is reinforced with a layer of canvas or wrapped cotton to resist the pressure of the gases in the hose. Finally there is a tough rubber outer casing that is abrasion-resistant. Welding hoses are coloured *black for oxygen* and *red for acetylene* or any other flammable gas that may be used. Table 11.1 lists the sizes of welding hose in normal use. Care must be taken not to damage welding hoses by allowing trolleys to run over them where they cross the workshop floor, not to damage them by dropping heavy weights on them and not to damage them by allowing the welding flame to come into contact with the hoses. Figure 11.8 shows a selection of hose fittings for connecting to the regulator at one end and the welding torch at the other. *Hose protectors* were dealt with in Section 1.17.4, Fig. 1.40: these prevent the gas flow being reversed in the event of the hoses being reversed or in the event of a *blow-back* from the welding torch. They should be fitted at the *torch end* of the hose. In addition, a *flashback arrester* should be fitted at the *outlet of each regulator* between the regulator and the hose. In the event of a flashback (backfire) the flame is blown back up one or other of the hoses with sufficient force that it closes the flashback arrester instantly and shuts off the supply from the cylinder. A hose protector alone is not sufficient to stop the force of a flashback from bursting a hose. Flashbacks are usually caused by insufficient cylinder pressure (contents too low) or the nozzle becoming blocked either by spatter reducing the nozzle orifice or by the nozzle being brought too close to the work.

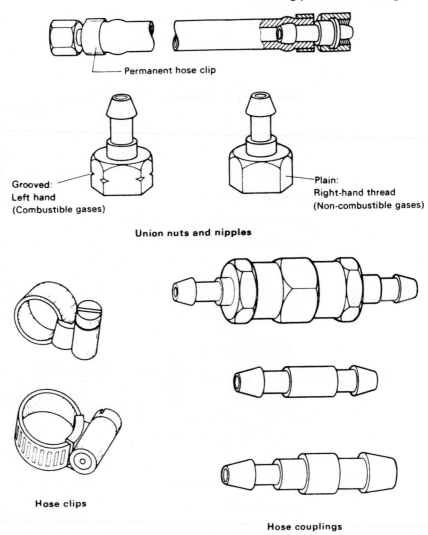

Permanent hose clip

Grooved:
Left hand
(Combustible gases)

Plain:
Right-hand thread
(Non-combustible gases)

Union nuts and nipples

Hose clips

Hose couplings

Figure 11.8 Welding hose fittings

11.2.4 Welding torches

Welding torches are specifically designed to mix and control the flow of the welding gases to the welding nozzle (tip). The torch also provides a means of holding and directing the welding nozzle. The essential features of a welding torch are shown in Fig. 11.9(a) and an enlarged section through the mixing chamber is shown in Fig. 11.9(b). The hose connections on the torch body are handed left or right so that only the correct hose can be attached and there can be no chance of incorrect connection. There are two control valves so that the operator can set the flame to suit the job in hand or turn the supply off completely. The colour coding of the cylinders and hoses is applied to the valve knobs for continuity and to avoid confusion by the operator. The oxygen valve is coloured blue or black and the acetylene valve is coloured red.

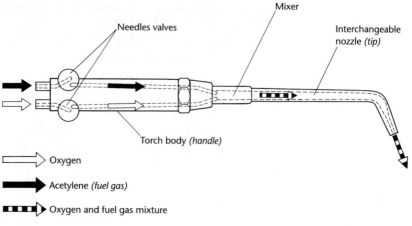

(a) High-pressure welding torch

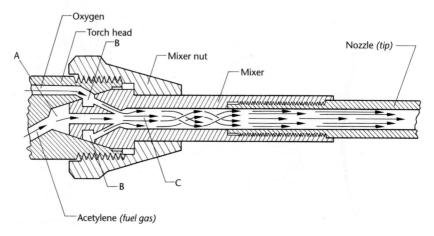

(b) Mixing chamber for a welding torch

Figure 11.9 Oxy-acetylene welding torch

After passing through the control valves the gases flow through metal tubes in the welding torch handle and are brought together by the *gas mixer* in the mixing chamber. The gas mixture then passes through the neck pipe until it emerges at the *nozzle* or 'tip'. The two gases enter at the points marked 'A' (acetylene) and 'B' (oxygen). The mixing of the gases commences at the point 'C' and continues throughout the chamber and through the *neck of the torch* before emerging from the *welding nozzle* where they are ignited to give the welding flame. The gas mixer should perform the following important functions:

- Mix the gases thoroughly for complete and proper combustion.
- Offer a first line of defence against flashbacks by stopping the flame travelling further back than the mixer and before the hose protector and flashback arrester are brought into action.
- Permit a wide range of nozzle sizes to be used with one size of mixer.

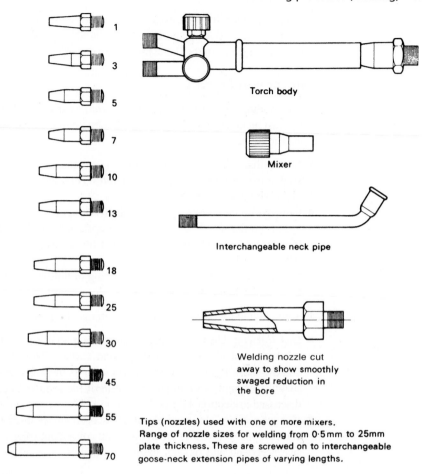

Torch body

Mixer

Interchangeable neck pipe

Welding nozzle cut away to show smoothly swaged reduction in the bore

Tips (nozzles) used with one or more mixers. Range of nozzle sizes for welding from 0·5 mm to 25 mm plate thickness. These are screwed on to interchangeable goose-neck extension pipes of varying lengths.

Figure 11.10 Welding torch and nozzle/mixer combinations

11.2.5 Welding nozzles

A welding nozzle (tip) is situated at the end of the welding torch neck where the gases emerge prior to ignition and combustion. A well-designed nozzle enables the operator to guide the flame and direct it with ease and efficiency. Nozzles are made from copper or copper alloy since these metals have a high thermal conductivity and this property reduces the possibility of the nozzle overheating and becoming damaged by the heat radiating from the weld zone. Welding torch nozzles are available in a wide variety of sizes, shapes and construction. However, there are essentially two main classes of nozzle and mixer combinations in normal use:

- A separate nozzle and mixer unit where each size of nozzle has a specific mixer which provides efficient combustion conditions and optimum heating efficiency.
- One or more mixers used for the entire range of nozzle sizes. The neck and nozzle combination is screwed into the appropriate mixer. Each size of mixer has a particular screw thread that will only accept the range of neck and nozzle combinations for which it is designed. This system is shown in Fig. 11.10.

Heat radiation from the weld zone is very great, particularly when using the larger sizes of nozzle on thick plate. Therefore, it is advisable to have the torch body at a comfortable distance from the flame by selecting a neck pipe of adequate length for the job.

Care and use of nozzles

Since welding torch nozzles are made from relatively soft copper or copper alloys, care must be taken to prevent them from becoming damaged. The following precautions should be observed:

- Ensure that the nozzle seat and threads are absolutely clean and free from damage so that they do not become scored and form a gas-tight seal when tightened onto the torch/neck assembly.
- Only use tip cleaners which are specially designed for the purpose to remove blockages from the nozzle orifice. Use a proprietary nozzle-cleaning compound diluted in water (usually 50 g of compound to 1 litre of water) when cleaning dirty nozzles. Allow to soak for at least two hours.
- Nozzles should never be used for moving or holding work and should be kept in a fitted case when not in use, never in a drawer with other tools.

Before moving on there are two further items of equipment to discuss, namely:

- *Gas lighters.* These are spark lighters employing a flint which presses against a hardened steel wheel with a knurled surface. The acetylene is turned on first and the spark lighter is operated. As soon as the acetylene is ignited and burning steadily, oxygen is gradually added until the required flame type is achieved, as discussed in Section 11.1.5.
- *Gas economizer.* An example of a gas economizer is shown in Fig. 11.11. As its name implies it provides an efficient and convenient means of conserving the expensive acetylene and oxygen gases. The unit is fastened to the welding bench in a

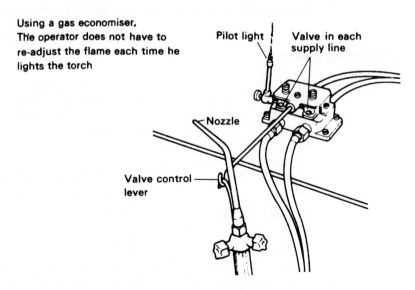

Using a gas economiser.
The operator does not have to re-adjust the flame each time he lights the torch

Pilot light

Valve in each supply line

Nozzle

Valve control lever

Figure 11.11 Gas economiser unit

convenient position. The two hoses from the regulators are connected to the economizer input and two more hoses connect the torch to the economizer outlet. When welding has to be interrupted the welding torch is hung on the economizer as shown and its weight closes the gas valves and turns off the supply to the torch. When welding is to be resumed the torch is lifted off the economizer, the gas supply is restored and the torch is re-lighted at the small pilot flame on the economizer unit.

11.2.6 The oxy-acetylene flame

The welding flame is produced by burning approximately equal volumes of acetylene and oxygen mixture at the nozzle of the welding torch. Figure 11.12 shows a typical welding flame.

This flame has the following zones:

1. The innermost cone of mixed but unburnt gases leaving the nozzle. It appears intensely white and clearly defined if the correct neutral setting of the gases has been set by the operator by adjustment of the torch gas valves. Any small excess of acetylene is indicated by a white 'feather' around the edge of this cone. Excess oxygen is indicated by the cone becoming shorter, more pointed and tinged with blue. Also the gases burn with a 'roaring' sound rather that the normal, gentle 'hiss'.
2. The very narrow stationary zone where the chemical reaction of the first stage of burning takes place, producing a sudden and substantial rise in temperature.
3. The *primary zone* is where the primary products of combustion are concentrated and appears dark to light blue. These products determine the chemical nature of the flame and may be *carburizing* if there is an excess of acetylene present, *oxidizing* if there is an excess of oxygen present and *neutral* if the oxygen and acetylene gases are correctly balanced.
4. Within the primary zone and approximately 3 mm from the tip of the clearly defined cone of unburnt gases there is a region of maximum temperature. This is the zone used for welding.
5. The outer envelope or 'plume' of the flame varies in colour from yellow to pale pink and lies beyond the previous zones. It is in this part of the flame that secondary combustion reactions take place to complete the combustion process. The final products of combustion are carbon dioxide and water vapour with the nitrogen from the surrounding air.

For oxy-acetylene welding it is essential to maintain a neutral flame. Except in special circumstances a reducing flame or an oxidizing flame will result in welds having unsatisfactory mechanical properties.

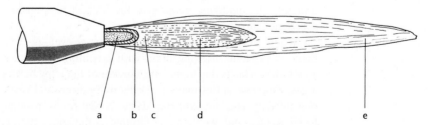

Figure 11.12 The structure of the oxy-acetylene welding flame

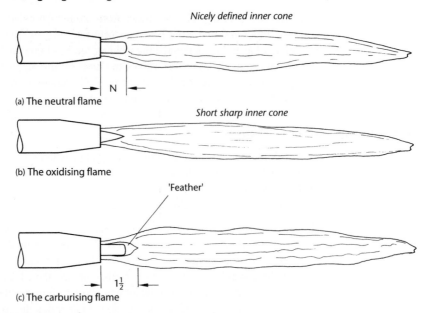

Nicely defined inner cone

(a) The neutral flame

Short sharp inner cone

(b) The oxidising flame

'Feather'

(c) The carburising flame

Figure 11.13 Oxy-acetylene welding flame conditions

The neutral flame

For most welding applications the *neutral flame* condition is used as shown in
Fig. 11.13(a). As has been previously stated this is produced when approximately
equal volumes of oxygen and acetylene are mixed in the welding torch. It is called
a neutral flame because it does not cause any chemical change in the metals being
joined. It neither oxidizes nor carburizes the parent and filler metals. It is easily recognized by its characteristic clearly defined white inner cone at the tip of the welding
torch nozzle. The neutral flame is most commonly used when welding low carbon and
stainless steels, cast iron, and copper.

The oxidizing flame

Figure 11.13(b) shows an *oxidizing flame*. The flame is first set to the neutral condition and the acetylene supply is slightly *reduced* by the welding torch control valve.
The oxidizing flame is recognized by the shorter and sharply pointed inner cone and
by a noisier (roaring) combustion process. The oxidizing flame is undesirable in
most cases as it oxidizes the molten metal in the weld pool, causing heavy scaling.
However, an oxidizing flame can be of advantage when welding some brasses and
bronzes. The oxidizing flame give the highest temperature providing the oxygen:
acetylene ratio doe not exceed 1.5:1.

When welding brasses, brazing, or braze-welding other metals an oxidizing
flame is essential. Melting brass with a neutral flame causes violent evolution of
zinc fumes. This is due to the zinc content of the alloy boiling and being given off as
a gas. Zinc lost in this manner results in a porous and weak welded joint. It can be
prevented by using an oxidizing flame. A trial run is made on an off-cut of the metal
to be welded and, as soon as 'fuming' commences the acetylene valve is turned
down until the brass melts without 'fuming' occurring, indicating that the correct
flame condition has been achieved. This is shown in Fig. 11.14.

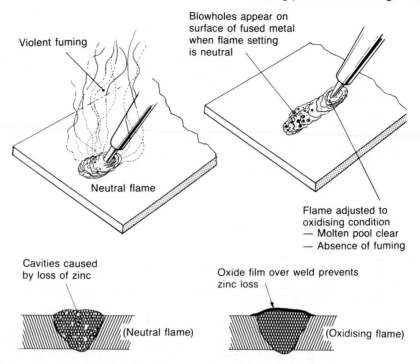

Note: Brazing filler rods usually contain about 0·25% SILICON,
which, during the brazing operation, produces a 'silica film'
on the surface of the liquid metal which prevents 'fuming',
and thus produces a sounder deposit.

Figure 11.14 The importance of an oxidising flame – brass welding

Reducing flame

When a *reducing flame* is required as shown in Fig. 11.13(c), the flame is first set
to the neutral condition and the acetylene supply is slightly *increased* by the control
valve on the welding torch. This flame is recognized by the 'feather' of incandescent
carbon particles between the inner cone and the outer envelope. A reducing flame is
used for carburizing (surface-hardening) ferrous metals and when 'flame brazing'
aluminium and its alloys.

11.2.7 Nozzle size

Welding torch nozzles are available in sizes ranging from 1 to 90. The size indicates the average amount of gas passing through the nozzle in cubic feet per hour.
Advances in gas shielded electric arc welding have overshadowed oxy-acetylene
welding. All forms of arc welding (see Section 11.2) are more cost-effective than
oxy-acetylene welding but, nevertheless, oxy-acetylene welding is still used extensively for repair and reclamation work or where the portability of oxy-acetylene
equipment to sites where a source of electricity is not available.

Table 11.2 relates the parent metal thickness to the nozzle size, the filler rod
diameter required and the rate of gas consumption for the *downhand* butt welding

Table 11.2 Downhand butt welds in steel

Thickness of metal	Diameter of filler rod (mm)	Diameter of filler rod (inch)	Joint edge preparation	Plate thickness	Nozzle size	Approximate consumption of each gas — Litres per hour	Approximate consumption of each gas — Cubic feet per hour	Welding technique
Less than 20 swg	1.2 to 1.6 for square each preparation	3/64 to 1/16		0.8	1	28	1	Leftward welding technique
			No filler rod	1.2	2	57	2	
				1.6	3	86	3	
20 swg to 3.2 mm (1/8 in)	1.6 to 3.2	1/16 to 1/8	0·8—3·2 mm Gap	2.4	5	140	5	
				3.2	7	200	7	
3.2 mm to 5.0 mm (1/8 to 3/16 in)	3.2 to 4.0	1/8 to 5/32	80° 1·6—3·2 mm Gap	4.0	10	280	10	
				5.0	13	370	13	
5.0 mm to 8.2 mm (3/16 to 5/16 in)	3.2 to 4.0	1/8 to 5/32	3·2—4·0 mm Gap	6.5	18	520	18	Rightward welding technique
				8.2	25	710	25	
8.2 mm to 15.2 mm (5/16 to 5/8 in)	4.0 to 6.5	5/32 to 1/4	60° 3·2—4·0 mm Gap	10.0	35	1000	35	
				13.0	45	1300	45	
				16.2	55	1600	55	
16.2 mm and over	6.5	1/4	60° 80° 3·2—4·0 mm Gap	19.0	60	1700	60	
				25.0	70	2000	70	
				Over 25.0	90	2500	90	

of steel. The sizes given in the table are for guidance only. Use the correct size of nozzle for the job in hand. Too large a nozzle will result in overheating the metal whilst too small a nozzle will result in lack of penetration and a weak joint. Also it will cause the *heat-affected zone* to travel back into the parent metal because of the length of time it takes to raise the joint edges to the welding temperature. This results in grain growth and weakness in the parent metal.

11.2.8 Leftward welding

When *leftward welding*, the flame is directed away from the finished weld and towards the unwelded part of the joint. The filler rod, when required, is directed towards the welded part of the joint. This technique is used on low-carbon steels when making:

- Flanged edge welds when the parent metal is less than 1.2 mm (20 swg) thick. These welds do not require a filler rod.
- Square-butt welds on sheet steel with unbevelled edges up to 3.2 mm (10 swg) thick.
- Vee-butt welds on steel plate with bevelled edges between 3.2 mm and 5 mm thick. The leftward method of welding is considered to be uneconomical when making joints in metal over 5 mm thick.

The techniques used in leftward welding are shown in Fig. 11.15.

Flanged-edge welding

This technique is shown in Fig. 11.15(a). No filler rod is required. The welding flame is manipulated with a steady semi-circular sideways movement and progressively forward only as fast as the edges of the parent metal are melted. The tip of the flame cone must be kept at 3 mm from the weld pool of molten metal.

Square-butt weld

This technique is shown in Fig. 11.15(b). As welding proceeds and the filler rod is melted, it should be fed forward to build up the molten pool of weld metal. The welding flame is guided progressively forwards with small sideways movements as fast as the sheet or plate edges are melted. A pear-shaped melted area (often referred to as an 'onion') should be maintained ahead of the weld pool. Care must be taken to ensure uniform melting of both the joint edges.

Vee-butt weld

This technique is shown in Fig. 11.15(c). Care must be taken to ensure that the tip of the welding flame cone never touches the weld metal or the filler rod. As the filler rod is melted it should be fed forward into the molten weld pool in order to build it up. The rod is then retracted slightly to enable the heat to fuse the bottom edges of the 'vee' in order to ensure *full penetration of the weld* to the bottom of the plate edges.

The side-to-side movement of the welding torch should only be sufficient to melt the sides of the 'vee'. The weld deposit is built up as the welding torch and the filler rod move progressively forward, filling the 'vee' to a level slightly higher than the edges of the parent metal. Table 11.3 summarizes the data for leftward welding.

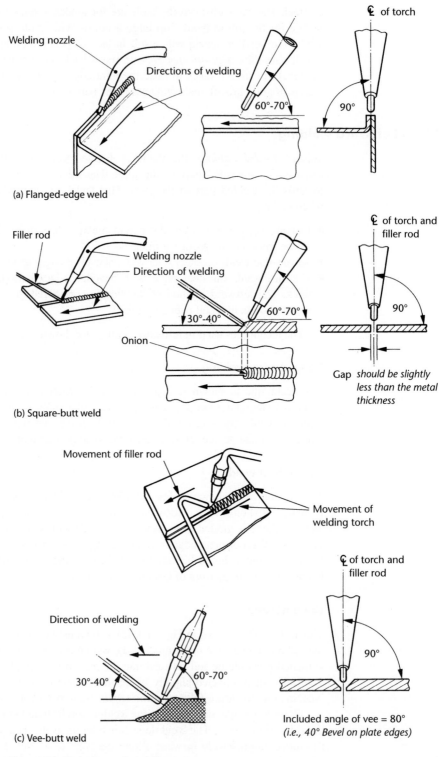

(a) Flanged-edge weld

(b) Square-butt weld

(c) Vee-butt weld

Figure 11.15 Leftward welding

Table 11.3 Welding speeds and data for leftward welding

Thickness of metal		Edge preparation	Gap		Diameter of filler rod		Power of torch (gas consumption)		Rate of welding		Filler rod used	
mm	inch		mm	inch	mm	inch	litres/hr	cubic ft/hr	metres/hr	feet/hr	per metre (m)	per foot (ft)
0.8	1/32	Square	–	–	1.6	1/16	28–57	1–2	6.0–7.6	20–25	0.3	1
1.8	1/16	Square	1.6	1/16	1.6	1/16	57–86	2–3	7.6–9.0	25–30	0.53	1.75
2.4	3/32	Square	2.4	3/32	1.6	1/16	86–140	3–5	6.0–7.6	20–25	0.84	2.75
3.2	1/8	Square	3.2	1/8	2.4	3/32	140–200	5–7	5.4–6.0	18–20	0.50	1.65
4.0	5/32	80° Vee	3.2	1/8	3.2	1/8	200–280	7–10	4.6–5.4	15–18	0.64	2.1
5.0	3/16	80° Vee	3.2	1/8	4.0	5/32	280–370	10–13	3.6–4.6	12–15	1.40	4.8

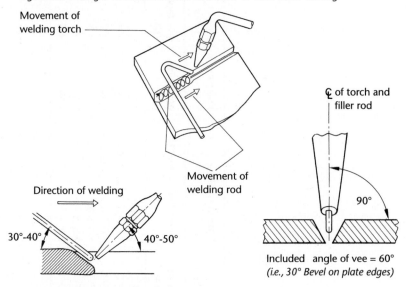

Rightward welding is sometimes termed 'backward' or 'back-hand' welding

Movement of welding torch

Ç of torch and filler rod

Direction of welding

Movement of welding rod

30°-40°   40°-50°

90°

Included   angle of vee = 60°
(i.e., 30° Bevel on plate edges)

Figure 11.16 Rightward welding

11.2.9 Rightward welding

As its name implies, this technique is the reverse of the one just described. For a right-handed operator the weld commences at the left-hand end of the joint and moves towards the right-hand end. The torch flame, in rightward welding, is directed *towards the metal being deposited*, unlike the leftward method of welding where the flame is directed away from the metal being deposited. Rightward welding is used as follows for steel plates which exceed 8 mm in thickness:

- Square-butt welds on unbevelled steel plates between 5 mm and 8 mm in thickness (inclusive).
- Vee-butt welds on bevelled steel plates over 8 mm in thickness. The angles of the welding torch and the filler rod are shown in Fig. 11.16.

The welding torch is moved progressively to the right along the joint, with the welding flame cone being kept just clear of the filler rod and the deposited weld metal. By comparison with the leftward method of welding (Fig. 11.15(c)) it can be seen that the flame is deeper in the 'vee'. The filler rod is given an elliptical looping movement as it travels progressively to the right.

The welding torch and the filler rod must be kept in the same vertical plane as the centre line of the weld otherwise unequal fusion of the two sides of the weld will occur. When the weld is completed, examination of the back of the joint should show an '*under bead*' that is perfectly straight and even. It is the appearance of this 'under bead' that indicates the quality of the weld. Table 11.4 summarizes the data for rightward welding.

The advantages of rightward welding may be summarized as follows:

- No bevel is required for plates up to 8 mm in thickness and this saves in the cost of preparation and reduces the amount of filler material required.

Table 11.4 Welding speeds and data for rightward welding

Thickness of metal		Edge preparation	Gap		Diameter of filler rod		Power of torch (gas consumption)		Rate of welding		Filler rod used	
mm	inch		mm	inch	mm	inch	litres/hr	cubic ft/hr	metres/hr	feet/hr	per metre (m)	per foot (ft)
5.0	3/16	Square	2.4	3/32	2.4	3/32	370–520	13–18	3.6–4.6	12–15	1.03	3.4
6.5	1/4	Square	3.2	1/8	3.2	1/8	520–570	18–20	3.0–3.6	10–12	1.03	3.4
8.2	5/16	Square	4.0	5/32	4.0	5/32	710–860	25–30	2.1–2.4	7–8	1.03	3.4
10.0	3/8	60° Vee	3.2	1/8	5.0	3/16	1000–1300	35–45	1.8–2.1	6–7	1.22	4.6
13.0	1/2	60° Vee	3.2	1/8	6.5	1/4	1300–1400	45–50	1.3–1.5	4.5–5	1.69	4.75
16.2	5/8	60° Vee	3.2	1/8	6.5	1/4	1600–1700	55–60	1.1–1.3	3.75–4.25	2.05	6.75
19.0	3/4	60° Vee (Top) 80° Vee (Bottom)	4.0	5/32	6.5	1/4	1700–2000	60–70	0.9–1.0	3–3.25	2.96	9.75
25.0	1	60° Vee (Top) 80° Vee (Bottom)	4.0	5/32	6.5	1/4	2000–2500	70–90	0.6–0.7	2–2.25	5.08	16.75

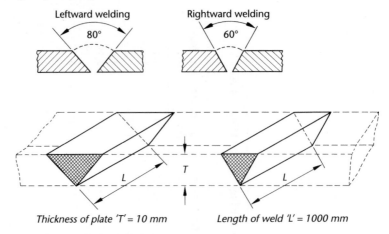

Thickness of plate 'T' = 10 mm   *Length of weld 'L' = 1000 mm*

Figure 11.17 Amounts of deposited metal

- When bevelling of the joint edges becomes necessary, the included angle of the 'vee' needs only to be 60° again, reducing the amount of filler material required compared with 80° 'vee' needed for leftward welding as shown in Fig. 11.17.
- Larger welding nozzles can be used when rightward welding, resulting in higher welding speeds. With the use of more powerful nozzles, the force of the flame 'holds back' the molten metal in the weld pool and allows more metal to be deposited so that welds can be made in plates of up to 10 mm in thickness at one pass.
- The operator's view of the weld pool and the sides and bottom of the 'vee' is unobstructed. This enables the operator to provide accurate control the molten metal and ensures that full fusion of the joint (parent metal) edges, particularly the bottom edges, is always maintained, thus ensuring an adequate and continuous bead of penetration.
- The quality and appearance of the weld are better than that obtained when leftward welding. This is due to the deposited metal being protected by the envelope of the flame which retards the rate of cooling.
- Compared with leftward welding, the smaller volume of deposited metal when using the rightward technique reduces shrinkage and distortion. The heat of the welding flame, when rightward welding, is localized in the joint and is not allowed to spread across the parent metal either side of the joint. By contrast, the welding torch movement when leftward welding causes a greater spread of heat, causing grain growth and weakness around the joint. Further, the amount of distortion depends upon the amount of heat put in; therefore, it is important that the heat of welding is confined as far as possible to the weld seam itself.
- The cost of welding is lower when using the rightward technique despite the use of larger and more powerful welding nozzles. This due to the smaller amount of filler material required and the greater speed with which the weld metal is deposited.

11.2.10 Further welding techniques

So far only butt welds in the downhand position have been considered. Figure 11.18 shows a selection of other common joints that may also be made in the downhand position using either the leftward or the rightward technique depending upon the metal thickness. The illustrations in the figure are all shown using the leftward technique.

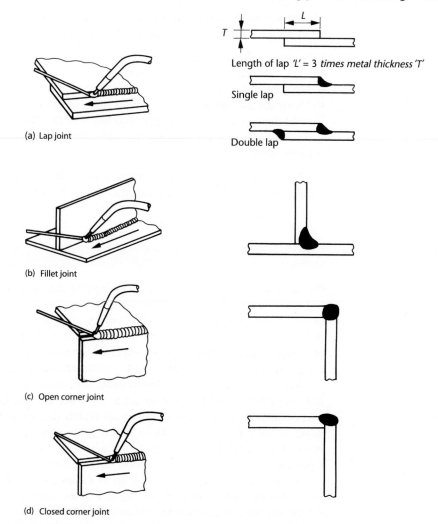

(a) Lap joint

Length of lap 'L' = 3 *times metal thickness* 'T'

Single lap

Double lap

(b) Fillet joint

(c) Open corner joint

(d) Closed corner joint

Note: All the joints in the above examples are being produced by the leftward technique. The rightward technique can also be used for all these joints.

Figure 11.18 Types of joint

11.2.11 Oxides in welding

The removal and dispersion of the oxide film which forms when metals are heated is one of the traditional problems associated with welding. Difficulties in welding occur when:

- The surface oxide forms a tenacious film.
- The oxide film has a melting point very much higher than that of the parent metal.
- The oxide film forms very rapidly.

All these factors are present when welding or brazing aluminium but it is the tenacity of the oxide film that is the main source of trouble, whatever the metal being welded. One of the most important factors in the weld quality is the removal of oxides from the surface of the parent metals being joined. Unless the oxides are removed, the following undesirable conditions may result:

- Fusion may be difficult.
- Inclusions may be present in the weld.
- The joint will be weakened.
- The oxides will not flow from the weld zone but remain to become entrapped in the metal during solidification so as to interfere with the addition of filler material. This condition generally occurs when the oxide film has a higher melting point than the parent metal.

Readers wishing to obtain further information about oxy-fuel gas welding should consult Trade File 5.7 published by the British Oxygen Company.

11.3 Manual metal-arc welding

Providing access is available to a mains supply of electricity, manual metal-arc welding is very much cheaper than oxy-acetylene welding. It also produces higher temperatures and greater amounts of heat energy, enabling much thicker sections to be welded successfully. For site work, welding generators (see Section 1.19, Fig. 1.50) are required which, although more expensive in initial cost, are sill cheaper to run, especially where heavy sections such as girders have to be welded.

The 'arc' is produced by a low-voltage, high-amperage current jumping the air gap between the electrode and the joint to be welded. The heat of the electric arc is concentrated on the edges of the parent metals that are to be joined, causing the edges of the parent metal to melt. Whilst these edges are still molten, additional metal in the molten state is transferred across the arc from a suitable electrode. Upon cooling, this molten mass of weld metal solidifies to produce a strong joint.

As soon as the arc is struck the tip of the electrode commences to melt, thus increasing the gap between the electrode and the work. Therefore, it is necessary for the operator to advance the electrode continuously towards the joint in order to maintain a constant arc gap (length) of approximately 3 mm during the welding operation. The electrode is also moved along the joint to be welded with a uniform velocity. This requires considerable skill that is only acquired with practice.

11.3.1 The electrode

Most electrodes used in the manual arc-welding process are *coated* electrodes; that is, the coated electrode consists of a wire core of suitable composition surrounded by a concentric covering of a flux and/or other materials which will melt uniformly along with the core wire to form a partly vaporized and partly molten shield around the arc stream. This shield protects the arc from contamination by atmospheric gases. The liquid slag that is produced performs three important functions:

1. It protects the solidifying weld metal from atmospheric contamination.
2. It prevents the weld metal cooling too rapidly.
3. It controls the contour of the completed weld.

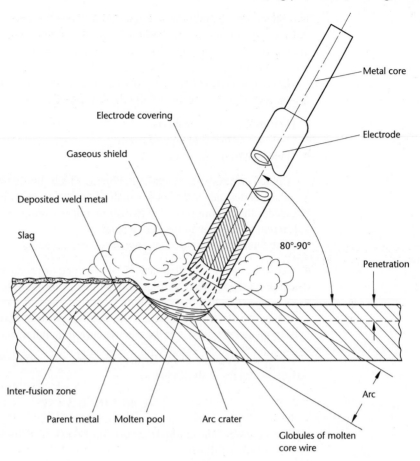

Figure 11.19 Basic features of arc-welding

The function of an electrode is more than acting simply as a conductor for the electric-arc current – it also acts as a filler rod. The core wire melts in the arc and tiny globules of molten metal transferred explosively across the arc into the molten weld pool (the arc crater in the parent metal) during welding. These tiny globules of molten metal are not transferred by the force of gravity (otherwise overhead welding would not be possible) but are explosively transferred across the arc to form the *arc stream*.

The chemical coating surrounding the core wire melts or burns in the heat of the arc. Since it melts at a slightly higher temperature than the metal core it protrudes a little beyond the core and directs the arc. This extension also prevents sideways arcing when welding in deep grooves. The arc stream and other basic features of manual metal-arc welding are shown in Fig. 11.19.

11.3.2 Electrode coatings

Before the introduction of coated electrodes, bare wire electrodes were used. They were much more difficult to use and, because they have no flux, no protective shield is produced, resulting in an arc that is less powerful and which tends to 'short out' more easily. Because of the unstable arc associated with uncoated electrodes, they are unsuitable for use with alternating-current welding sets but can be used satisfactorily with direct-current welding sets using electrode negative polarity.

Let's now consider the *coated electrode* in greater detail. Coated electrodes are the subject of British Standard 499. It classifies electrodes according to:

- Their mechanical properties.
- The composition and type of coating.
- Their operating conditions.

Coated electrodes are supplied in boxes which should be marked with the British Standard code, indicating their suitability for the job in hand. On no account use electrodes that do not carry the British Standard coding or carry a coding for use on a different type of job to the one in hand.

The Standard is divided into two categories, namely:

- *A compulsory section* that deals with the type of product or process, the mechanical properties of the electrode and the flux covering of the electrode.
- *An optional section* that deals with the more practical aspects of the electrode's use such as the electrode efficiency, the welding position best suited for the electrode, the type and magnitude of the electric current required and any special conditions.

A typical example of the British Standard 499 coding that should appear on the box of electrodes is shown below:

E 46 3 INi B 5 4 H5

Let's consider the compulsory section first (E to B inclusive in the figure). This is interpreted as follows:

- The first letter 'E' indicates the product or process being used.
- The number 46 indicates the strength and elongation properties of the electrode material.
- The number 3 indicates the impact strength of the electrode material.
- The letters 'INi' indicate the chemical composition of the electrode material.
- The letter 'B' indicates the flux covering (this will be discussed later in this section).

Now let's consider the optional section (5 to H5 inclusive):

- The number 5 indicates the amount of weld metal deposited per electrodes and the type of welding current needed.
- The number 4 indicates the welding position.
- Finally 'H5' which indicates the hydrogen content of the weld metal.

Let's now consider the electrode coatings available. The types most commonly used are:

- *Rutile* (code **R**), which would be used to coat an electrode for general-purpose welding. The residual slag is easy to remove. This may be subdivided into **RR** (a thick coating), **RC** (a rutile cellulosic coating), **RA** (a rutile acid coating), and **RB** (a rutile basic coating).

- *Celluosic* (code **C**) is an electrode coating containing cellulose. This produces a forceful arc with deep weld penetration.
- *Basic* (code **B**) is an electrode coating that ensures a very low hydrogen content in the weld metal and is used where a high-strength weld is required.

Care must be taken in the storage of coated electrodes since the coatings tend to be hygroscopic; that is, they tend to absorb moisture from the atmosphere. They must be kept in a dry, well-ventilated store, they must be handled with care to avoid damage to the coating, and new packets should not be opened until they are required. Let's now summarize the advantages of using coated electrodes.

1. The coating facilitates the striking of the electric arc and enables it to burn stably, thus making the electrode suitable for use with an *alternating current*.
2. It serves as an insulator for the core wire and prevents stray arcs occurring.
3. It provides a flux for the molten weld pool from which it removes impurities and forms a protective slag which can be readily chipped off when cooled and solidified.
4. It stabilizes the arc and directs the globules of molten filler metal from the core of the electrode into the weld pool.
5. It provides a protective non-oxidizing (reducing) gas shield around the arc to keep the oxygen and the nitrogen in the air away from the molten metal.
6. It speeds up the rate of melting (metal deposition) and, therefore, increases the speed of welding.
7. Additions to the composition of the coating can be made at the time of manufacture in order to replace any alloying elements in the electrode core, or in the parent metals being joined, that are likely to be lost during the welding process.
8. It improves the penetration of the weld and therefore ensures a sound joint.
9. The coating can be formulated to increase or decrease the fluidity of the protective slag. For example, electrodes designed for overhead welding can be made with a coating that provides a less fluid slag.

11.3.3 Welding with direct current (d.c.)

The equipment used for welding with a direct current (d.c.) operates with an open-circuit potential (voltage) that is much lower than that used for alternating-current arc-welding. With a direct current the electron current *always flows in the same direction*. This causes the positive pole to become hotter than the negative pole. In fact, two thirds of the heat is generated at the positive pole and only one third of the heat is generated at the negative pole. This is very important when welding since it enables the welding operative to vary the heat according to the job in hand as shown in Fig. 11.20. The polarity of the electrode is most important when welding with d.c. and the manufacturer's recommendations must be adhered to rigidly. In order to prevent confusion, electrodes that are only to be connected to the positive side of the supply are termed '*electrode positive*'.

11.3.4 Welding with alternating current (a.c.)

With alternating current (a.c.) the electron current is changing in magnitude and direction 50 times per second in the UK and, therefore, there are no positive and negative sides of the supply in the ordinary sense; therefore the electrodes for use

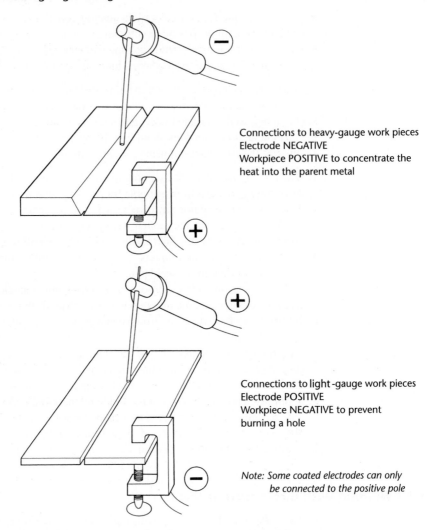

Connections to heavy-gauge work pieces
Electrode NEGATIVE
Workpiece POSITIVE to concentrate the
heat into the parent metal

Connections to light-gauge work pieces
Electrode POSITIVE
Workpiece NEGATIVE to prevent
burning a hole

*Note: Some coated electrodes can only
be connected to the positive pole*

Figure 11.20 Connections for direct-current arc-welding

with alternating current do not have a specific a polarity and the *same amount of heat is produced at the electrode and at the joint.* The arc is extinguished every time the current changes direction but re-establishes itself because the arc atmosphere in the weld zone is heavily ionized. Alternating-current welding equipment was discussed in greater detail in Section 1.18.

11.3.5 Welding current and voltage values

All electrodes are designed for use with a specific value of welding current and, for this reason, the values quoted by the electrode manufacturers should always be strictly adhered to. Any alteration will affect the behaviour of the electrode, the appearance of the weld and possibly the properties of the deposited metal. The average welding currents when joining steel plates are listed in Table 11.5.

Table 11.5 Welding currents (a.c.)

Minimum thickness (mild steel plate)			Welding current value	Diameter of electrode	
Mm	s.w.g.	inch	Amps	mm	inch
1.62	16		40–60	1.6	1/16
2.03	14		60–80	2.4	3/32
2.64	12		100	3.2	1/8
3.18		1/8	125	3.2	1/8
3.25	10		125	3.2	1/8
4.06	8		160	4.8	3/16
4.76		3/16	190	4.8	3/16
4.88	6		190	4.8	3/16
5.89	4		230	6.3	1/4
6.35		1/4	250	6.4	1/4
7.01	2		275–300	7.9	5/16
8.23	0		300–400	7.9	5/16
8.84	0		400–600	9.5	3/8
9.53		3/8	400–600	9.5	3/8

Note: The diameter of the electrode is the size of the core wire.

There are three voltage conditions that the operator should be familiar with when arc-welding.

1. The *open circuit voltage* or electromotive force (emf). This is the maximum voltage available from the welding set under zero current conditions. This is approximately 40 to 100 volts.
2. The *striking voltage*. This is the voltage necessary to strike the arc, and its magnitude will depend upon the welding equipment and the type of electrode being used.
3. The *welding or arc voltage*. This is the electrical potential that produces the welding current once the arc has been struck. It is the lowest of the three voltages (approximately 10 to 50 volts) and is the lowest value that will maintain a stable arc.

11.3.6 Welding techniques

Striking the arc

Before welding can commence it is necessary to *strike the arc*. To do this, the electrode must first touch the work briefly and then be withdrawn to the correct arc distance. There are two methods for striking the arc:

- Use of a scratching motion.
- Use of an up-and-down motion.

At first the electrode may stick to the job or the arc may not be maintained if the electrode is withdrawn too far. Regular practice will enable you to overcome these difficulties.

Electrode angles

There are two angles to consider when welding and these are shown in Fig. 11.21. These angles are known as the *tilt angle* and the *slope angle*. The *tilt angle* is the

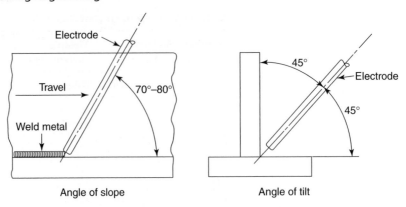

Figure 11.21 Electrode angles

angle between the electrode and the work in a plane perpendicular to the direction of electrode travel along the joint. The *slope angle* is the angle between the electrode and the work in a plane parallel to the joint. Figure 11.21(a) shows how these angles are applied to a 'T' fillet joint.

Figure 11.22 shows how the slope angle affects the quality of the weld. The nearer the electrode slope angle approaches the vertical, the greater will be the depth of penetration since the full force of the arc is directed and concentrated on to the parent metal. In this position the force of the arc stream tends to drive the molten metal from the edge of the weld pool outwards and, in doing so, produces undesirable *undercutting* of the weld profile. Further, with the electrode in the upright position, the slag will build up and surround the weld metal deposited as shown in Fig. 11.22(a). This causes problems as soon as the electrode is moved along the joint line as slag will become entrapped in the weld. Also it is difficult for the operator to see and control the weld pool.

When the electrode is inclined at the correct angle as shown in Fig. 11.22(b), the direction and force of the arc stream causes the slag to form away from the molten weld pool and ahead of the arc, thus eliminating the problem of slag entrapment. Further, the operator can easily see the weld pool and control the deposition of the weld metal and the slag. The inclination of the electrode will vary with the job in hand but is usually about 60°.

If the angle of inclination is too low, as shown in Fig. 11.22(c), there will be lack of penetration, resulting in a weak weld. This will also be accompanied by excessive spatter and a poor 'fish-tail' weld profile.

One factor that influences the choice of electrode angle is the class of electrode being used. These may have coatings that produce a very fluid slag whilst others have coatings that form a highly viscous slag. A very fluid slag is difficult to control and there is the ever-present danger of slag entrapment as the angle of inclination of the electrode approaches the vertical. Between the extremes of flat and upright positions there is always one angle that provides the optimum welding conditions of:

- Adequate penetration.
- Correct weld profile.
- Correct width of weld bead.
- Minimum spatter.
- Adequate slag control.

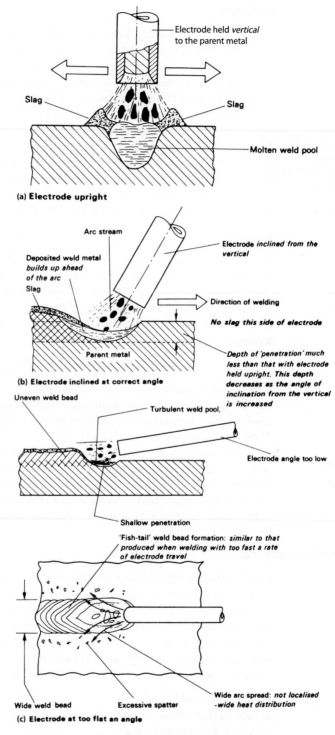

Figure 11.22 Effect of variation in electrode angle

Arc length

The *arc gap* is the distance, measured in millimetres, between the tip of the electrode and the molten weld pool during welding. Considerable skill is required to maintain the correct arc length during welding as the electrode melts away. Only constant practice enables the welder to attain this skill. A weld made with an arc gap of the correct length will enable the electrode metal to be deposited in a stream of small molten particles forming an even bead with a fine rippled appearance and good penetration.

If the *arc length is too great* the arc is noisy (crackles) and the weld metal is deposited in large uneven globules resulting in a poor weld with little penetration and low strength. On the other hand if the arc length is too short it is difficult to maintain an even weld bead and the electrode will keep sticking to the work. Again a weld that is unsatisfactory in appearance and poor in mechanical strength will result, as shown in Fig. 11.26(c).

> Controlling the arc length is very important since it affects the quality of the finished weld

Weaving

Weaving is the side-to-side movement of the electrode during welding and is required to spread the molten metal across the joint width. There are a number of different weave patterns that can be used and the maximum weave width for a given size of electrode is two to three times the electrode diameter. A larger weave pattern would produce a weld of coarse appearance and lack of fusion, leading to poor mechanical properties.

Stopping and re-starting

Stopping requires practice to prevent metal being lost at the end of the weld. This is not only unsightly but could result in the weld breaking as a result of fatigue failure at a later date. To prevent loss of metal the weld crater should be built up at the end of the weld as shown in Fig. 11.23(a).

Re-starting is necessary after stopping to change an electrode. The skill required here is not only in striking the arc in the correct place but also in joining up with the existing weld. Re-strike the arc with the new electrode at the edge of the existing weld crater and move the electrode quickly back over the crater as shown in Fig. 11.23(b). Then resume welding in the correct direction of travel at the normal welding speed.

11.3.7 Welding defects

British Standard BS4872:Part 1:1982 should be consulted for further information on weld quality since faulty welding can lead to catastrophic failure, leading to injury and loss of life if, for example, a bridge collapses or a pressure vessel seam fails. Some of the more important factors affecting weld quality are considered here. Figure 11.24(a) shows the more important terms used to describe the weld profiles. The weld metal and parent metal affected by the welding process may be divided into three distinct zones as shown in Fig. 11.24(b):

1. The actual weld metal zone.
2. The weld penetration zone. This consists of the parent metal that has been fused during the welding process.
3. The heat-affected zone where the parent metal, although not fused, has been affected by the heat generated by the welding process.

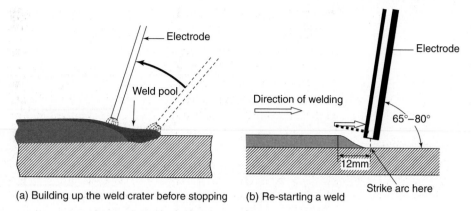

(a) Building up the weld crater before stopping   (b) Re-starting a weld

Figure 11.23 Stopping and restarting a weld

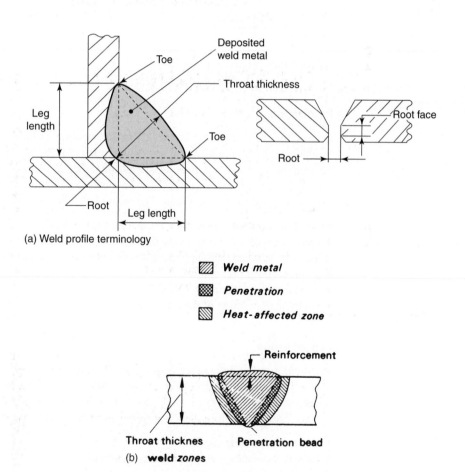

(a) Weld profile terminology

(b) **weld zones**

Figure 11.24 A Weld profile terminology (a); weld terminology (b)

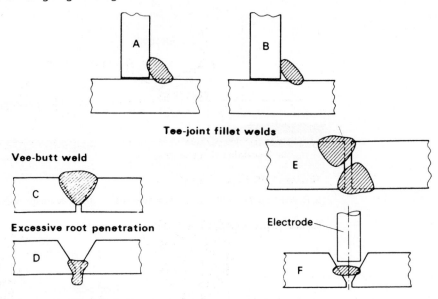

Figure 11.25 Incorrect penetration

Penetration

One of the most common causes of faulty welding is lack of penetration. Some examples of lack of penetration are shown in Fig. 11.25. Where there is lack of penetration the joint will not meet its strength specification and, in all probability, it will fail in service. It is essential for the weld metal to penetrate into the root of the weld in order to produce a strong joint. The most likely causes of lack of penetration are:

- Too low a welding current for the size of the electrode.
- Incorrect polarity if employing direct-current welding.
- The gap between the parent metals is too small.
- Incorrect edge preparation.
- Too large a root face.
- Faulty manipulation of the electrode.

Lack of fusion

The weld metal and the parent metal must be fused together to form a homogeneous whole. The weld toes must blend smoothly with the parent metal and there must be no undercutting. Examples of lack of fusion are shown in Fig. 11.26. The main causes of lack of fusion result from too low a temperature in the weld zone caused by:

- Too low a welding current.
- Incorrect polarity if employing direct current.
- Too fast a welding speed.
- Faulty manipulation of the electrode.
- Failure to clean the parent metal joint edges of oxides, dirt, grease and other foreign matter.

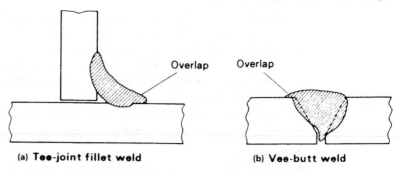

(a) **Tee-joint fillet weld**   (b) **Vee-butt weld**

Figure 11.26 Lack of fusion

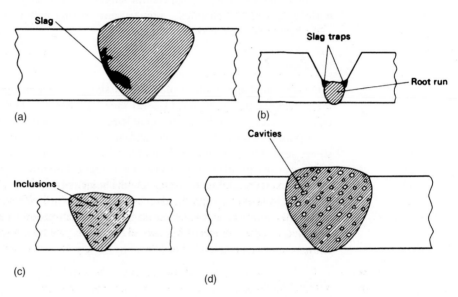

Figure 11.27 Inclusions and porosity

Slag inclusions and porosity

Slag inclusions (or simply 'inclusions') refer to slag or other non-metallic foreign matter trapped in the weld metal and causing a discontinuity leading to joint weakness and failure. The usual source of inclusions is the slag formed by the electrode coating, usually resulting from using an incorrect value of welding current – too high a value or too low a value. Other factors include the incorporation of bad tack-welds, too long an arc length, too high a speed of travel, lack of penetration, or the use of too large a diameter electrode. Dirty parent metal or joint edges still coated with mill-scale from the rolling process can also be a source of inclusions. Where multi-run welds are used for joining thick plates, lack of attention to the removal of the slag from a previous weld run can also lead to slag inclusions. Examples of inclusions are shown in Fig. 11.27(a) and (b), where it can be seen that slag is generally localized and concentrated, whereas foreign matter such as oxides from the edges of the parent metal is generally scattered throughout the weld in fine particles as shown in Fig. 11.27(c).

Porosity consists of small cavities caused by gas trapped in the weld metal. These cavities are commonly referred to as 'blow-holes' when they come out on the surface of the welded joint. Figure 11.27(d) shows a section through a porous weld. The most common cause of porosity is *excessive moisture absorbed in the electrode coating*. It is for this reason that electrodes must be carefully stored in a dry place. Always use dry electrodes and ensure the surfaces being welded are also dry. Porosity may be scattered uniformly throughout the weld, isolated in small patches, or concentrated at the weld root. Some other causes of porosity are:

- Rapid weld freezing – increase heat input.
- Oil, paint, or rust on the joint surfaces of the parent metal – clean these surfaces.
- Incorrect arc length, incorrect current setting, faulty electrode manipulation – use manufacturer's recommended control settings and correct your welding technique.
- Heavy galvanized coatings – remove zinc coating with hydrochloric acid in the vicinity of the weld before commencing.

Weld profile

The size and shape (profile) of the weld should be the same as that specified on the assembly drawing. Figure 11.28(a) shows a theoretical butt-weld profile. It is an equilateral triangle (weld legs of equal length) with adequate penetration.

Figure 11.28(b) shows some undesirable butt-weld profiles. Figure 11.28(c) shows some fillet- or mitre-weld profiles. These may be *concave* or *convex*. A weld with a *concave* profile is weaker than the theoretical fillet weld profile since the throat thickness is reduced despite the leg lengths being correct. If it is not possible to attain a near perfect triangular profile then it is better to have a convex weld than a concave weld. The convexity will not increase or decrease the design strength of the weld, it merely uses up more electrodes and, if excessive, can be unsightly. Figure 11.28(d) shows some undesirable fillet weld profiles.

Undercutting

Undercutting is caused by the burning away of the parent metal edges so as to reduce the joint thickness where the weld bead joins the parent metal (toes of the weld). This results in weakening of the joint and should be avoided if possible. In multi-run welds it can also result in slag entrapment. Examples of undercutting are shown in Fig. 11.29. British Standard 4872: part 1 recommends that any undercut shall not exceed 10% of the parent metal thickness or 1 mm, whichever is the lesser. The causes of undercutting are:

- Excessive welding speed.
- Excessive current for the size of the electrode.
- Incorrect electrode angles.
- Incorrect type of electrode.
- Excessive weaving of the electrode.

Spatter

Excessive spatter does not in itself affect the strength of the joint but, being a result of poor welding technique and incorrect current setting for the electrode used it can

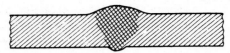

(A) DESIRABLE BUTT-WELD PROFILE

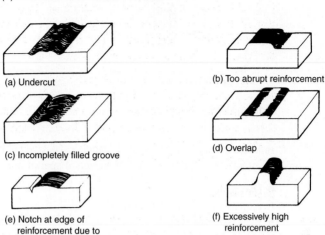

(a) Undercut

(b) Too abrupt reinforcement

(c) Incompletely filled groove

(d) Overlap

(e) Notch at edge of
    reinforcement due to
    incompletely filled groove

(f) Excessively high
    reinforcement

(B) UNDESIRABLE BUTT-WELD PROFILES

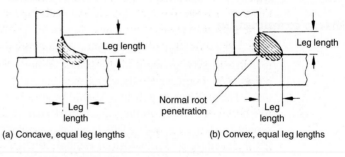

(a) Concave, equal leg lengths

(b) Convex, equal leg lengths

(C) DESIRABLE FILLET-WELD PROFILES

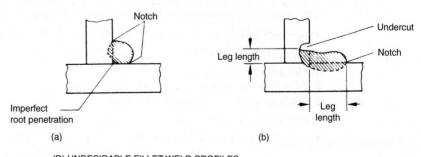

(a)

(b)

(D) UNDESIRABLE FILLET-WELD PROFILES

Figure 11.28 Weld profiles

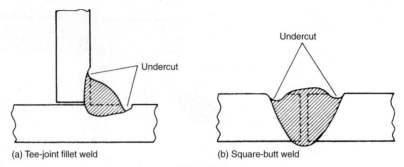

(a) Tee-joint fillet weld   (b) Square-butt weld

Figure 11.29 Undercutting

indicate a poor-quality weld. Further, excessive spatter is unsightly and shows lack of attention to detail. The main causes of spatter are:

- Damp electrode coating.
- Excessive welding current.
- Excessive arc-length.
- The electrode being presented to the weld at too flat a slope angle.

11.4 Workshop testing of welds

11.4.1 Macro examination

This is the examination of a section through a weld with the naked eye or with a magnifying glass giving a magnification of not greater than ×10. The specimen is polished on progressively finer grades of emery cloth on a flat surface such as plate glass. The direction of polishing should be at right angles for each successive change of emery cloth. When all the scratch marks have been removed the polished surface should be etched in 'nitol'. This is a commercial etchant consisting of a solution of 2% to 5% nitric acid in industrial alcohol. When the grain structure of the weld can be clearly seen the specimen should be washed in water, rinsed in acetone and dried in a current of warm air. A thin coating of clear lacquer can be used to preserve the specimen against atmospheric corrosion. Macro-examination is used to check for:

- Incorrect profile.
- Undercutting.
- Slag inclusions.
- Porosity and cracks.
- Poor root penetration and lack of fusion.

11.4.2 Impact (nick) test

A specimen approximately 15 mm wide is cut from the weld at right angles to the joint. A saw cut some 3 mm to 5 mm deep is made in the weld along the centre line of the weld. The specimen is held in a vice and given a sharp blow with a hammer. If the weld is sound it should bend, but any brittleness or weakness will cause it to fracture.

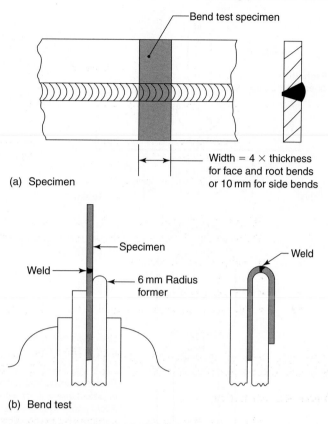

(a) Specimen

Bend test specimen

Width = 4 × thickness
for face and root bends
or 10 mm for side bends

Specimen

Weld

6 mm Radius
former

Weld

(b) Bend test

Figure 11.30 Bend testing of welds (thin metal)

11.4.3 Bend tests

Free bend test for thin material

This is suitable for specimens cut from sheet metal up to 1.6 mm thick. The specimen is cut as shown in Fig. 11.30(a) and bent over a former having a 6 mm radius as shown in Fig. 11.30(b). The specimen should be positioned in the vice so that the joint lies over the centre of the former when bending is complete as shown. Any weld reinforcement should be removed from the face of the weld for oxy-acetylene welds. For arc welds, both face and root reinforcement should be removed. No cracking should be evident if the weld is sound.

Guided bend test for material thicker than 1.6 mm

The correct procedure for preparing test welds is fully explained in British Standard 1295. For the purpose of bend tests, two specimens are cut from the test weld. One specimen is tested with the *face of the weld* in tension and the other specimen is tested with the *root of the weld* in tension. The procedure for carrying out the test is shown in Fig. 11.31.

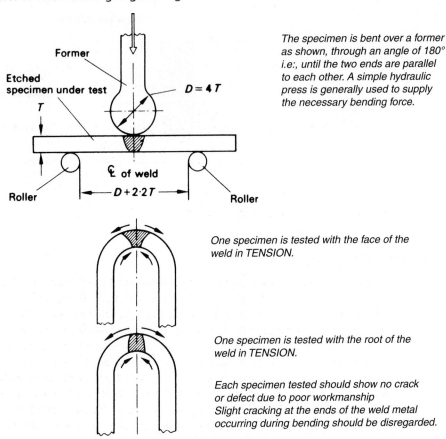

The specimen is bent over a former as shown, through an angle of 180° i.e:, until the two ends are parallel to each other. A simple hydraulic press is generally used to supply the necessary bending force.

One specimen is tested with the face of the weld in TENSION.

One specimen is tested with the root of the weld in TENSION.

Each specimen tested should show no crack or defect due to poor workmanship Slight cracking at the ends of the weld metal occurring during bending should be disregarded.

Figure 11.31 Bend testing of welds (thick metal)

The upper and lower surfaces of the weld may be filed, ground or machined flush with the surface of the parent metal. The direction of cutting for any of these processes should be along the specimen and across the weld. The sharp corners of the test piece should be rounded to a radius not exceeding one-tenth of the metal thickness. The centre line of the weld must be accurately aligned with the centre line of the former as shown, and the specimen is bent through 180°; that is, until the two tails are parallel to each other. A hydraulic press is normally used to supply the required bending force.

11.5 Miscellaneous fusion welding processes

In addition to the use of coated electrodes for arc welding, the following processes are also used:

- Submerged arc welding.
- Tungsten inert gas (TIG) welding.
- Metal inert gas (MIG) welding.

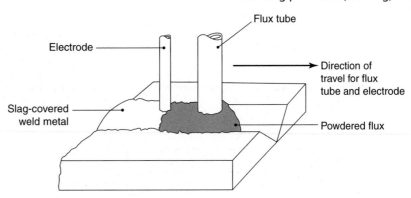

Figure 11.32 Submerged arc welding

11.5.1 Submerged arc welding

This is an automatic process used for arc welding thick plate where very heavy welding currents are required. The electrode is a bare wire fed from a coil carried on the welding machine. The end of this wire electrode is submerged under powdered flux, thus preventing splatter and ultraviolet radiation, which could be a hazard for persons working nearby. The principle of this process is shown in Fig. 11.32. The tube delivers the powdered flux to the joint from a hopper and surplus flux is removed by a vacuum cleaning unit (not shown) and recycled to the hopper. The wire electrode is fed into the joint automatically to keep the arc constant. Much of the flux melts and rises to the top of the weld pool as a protective slag. The automatic flux supply and electrode feed mechanism is traversed along the joint by a power driven 'tractor'. The process is most suitable for straight line welding of heavy plates and is widely used in ship construction and other large-scale applications.

11.5.2 Arc welding with shielding gases

As an alternative to using fluxes and coated electrodes, shielding gases can be used to prevent oxidation of the weld pool. Although processes using shielding gases have higher material costs than those for manual metallic arc (stick) welding, labour costs are reduced as descaling is no longer required.

Tungsten inert gas (TIG) welding

This employs a tungsten electrode which is non-consumable. The atmospheric gases of oxygen and nitrogen are excluded from the weld pool by a blanket of inert gas, such as argon or helium, which will not react with the molten metal. The principle of TIG welding is shown in Fig. 11.33. It can be seen that the arc is struck between the tungsten electrode and the parent metal and, unlike other electric arc welding processes a separate filler rod is used in a similar way to oxy-acetylene welding.

Metal inert gas (MIG) welding

This is a semi-automatic process: although the welding gun has to be guided by hand, the electrode, which is a bare wire, is fed continuously from a drum into the

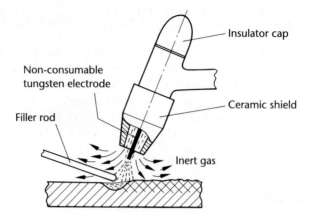

*The atmosphere is excluded from the weld by shielding with an inert gas.
No chemical reactions take place*

Figure 11.33 Protection of the weld – TIG process

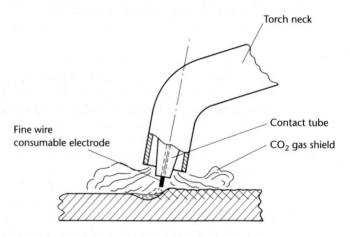

Figure 11.34 Protection of the weld – MIG process

welding gun by means of an *automatic electrode wire drive unit* which 'senses' the welding current and voltage across the arc in order to maintain a constant length of arc. The shielding gases used may be argon, helium or carbon monoxide. The use of carbon dioxide considerably reduces the operating costs of the process and when this latter gas is used the process is often called 'CO_2 welding'.

Unfortunately, carbon dioxide is not an inert gas and cannot be used with some non-ferrous metals and alloys. When the gas passes through the arc it breaks down into oxygen and carbon monoxide. The effect of these gases is to materially alter the strength characteristics of the finished joint and can lead to fatigue failure in service. It is particularly important to ensure that the liberated oxygen does not contaminate the deposited weld metal; for this reason deoxidizers are incorporated into the welding wire. They react with the liberated oxygen to form a thin and neat protective layer of slag on the surface of the completed weld. This slag does not have to be removed. The principle of the MIG welding process is shown in Fig. 11.34.

11.6 Workholding devices for fusion welding

When assembling components prior to welding them together, many different types of clamping devices are required. G-clamps are most commonly used but care must be taken to prevent their threads being damaged by weld spatter by coating the threads with anti-spatter compound. Special G-clamps for welding are available which have shielded threads. Where the components to be welded together are assembled in a welding fixture, quick-release toggle clamps or cam-action clamps may be used to speed up the assembly process.

11.6.1 Strong-backs

When parts are set up prior to welding, it is important that the plate edges and particularly the root faces and root gaps are correctly aligned. One device which is widely used to achieve this alignment is the 'strong-back'. This consists of a rigid piece of plate or angle section that is tack-welded to one side of the joint and a wedge or a bolt and a dog is used to achieve alignment as shown in Fig. 11.35.

11.6.2 Cleats

Angle cleats are used to push or pull (draw) the plate edges together when assembling long and circular weld seams. When cleats are used for pulling plate edges together they are often called 'draw-lugs'. The use of cleats is shown in Fig. 11.36.

11.6.3 Glands

These are simple bridge pieces made from stout off-cuts and often used in conjunction with cleats to align the edges of an assembly. Whereas *cleats* are used to maintain a constant gap width between the joint faces, *glands* are used to provide lateral alignment as shown in Fig. 11.37. Dog-bolts or gland-screws are positioned on one side of the joint a tack welded to the plate. The glands or bridging pieces are placed in position and held by means of nuts and washers as shown in Fig. 11.37. The gland nuts are tightened and adjusted until the desired alignment of the joint has been achieved. After completion of the welding operation the glands and cleats are carefully removed together with their tack-bolts and retained for future use.

11.6.4 Dogs and wood blocks

Stiffeners such as angle bars can be clamped in position prior to welding by the use of clamping dogs and wooden blocks. A bolt is tack-welded adjacent to the required location of the stiffener and the assembly is held in position as shown in Fig. 11.38. After welding is complete the bolt is removed by breaking the tack welds and the surface is cleaned up with an angle grinder. In fact this cleaning-up process applies to any assembly in which tack-welding is used to secure temporary clamping and location devices.

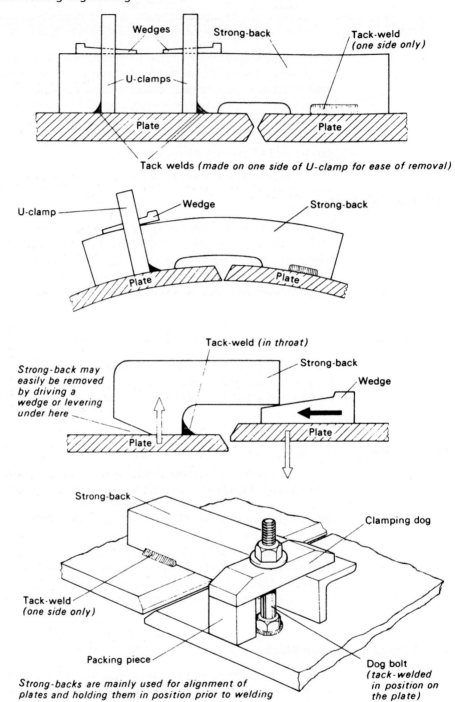

Figure 11.35 Use of strong-backs for welding

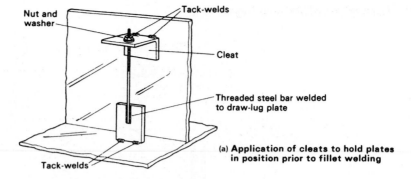

Nut and washer

Tack-welds

Cleat

Threaded steel bar welded to draw-lug plate

Tack-welds

(a) **Application of cleats to hold plates in position prior to fillet welding**

When cleats are used as an aid to holding work for welding, they may be easily removed by tapping with a hammer to bend them over and break the tack-welds as shown by the arrows

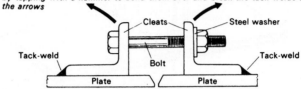

Cleats

Steel washer

Tack-weld

Bolt

Tack-weld

Plate

Plate

(b) **Application of cleats as draw-lugs for closing a butt weld**

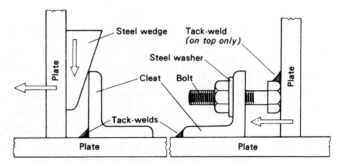

Steel wedge

Tack-weld *(on top only)*

Steel washer

Plate

Cleat

Bolt

Plate

Tack-welds

Plate

Plate

1. Pushing the plates into position   *2. Pulling the plates into position*

(c) **Application of cleats for assembling plates in correct position for fillet welding**

Figure 11.36 Use of cleats for holding work for welding

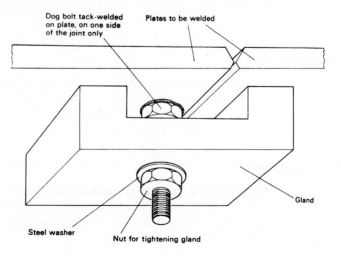

Dog bolt tack-welded on plate, on one side of the joint only

Plates to be welded

Gland

Steel washer

Nut for tightening gland

Figure 11.37 Use of glands

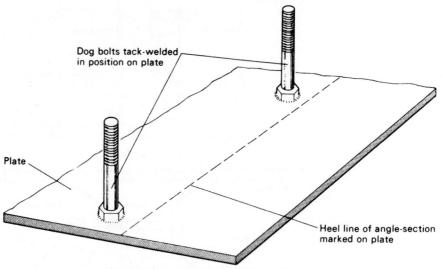

Dog bolts tack-welded in position on plate

Plate

Heel line of angle-section marked on plate

Dog bolts tack-welded in position adjacent to required position of stiffener to be welded to plate

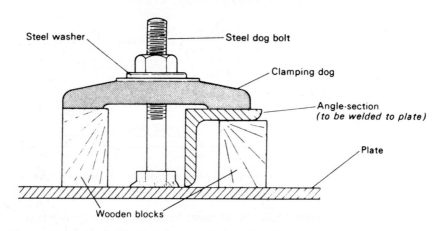

Steel washer

Steel dog bolt

Clamping dog

Angle-section *(to be welded to plate)*

Plate

Wooden blocks

Figure 11.38 Application of the use of clamping dogs

11.6.5 Bridge pieces

Bridge pieces are often referred to as U-clamps and are used together with wedges for clamping sections and plates together as shown in Fig. 11.39. They are also used for flat plates which are out of alignment.

11.6.6 Chain and bar

When welding longitudinal seams on cylindrical vessels, difficulty is often experienced with misaligned joint edges, particularly if the plate has not been correctly rolled. This problem can be overcome by the use of a chain and bar as shown in Fig. 11.40.

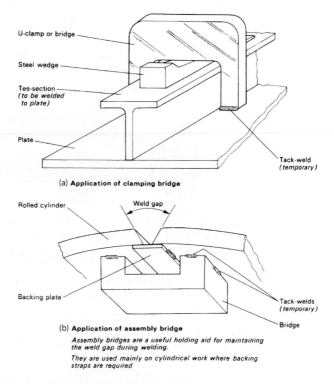

U-clamp or bridge

Steel wedge

Tee-section
(to be welded
to plate)

Plate

Tack-weld
(temporary)

(a) **Application of clamping bridge**

Rolled cylinder

Weld gap

Backing plate

Tack-welds
(temporary)

Bridge

(b) **Application of assembly bridge**

*Assembly bridges are a useful holding aid for maintaining
the weld gap during welding.*

*They are used mainly on cylindrical work where backing
straps are required*

Figure 11.39 Use of bridge pieces

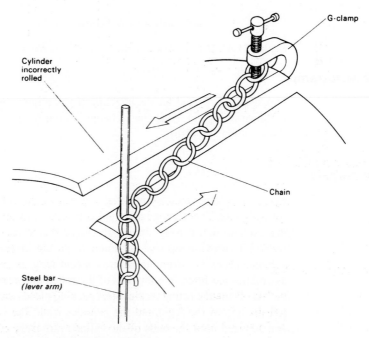

G-clamp

Cylinder
incorrectly
rolled

Chain

Steel bar
(lever arm)

Figure 11.40 Use of chain and bar

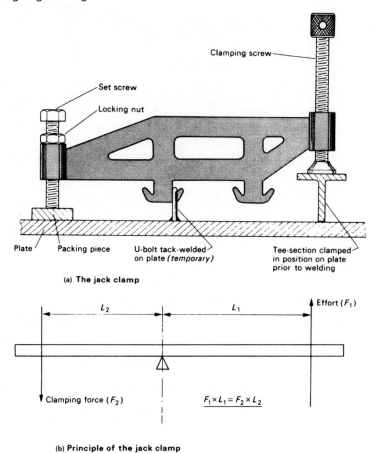

(a) The jack clamp

(b) Principle of the jack clamp

Figure 11.41 Use of jack clamps for holding work

11.6.7 Jack clamps

Jack clamps are used for the attachment and alignment of heavy plate. An example is shown in Fig. 11.41(a). Jack clamps exploit the principle of moments (levers) as shown in Fig. 11.41(b).

11.6.8 Spiders

Large cylindrical vessels are often constructed from two or more sections of roll-formed plate. The sections are welded together circumferentially and it is essential that each section is truly circular to ensure the correct alignment of the joint being welded. To provide this true alignment of the circumferential joints, spiders are used at both ends of the section to ensure concentricity, as shown in Fig. 11.42. The spider assemblies are fitted in both ends of the rolled and seam-welded steel cylinder with the aid of suitable lifting tackle. Steel packing pieces are place between the set screws (on the end of the legs) and the cylinder wall. The set screws are then tightened and adjusted until the ends of the cylinder are truly circular. Once the contour has been checked by suitable templates, location plates are tack-welded to the legs and the inside of the cylinder wall.

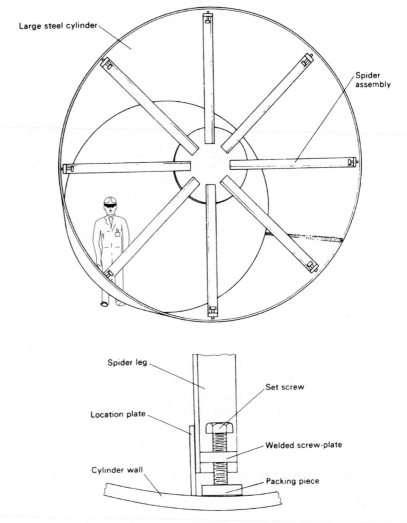

Figure 11.42 Use of spider assemblies for holding work

11.6.9 Magnetic clamps

Magnetic clamps of the types shown in Fig. 11.43 are useful for holding components together at an angle, particularly if they are made from thin gauge material and likely to be distorted by the clamping devices previously described. Care must be taken to keep such magnetic clamping devices cool as heating will destroy their magnetism.

11.7 Resistance welding

The welding processes discussed so far in this chapter have all been *fusion processes*; that is, the filler metal and the joint edges of the parent metal have been melted and allowed to run together. In resistance welding the metal is raised to *just below its melting point* and the weld is completed by *applying pressure* as in forge welding.

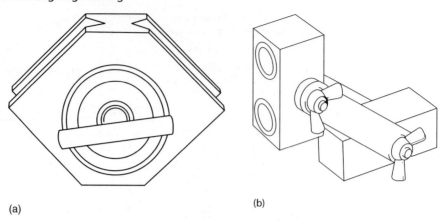

(a)

(b)

Figure 11.43 Magnetic clamps. (a) magnetic holder; (b) adjustable magnetic links

11.7.1 Spot welding

This is the most common of the resistance welding processes. It is much quicker than riveting or fusion welding as a technique for joining sheet metal components and, since no holes are drilled in the components being joined, they are not weakened. Also, since the process is largely automated, less skill is required, resulting in lower labour costs. The joint is produced by making a series of spot welds side by side at regular intervals. Such joints are not fluid-tight and have to be sealed to prevent leakage and/or corrosion. Apart from ensuring both the joint faces are clean and free from corrosion, no special joint preparation is required.

The temperature of the components being joined is raised locally by the passage of a heavy current, at low voltage, through the components as shown in Figs. 11.44(a) and 11.44(b). When an electric current flows through a conductor offering resistance to the flow of current, the electrical energy is partially converted into heat energy and the temperature of the conducting metal is raised. *Resistance welding uses this principle*. Resistance to the current flow occurs between the two surfaces of the components being joined, over the cross-section of the electrodes. Sufficient heat is generated at this 'spot' to raise the components to the welding temperature. The current is then switched off and the weld is completed by squeezing the components tightly together between the electrodes. The electrodes are made from copper since this metal has high thermal and electrical conductivity. They are also water-cooled on the larger production machines to prevent them overheating. The complete cycle of events is controlled by an automatic timer.

No additional filler metal is required to make a resistance-welded joint. Since the process is akin to forge welding, the grain structure of the metal in the weld zone is in the wrought condition, rather than the 'as-cast' condition associated with fusion welding. This makes resistance welds stronger and more ductile than fusion welds of similar cross-sectional area. Because of the relatively high initial capital outlay and the lead time in programming the automatic control gear, resistance-welding processes are more suitable than fusion welding for medium- and large-volume production.

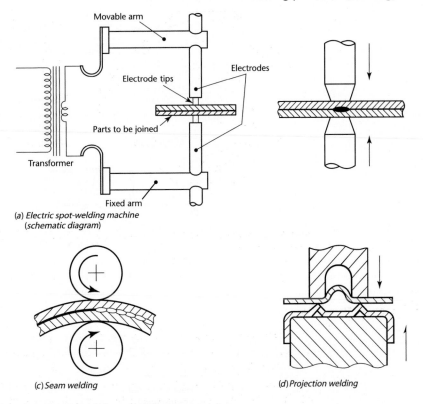

(a) *Electric spot-welding machine (schematic diagram)*

(c) *Seam welding*

(d) *Projection welding*

Figure 11.44 Principles of resistance welding

11.7.2 Seam welding

Seam welding is similar to spot welding except that the components to be joined are gripped between revolving, circular copper rollers as shown in Fig. 11.44(c). The welding current is applied in a series of pulses resulting in a corresponding series of overlapping spot welds being made along the seam. This method of spot welding is used for the volume production of fluid containers and vehicle fuel tanks.

11.7.3 Projection welding

In this process the electrodes act as locations for holding the parts to be joined and are, therefore, job-specific. The joint is so designed that projections are preformed on one of the parts to be joined, as shown in Fig. 11.44(d). Projection welding enables the welding pressure and the heated weld zone to be localized at predetermined points. This technique is largely used for small, precision components that need to be accurately located.

11.7.4 Butt welding

The principles of resistance butt-welding are shown in Fig. 11.45. In this example the two ends of the rods are brought together with just sufficient force to ensure the current can flow without arcing. The resistance of the joint interface ensures that local heating will take place on the passage of a heavy electric current at low voltage.

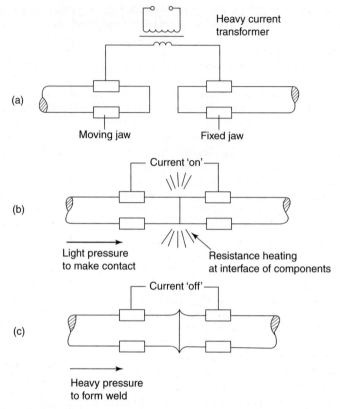

Figure 11.45 Resistance butt-welding

When the metal in the joint zone has reached its welding temperature, the current is switched off and the axial force on the joint is increased to complete the weld.

As for all resistance-welding processes, a sound weld is achieved by raising the temperature of the metal to *just above* the pressure welding temperature. Should the surfaces of the components at the joint interface commence to melt, this molten metal is displaced by the welding force until metal at the correct temperature for pressure welding is reached in the joint substrate and the weld is completed.

11.8 Further welding processes

Those readers who require knowledge of the following welding processes should refer to Appendix A for useful websites:

- Plasma arc welding.
- Electron beam welding.
- Laser beam welding.
- Automated welding.

Exercises

11.1 Oxy-acetylene welding
  a) List the essential differences between brazing and fusion welding.
  b) With the aid of diagrams explain the essential differences between oxy-acetylene welding and manual metal-arc welding.
  c) With the aid of sketches describe how oxygen and acetylene cylinders can be identified, and name the safety feature that prevents the welding equipment being connected to the wrong cylinder.
  d) List the advantages and limitations of a welding gas manifold system compared with individual cylinders at each work station.
  e) Explain the purpose of the pressure regulators that have to be mounted on the cylinders.
   i) Explain why such regulators have two pressure gauges.
   ii) Describe the correct procedure to be followed when mounting a regulator on a cylinder.
  f) Describe the requirements of welding hoses, how they can be identified and the purpose of a hose protector and a flash-back arrestor.
  g) With the aid of sketches describe the essential features of an oxy-acetylene welding torch.
  h) With the aid of sketches explain the difference between a reducing flame, an oxidizing flame and a neutral flame, indicating the hottest part of the flame. State which flame is usually used for welding plain carbon steels.
  i) With the aid of sketches describe what is meant by 'leftward' and 'rightward' welding and list their relative advantages and limitations.
  j) With the aid of sketches describe the differences between flanged-edge welding, square-butt welding and vee-butt welding. Under what circumstances would these techniques be used?
  k) Discuss the problems associated with oxidation in welded joints, paying particular attention to the difficulties that will occur in the welding process and the effect on the strength of the weld.

11.2 Manual metal-arc welding
  a) Describe the purpose of the electrode and why it is normally coated and what the purpose of this coating is when 'stick' welding.
  b) A typical standard coding, which should be plainly marked on a box of electrodes, could be **E 50 Ini R 6 4 H6.** What does this mean?
  c) With the aid of diagrams explain the difference between arc-welding with direct current and arc-welding with alternating current.
  d) Explain the difference between 'open circuit voltage', 'striking voltage' and the 'welding or arc voltage'.
  e) With the aid of sketches show the difference between the 'tilt' and 'slope' angles of the electrode. Describe how these angles affect the quality of the welded joint.
  f) With the aid of sketches explain what is meant by the 'arc length' and how this length affects the quality of the welded joint.
  g) Explain what is meant by the 'heat-affected zone' and how this affects the strength of the parent metal adjacent to the weld pool.
  h) Describe the edge preparation required before welding and how this influences the quality of the weld.
  i) Explain the difference between single-run and multi-run welding. Where should the latter process be used and what precautions should be taken to ensure a sound weld?

11.3 Welding defects
  a) A common fault when welding is lack of penetration. Explain the main causes of this fault and how it affects the quality of the weld.
  b) Explain what is meant by lack of fusion and state the possible causes of this fault.
  c) With the aid of sketches discuss the causes and the effects of slag inclusions and porosity. How is porosity affected by the conditions under which electrodes are stored?
  d) With the aid of sketches explain what is meant by the 'weld profile' and explain how this affects the quality of the joint.
  e) With the aid of sketches explain what is meant by the 'undercutting' and explain how this affects the quality of the joint.
  f) Discuss the causes of 'spatter'.
  g) With the aid of sketches describe:
   i) an impact test on a welded joint that may be performed under workshop conditions;
   ii) a guided bend test for material over 1.6 mm thick.
  h) With the aid of sketches describe what is meant by submerged arc welding and state under what conditions it could be used.
  i) Compare the differences between tungsten inert gas (TIG) welding and metal inert gas (MIG) welding and give examples of typical applications where it would be appropriate to use these processes.

11.4 Work holding devices

  a) With the aid of sketches describe the following workholding devices and give examples where they would be used:

  i) strong-backs

  ii) cleats

  iii) glands.

  b) With the aid of sketches describe the following workholding devices and give examples where they would be used:

  i) dogs

  ii) bridge pieces

  iii) magnetic clamps.

  c) With the aid of sketches describe the following workholding devices and give examples where they would be used:

  i) spiders

  ii) jack clamps

  iii) chain and bar.

11.5 Resistance welding

  a) Describe the difference between resistance welding and fusion welding.

  b) With the aid of sketches describe the following resistance welding processes and give examples where they would be used:

  i) spot welding

  ii) seam welding

  iii) projection welding

  iv) butt welding.

12 Composite fabrication processes and adhesive bonding

When you have read this chapter you should be able to:

- Identify and describe laminate materials
- Identify and describe fibre reinforced materials
- Identify and describe particle hardened and reinforced materials
- Identify and describe sintered materials including 'cermets'
- Identify and describe 'Whiskers' and the principles of dispersion hardening
- Describe the processes used in heat bending and forming sheet plastics
- Describe the process of heat welding sheet plastics
- Describe the process of solvent welding sheet plastics
- Describe the process of friction welding plastic rods and tubes
- Describe the process of resistance welding plastics
- Describe the process of induction welding plastics
- Describe the process of dielectric welding plastics
- Describe the process of ultrasonic welding plastics
- Identify and describe the various techniques available for the assembly of metals and non-metals by adhesive bonding

12.1 Introduction

In its simplest form a *composite material* consists of two dissimilar materials in which one material forms a matrix to bond together the other (reinforcement) material. The matrix and reinforcement are chosen so that their mechanical properties complement each other, whilst their deficiencies are neutralized. For example, in a GRP moulding, the polyester resin is the matrix that binds together the glass fibre reinforcement.

In a composite material the *reinforcement* material can be in the form of rods, strands, fibres or particles and it is bonded together with the other *matrix* material(s). For example, the fibres may have some of the highest modulii and greatest strengths available in tension, but little resistance to bending and compressive forces. On the other hand, the matrix can be chosen to have high resistance to bending and compressive forces. When used together these two different types of material produce a composite material with high tensile and compressive strengths with a high resistance to bending.

12.2 Lamination

Figure 12.1 shows a natural composite material – natural wood. The figure shows how the *tubular cellulose fibres* forming the 'reinforcement' are bonded together by the relatively weak *lignin* 'matrix'.

Brittle materials, such as concrete and ceramics, are strong in compression but weak in tension since they are susceptible to crack propagation, as shown in Fig. 12.2(a). The initial crack can be prevented from spreading or 'running' by the addition of a reinforcement material to form a 'composite'. Two ways in which a crack can be prevented from 'running' are by:

- The use of lamination as shown in Fig. 12.2(b).
- The use of reinforcing fibres which hold the brittle matrix in compression, so an external tensile load cannot open up any surface cracks and discontinuities. This is shown in Fig. 12.2(c).

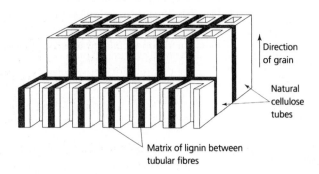

Direction of grain

Natural cellulose tubes

Matrix of lignin between tubular fibres

Figure 12.1 Wood a natural composite

12.2.1 Plywood

Wood, like all materials reinforced by parallel fibres, has highly directional strength characteristics. Figure 12.3(a) shows a plank being bent in a plane *perpendicular* to the lay of the grain. Loading the plank in this way exploits its greatest strength characteristics. Figure 12.3(b) shows what would happen if a piece of wood is loaded so that bending occurs *parallel* to the lay of the grain. The wood breaks easily since the lignin bond is relatively weak compared with the natural cellulose reinforcement fibres.

Plywood overcomes this problem and is a man-made composite which exploits the directivity of the strength of natural wood. Figure 12.4 shows how plywood is built up from veneers (thin sheets of wood) bonded together by a high strength and water resistant, synthetic resin adhesive. When laying up the veneers, care is taken to ensure that the grain of each successive layer is perpendicular (at right-angles) to the preceding layer. When the correct number of veneers has been laid up, the adhesive component of the composite is cured under pressure in a hydraulic press by the application of heat.

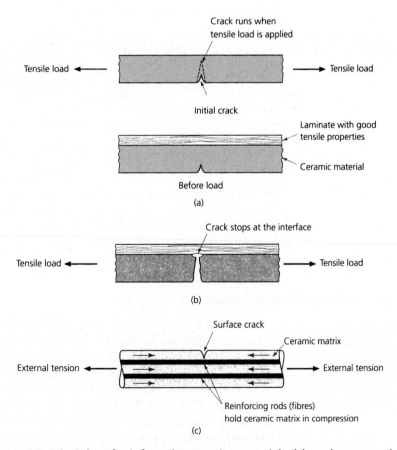

Figure 12.2 Principles of reinforced composite materials: (a) crack propagation in a non-reinforced ceramic material; (b) behaviour of a laminated composite when in tension; (c) fibre reinforcement

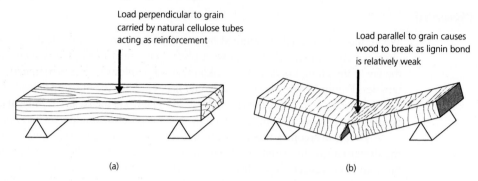

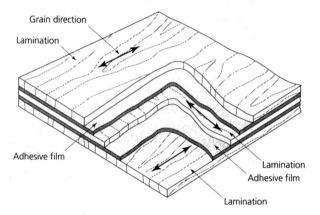

Figure 12.3 Effect of the direction of grain on the strength of wood

Figure 12.4 Structure of plywood

The resulting laminate material has uniform properties in all directions. However, if it has an even number of 'plies' (layers) it only has half the strength of a solid piece of timber when the load is applied uniaxially as shown in Fig. 12.5. In practice, an odd number of plies are used so that the grain in the outer laminations lies in the same direction.

12.2.2 Laminated plastic (Tufnol)

The thickness of the finished sheet is determined by the number of layers of impregnated reinforcement in the laminate

Fibrous material such as paper, woven cotton and woollen cloth, and woven glass fibre cloth can be used to reinforce phenolic and epoxy resins. Sheets of these fibrous reinforcing materials are impregnated with the resin and they are then laid up between highly polished metal plates in hydraulic presses.

Each layer of reinforcement is rotated through 90° of arc so as to ensure uniformity of mechanical properties. The laminates are then heated under pressure until they become solid sheets, rods or tubes.

Laminated composites can be machined dry with ordinary engineering tools using a low value of rake angle and fairly high cutting speeds. However, the material is rather abrasive and carbide tooling can be used to advantage, particularly when machining this material on a production basis. Although not toxic, the dust

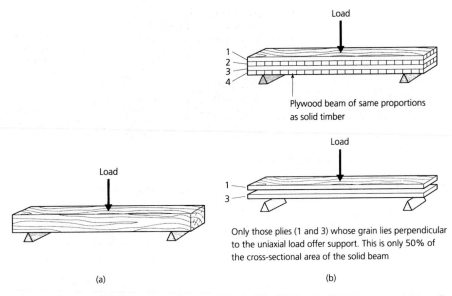

Plywood beam of same proportions as solid timber

Only those plies (1 and 3) whose grain lies perpendicular to the uniaxial load offer support. This is only 50% of the cross-sectional area of the solid beam

(a)   (b)

Figure 12.5 Uniaxial loading of (a) solid timber (100% support for uniaxial load) and (b) plywood (50% support)

and fumes generated during machining is unpleasant and adequate extraction should be provided. Care must be taken in the design and use of plastic laminates because of the 'grain' of the material which makes it behave in a similar manner to plywood.

'Tufnol' composites are widely used for making bearings, gears and other engineering products which have to operate in hostile environments or where adequate lubrication is often not possible as in food processing and office machinery. In the electronic industry, copper clad sheet tufnol is used for making printed circuit boards. It is also used for making insulators for heavy duty electrical equipment in the electrical power engineering industry.

12.3 Fibre reinforcement

This method of reinforcement can range from the glass fibres used in GRP plastic mouldings to the thick steel rods used to reinforced concrete.

12.3.1 Reinforced concrete

Concrete itself is a particle reinforced material. It consists of a mortar made from hydraulic cement and sand, reinforced with an *aggregate* of chipped stones as shown in Fig. 12.6. The stones are crushed so that they have a rough texture and sharp corners that will *key* into and bond with the mortar matrix. This basic concrete has a very high compressive strength but is very weak in tension. To improve its performance overall, metal reinforcing rods are added as shown in Fig. 12.7.

As well as the amount of reinforcement, the positioning of the reinforcement is also important. Since concrete is strong in compression but weak in tension, it

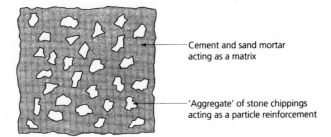

Cement and sand mortar
acting as a matrix

'Aggregate' of stone chippings
acting as a particle reinforcement

Figure 12.6 Structure of concrete

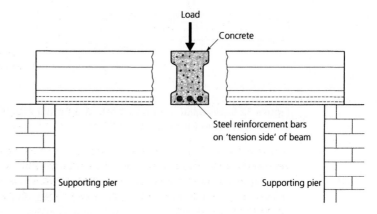

Figure 12.7 Reinforced concrete beam

would seem sensible to insert the reinforcement near the tension side of the beam as shown in Fig. 12.7. The tension side is the side of the beam that is on the outside of the 'bend' when the beam 'gives' under the effect of an externally applied load.

However, concrete beams often have to be craned into position because of their weight and this operation causes reverse stressing leading to fracture as shown in Fig. 12.8. For this reason, a small amount of additional reinforcement is introduced on what is normally the compression side of the beam. This is also shown in Fig. 12.8.

Unfortunately, the concrete is more rigid than the reinforcement. The steel reinforcement can 'give' without cracking, but the concrete cannot. Therefore, surface cracking can still occur. Although catastrophic failure is not likely to result directly from such surface cracking, the penetration of moisture can cause corrosion of the reinforcement and weakening of the structure. This is over come by *prestressing* the reinforcement so that the concrete is always in compression under normal load conditions.

12.3.2 Glass reinforced plastic (GRP)

This important composite material is produced when a plastic material, usually a polyester resin, is reinforced with glass fibre in strand or mat form. The resin is used to provide shape, colour and finish, whilst the glass fibres impart mechanical strength.

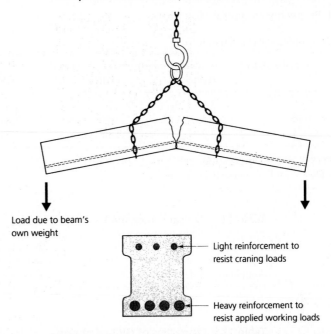

Figure 12.8 If a simple concrete beam is lifted by a crane the stresses are reversed due to the beam's own weight; the reinforcement designed to support the applied load cannot support this reversed stress; the beam fractures and the reinforcing bars bend – beams likely to be craned into position have additional reinforcement as shown

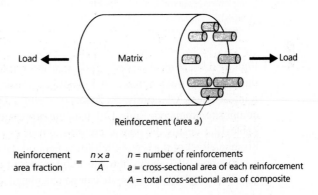

Reinforcement area fraction $= \dfrac{n \times a}{A}$

n = number of reinforcements
a = cross-sectional area of each reinforcement
A = total cross-sectional area of composite

Figure 12.9 Reinforced composite

When laid in parallel strands, the glass fibres provide uniaxial strength, whilst glass fibres woven into a fabric or in a random matt formation provide omniaxial strength.

The amount of reinforcement which can be used depends upon the orientation of the reinforcement. With long strands laid up parallel to each other the reinforcement area fraction can be as high as 0.9. The reinforcement area fraction is the cross-sectional area of reinforcement divided by the total cross-sectional area as shown in Fig. 12.9. With woven strand fabrics the reinforcement area fraction can be as

Table 12.1 Properties of typical glass fibres

Steel wire is included for comparison

Material	Relative density	Tensile strength (GPa)	Tensile modulus (GPa)	Specific strength (GPa)*	Specific modulus (GPa)[†]
E-Glass[†]	2.55	3.5	74	1.4	29
S-Glass	2.50	4.5	88	1.8	35
Steel wire	7.74	4.2	200	0.54	26

*Specific strength = (tensile strength)/(relative density)
[†]Specific modulus = (tensile modulus)/(relative density)

Table 12.2 Typical composition of e-glass fibres

Silicon dioxide	52–56%	Boron oxide	8–13%
Calcium oxide	16–25%	Sodium and potassium oxides	1%
Aluminium oxide	12–16%	Magnesium oxide	0–6%

Table 12.3 Properties of GRP composites

Material	Reinforcement (weight %)	Tensile strength (MPa)	Tensile modulus (GPa)
Chopped strand mat	10–45	45–180	15–15
Plain weave cloth	45–65	250–375	10–20
Long fibres (uniaxially loaded)	55–80	500–1200	25–50

high as 0.75. With chopped strand mat the reinforcement area fraction is substantially reduced but a figure of 0.5 should be considered the minimum for satisfactory reinforcement. Obviously, the lower the reinforcement area fraction, the weaker the composite produced. Table 12.1 lists the properties of some typical glass fibres, and Table 12.2 details the typical composition of e-glass fibres in particular. Table 12.3 lists the effect of the type of reinforcement on the strength of the composite.

Example 12.1 shows how the 'reinforcement area fraction', tensile modulus and tensile strength of a composite can be calculated.

Example 12.1

Given the following data for a fibre glass reinforced polyester component of rectangular cross-section in which the strands lie parallel to the direction of loading, calculate:

a) The matrix area fraction and the reinforcement area fraction.
b) The tensile modulus for the composite.
c) The tensile strength of the composite.

Average fibre diameter   = 0.005 mm
Number of fibres per strand   = 204
Number of strands   = 51470
Tensile modulus for polyester   = 4 GPa
Tensile modulus for glass fibre   = 75 GPa
Tensile strength of polyester   = 50 MPa
Tensile strength of glass fibre   = 1500 MPa
Cross sectional area of the component = 300 mm^2

a) Area of one filament of glass = [0.005^2 × π]/[4] = 0.00002 mm^2
   ∴ Area of one basic strand   = 0.00002 × 204 = 0.00408 mm^2
   ∴ Total reinforcement area   = 0.00408 × 51470 = 210 mm^2

Reinforcement area fraction = [total reinforcement area]/
                                   [total component area]
                                 = [210 mm^2]/[300 mm^2]
                                 = **0.7**

Matrix area = (total area) − (reinforcement area)
            = 300 mm^2 − 210 mm^2
            = 90 mm^2

Matrix Area fraction = [total matrix area]/
                           [total component area]
                         = [90 mm^2]/[300 mm^2]
                         = **0.3**

b) Modulus of composite = (modulus of matrix × modulus area fraction)
                          + (modulus of reinforcement
                               × reinforcement area fraction)
                          = (4 GPa × 0.3) + (75 GPa × 0.7)
                          = **53.7 GPa**

c) Strength of composite = (tensile strength of matrix × matrix area fraction)
                           + (tensile strength of reinforcement
                                × reinforcement area fraction)
                           = (50 MPa × 0.3) + (1500 MPa × 0.7)
                           = 15 MPa + 1050 MPa
                           = **1065 MPa**

The above example indicates the importance of the reinforcement in any composite material when the contribution of the reinforcement is compared with the contribution of the matrix. If the calculation is repeated for a range of reinforcement area fractions, the curves shown in Fig. 12.10 will be obtained. Figure 12.10(a) plots the tensile modulus against reinforcement area fraction, and Fig. 12.10(b) plots the tensile strength of the composite against the reinforcement area fraction.

When a GRP moulding is correctly 'laid up', the thick, viscous resin completely surrounds and adheres to the glass fibres of the 'chopped strand mat' or the woven fibre glass fabric (cloth). The resin sets to give a hard, rigid structure. To shorten the

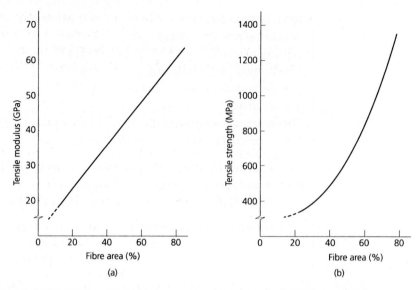

Figure 12.10 Effect of reinforcement area on a typical GRP composite

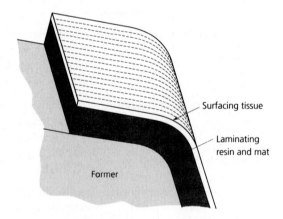

Figure 12.11 Composition of a GRP lay up

setting time a catalyst (accelerator) is often added to the resin immediately before moulding. To improve the appearance of the 'rough side' of the moulding (the side not in contact with the polished surface of the mould), a surfacing tissue is added made up from finer strands of mat as shown in Fig. 12.11. Glass reinforced plastic is widely used for products that require high strength to weight ratios such as boat hulls, car bodies, gliders, etc. It has the advantages of being resistant to most environments and it can be used to form more complex shapes than is possible in wood or sheet metal.

12.3.3 Carbon fibres

These have a higher elastic modulus and lower density than glass fibres, and can be used to reinforce composite materials having a higher strength to weight ratio.

Carbon fibres are used as a reinforcing material in polymeric materials for a wide range of light weight, high strength applications. For example, carbon fibres are used in the manufactures of gas turbine fan blades, racing car body panels, high performance tennis racket frames and high performance golf club shafts.

12.4 Particle reinforcement

In tough pitch copper particles of copper oxide interfere with the flow of the slip planes in the metal reducing its ductility but greatly increasing its strength compared with high-purity copper. This is, in fact, a simple form of particle reinforcement. Similarly the stone chippings in concrete also provide a simple form of particle reinforcement.

12.4.1 Cermets

These are materials which combine the ductility and toughness of a metal matrix with the hardness and compressive strength of ceramic particles to provide *particle reinforcement*. Certain metal oxides and carbides can be bonded together and 'sintered' into a metal powder matrix to form important composite materials called *cermets*. This is short for *ceramic* reinforced *metals*. The properties of the resulting composite materials enable extremely hard and abrasion resistant cutting tools to be produced. Such tools enable hard, tough and high strength materials to be machined at production rates and with surface finishes not previously possible.

The mixture of metal and ceramic powders is compressed in dies to form 'powder compacts'. These are then *sintered* in a furnace. Sintering consists of heating the compact to between 60 and 90% of the melting temperature of the majority material present in the compact. The process is carried out automatically in a conveyor type furnace under reducing atmosphere conditions to prevent oxidation. Some typical cermets are listed in Table 12.4.

12.4.2 'Whiskers' and dispersion hardening

'Whiskers' are high aspect ratio (hair like) crystals grown with very high strengths. They may have a diameter of only 0.5 to 2.0 microns, yet have a length of up to 20 millimetres. The properties of some whiskers are shown in Table 12.5. The boron and carbon whiskers are used to harden and reinforce polymeric materials, whilst whiskers of alumina are used to harden and reinforce the metal nickel to produce materials that retain their strength at high temperatures.

Alumina (aluminium oxide) can also be used in spheroidal particle form to increase the strength of composites rather than their hardness. Aluminium metal is ground into a fine powder in an oxygen rich atmosphere. This causes aluminium oxide (alumina) to form on the surface of the particles. After sintering, the material produced consists of 6 per cent alumina in a matrix of aluminium, and is called *sintered aluminium* powder or SAP. Figure 12.12 shows how the strength of SAP compares with duralumin. Although the SAP is not so strong at room temperature, it retains its strength at very much higher temperatures.

Table 12.4 Typical cermets

Type	Ceramic particles	Metal matrix	Applications
Borides	Titanium boride Molybdenum boride Chromium boride	Cobalt/nickel Chromium/nickel Nickel	Mostly cutting tool tips
Carbides	Tungsten carbide Titanium carbide Molybdenum carbide Silicon carbide	Cobalt Cobalt or tungsten Cobalt Cobalt	Mostly cutting tool tips and abrasives
Oxides	Aluminium oxide Magnesium oxide Chromium oxide	Cobalt/chromium Cobalt/nickel Chromium	Disposable tool tips; refractory sintered components, e.g. spark plug bodies, rocket and jet engine parts

Table 12.5 Properties same reinforcement 'whiskers'

Material	Relative density	Tensile strength (GPa)	Tensile modulus (GPa)	Specific strength* (GPa)	Specific modulus† (GPa)
Alumina	3.96	21.0	430	5.3	110
Boron carbide	2.52	14.0	490	5.6	190
Carbon (graphite)	1.66	20.0	710	12.0	430

*Specific strength = (tensile strength)/(relative density)

†Specific modulus = (tensile modulus)/(relative density)

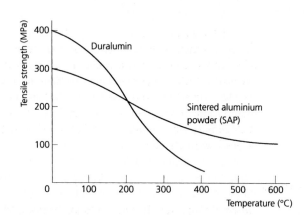

Figure 12.12 Comparative effects of temperature on duralumin and sintered aluminium powder

12.4.3 Rubber

Naturally occurring rubber (latex) comes from the sap of the tree called *Hervea Brasiliensis*. This material is of little use to engineers since it becomes tacky at elevated temperatures, has a very low tensile strength at room temperature, and is attacked by oils and quickly perishes in the presence of the ultraviolet rays of sunlight.

However, when compounded with other substances and vulcanized, it changes to become a useful and versatile material. The reinforcing filler material is *carbon black*. The latex and the carbon black are mixed together with small amounts of chemicals such as sulphur which speed up the vulcanizing process and extend the working life of the product by improving its ageing characteristics. Although rubber is a naturally occurring *elastomer*, this term is mainly used for the synthetic rubbers that are an important sector of the polymer industry.

Typical products made from natural rubber are vehicle tyres, inner tubes, footwear, respirator masks, and gloves

Reclaimed rubber is widely used for lightly stressed applications were strength is not of overall importance. The use of carbon black to improve the properties of the rubber is, again, an example of a particle reinforced composite material.

12.5 The manipulation and fabrication of polymers (plastics)

In addition to the processes already described, sheet polymers can be hot formed into a simple or complex shapes and remain in that shape after cooling.

12.5.1 Heat bending

Simple, straight bends in thermoplastic materials such as polyvinyl chloride (PVC), polyethylene (PE) and acrylics (Perspex) follow the techniques used for sheet metal working. The only difference being that the plastic material requires to be heated before bending and bending must take place whilst the plastic sheet is still hot. For this reason the bending jig must be faced with materials having a low thermal conductivity such as wood or Tufnol.

Electric strip heaters may be used to ensure that heating is localized along the line of the bend. This makes the plastic sheet easier to handle. Rapid cooling is required immediately after bending is complete to avoid loss of shape and degradation of the plastic. Thick sheet should be heated on both sides prior to bending because of the low thermal conductivity of plastic materials.

Other heat bending techniques such as vacuum forming, blow forming and pressing are outlined in Fig. 12.13. In all these examples the thermoplastic sheet such as polyvinyl chloride (PVC), polyethylene (PE) and acrylics (Perspex) is preheated before forming.

12.5.2 Heat welding

Heat welding can only be used to join thermoplastic materials since only these plastic materials soften upon heating. The temperature of an oxy-fuel-gas welding torch is much too high for plastic welding and would destroy the material being joined.

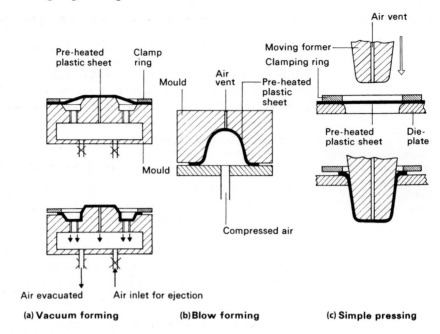

Figure 12.13 Forming plastic sheet materials

The low thermal conductivity and softening temperatures of thermoplastic materials necessitates the use of a low welding temperature so that the heat can penetrate into the body of the plastic before the surface degrades.

Heat is normally applied to the joint by a welding 'gun'. Air or nitrogen gas is heated in the 'gun' and directed as a jet of hot gas into the weld zone. Figure 12.14(a) shows the principle of an electrically heated welding gun, whilst Fig. 12.4(b) shows the principle of a gas heated welding gun. The easiest plastics to weld are polyvinyl chloride (PVC), polyethylene (PE) and acrylics (Perspex) as they have a wide softening range. The basic technique is to apply a jet of heated air or nitrogen into the joint so that the edges of the parent plastic sheet are softened. Filler material, in the form of a rod of the same material as that being welded, is added into the joint in much the same way as when welding metals except that the joint edges and filler rod are softened but not melted. Some degradation inevitably occurs, so the strength of the joint is slightly below that of the surrounding material.

Figure 12.15 shows examples of edge preparation for a range of joints in thermoplastic materials. A small root gap should be provided, no feather edges should be left, and the weld bead (reinforcement) should not be removed as it can increase the joint strength by up to 20 per cent. The technique of welding sheet plastic is shown in Fig. 12.16.

12.5.3 Solvent welding

Again, this process can only be applied to thermoplastic materials since, once cured, thermosets cannot be softened by solvents. When solvent welding, those edges or surfaces that are to be joined are softened by use of a suitable solvent 'cement'

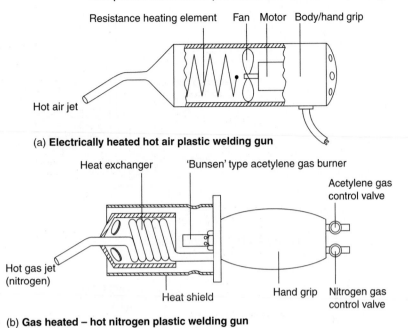

(a) **Electrically heated hot air plastic welding gun**

(b) **Gas heated – hot nitrogen plastic welding gun**

Figure 12.14 Plastic-welding guns: (a) electrically heated hot air plastic-welding gun; (b) gas-heated hot nitrogen plastic-welding gun

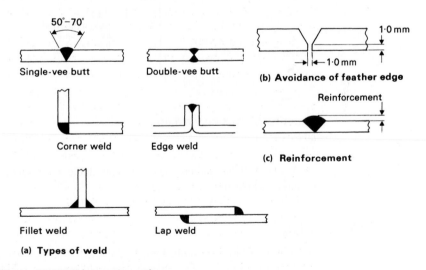

Figure 12.15 Edge preparation

To ensure rapid evaporation volatile solvents are required

instead of by the use of heat. The surfaces are pressed together after application of the solvent until evaporation is complete. The solvent has to be chosen to suit the material being joined. Often the solvent contains a small quantity of the parent material already dissolved in it to give the cement *gap-filling* properties.

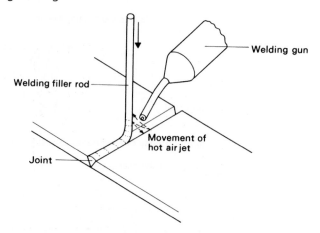

Figure 12.16 Welding plastics

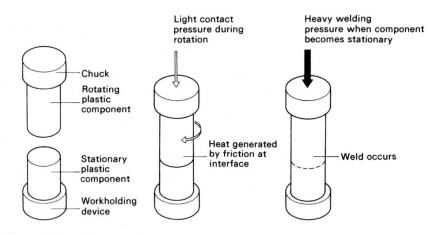

Figure 12.17 Friction welding

Many of these solvents give off flammable and toxic fumes and great care is required in their use.

12.5.4 Friction welding

When two surfaces are rubbed together without a lubricant, the friction between them results in mechanical energy being converted into heat energy at the interface. This effect is exploited in friction welding (also called spin welding). The components to be joined are pressed lightly together. One of the components rotates and one remains stationary. To avoid excessive speed being required for small diameter components, the two components can both be rotated but in opposite directions. As soon as the welding temperature has been reached, rotation ceases and the axial pressure is increased to complete the weld. The principle of friction welding is shown in Fig. 12.17.

Friction welding has one major disadvantage: the heating effect is not uniform across the joint face. This is because as the surface speed diminished towards the

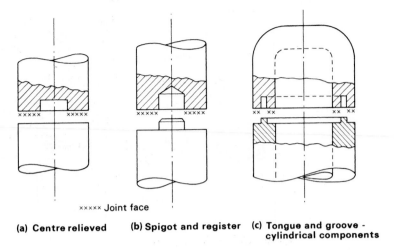

xxxxx Joint face

(a) Centre relieved   (b) Spigot and register   (c) Tongue and groove -
cylindrical components

Figure 12.18 Types of friction weld

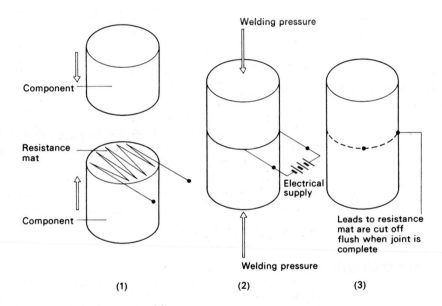

Figure 12.19 Resistance welding

centre of the rod. At the very centre the surface speed is zero and no heating takes place. Therefore it is best suited to tubular components as shown in Fig. 12.18.

12.5.5 Resistance (hot wire) heating

Resistance welding, however, is suitable for solid rods since the heating effect can be virtually uniform. A mat of resistance wire is placed in the joint as shown in Fig. 12.19. The two components are brought together and an electric current is passed through the resistance mat. As soon as the joint faces have softened, the current is disconnected and the welding pressure is applied. The resistance mat remains permanently in the joint and the ends of the wires are cut off flush.

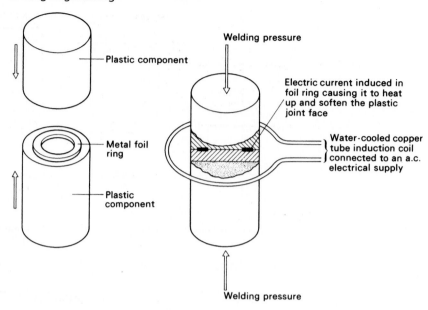

Figure 12.20 Induction welding

12.5.6 Induction heating

Instead of a resistance mat being placed in the joint, a ring of metal foil is used. The plastic components complete with the foil are passed through an induction coil. The induction coil acts as the primary winding of a transformer and the foil disc acts as the secondary winding of a transformer. An electric current is induced in the foil disc which becomes hot, softening the plastic material at the joint face. Again pressure is applied to complete the joint. This is shown in Fig. 12.20. The advantage of this technique is that no connections have to be made and no trimming is required after the joint is complete. It lends itself to quantity production.

12.5.7 Dielectric welding

This technique exploits the insulating properties of plastic materials. In this process the plastic to be joined forms the dielectric of a capacitor by being placed between two electrodes. The electrodes are connected to a high frequency alternating current generator operating at about 30 MHz. The dielectric losses that occur in some thermoplastics at this frequency can cause internal heating in the zone between the electrodes. Again the joint is completed by applying pressure once the welding temperature has been reached. This technique overcomes the degrading effects of externally heating plastic materials in the presence of atmospheric oxygen. The principle of this process is shown in Fig. 12.21. The following plastic materials cannot be welded by dielectric heating:

- Polyethylene (PE).
- Polypropylene (PP).
- Polycarbonate (PC).

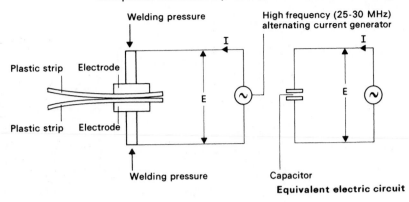

Figure 12.21 Dielectric welding

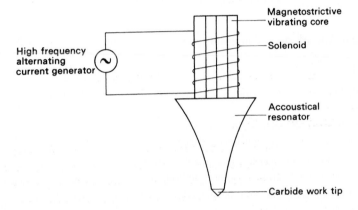

Figure 12.22 Ultrasonic tool

- Polytetrafluoroethylene (Teflon – PTFE).
- Polystyrene (PS).

12.5.8 Ultrasonic welding

This process uses acoustical energy at a frequency of 20 kHz or higher. That is, frequencies above the range of normal human hearing. Hence the name 'ultrasound' or 'ultrasonic'. Figure 12.22 shows the principle of an ultrasonic welding tool. *Magnetostriction* is the property of the core material to expand and contract in sympathy with the electromagnetic field produced in the solenoid (coil) by the high frequency alternating current supply.

To concentrate the heating effect of the high speed vibration, the weld zone must be prepared as shown in Fig. 12.23. That is the joint must contain conical *energy directors*. These restrict the energy required and localize the heat generated. The small volume of softened plastic spreads out under pressure to complete the joint. The generator is switched off and the pressure is held on for about one second for the joint to set. This process can be applied to most thermoplastics, except for the vinyls and the cellulosics.

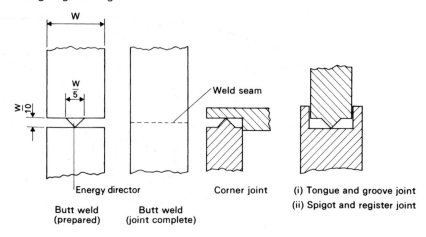

Figure 12.23 Weld preparation

Table 12.6 Advantage and limitations of bonded joints

Advantages	Limitations
The ability to join dissimilar materials, and materials of widely different thicknesses.	The bonding process is more complex than mechanical and thermal processes, i.e. the need for surface preparation, temperature and humidity control of the working atmosphere, ventilation and health problems caused by the adhesives and their solvents. The length of time that the assembly must be jigged up whilst setting (curing) takes place.
The ability to join components of difficult shape that would restrict the application of welding or riveting equipment.	
Smooth finish to the joint which will be free from voids and protrusions such as weld beads, rivet and bolt heads, etc.	
Uniform distribution of stress over entire area of joint. This reduces the chances of the joint failing in fatigue.	Inspection of the joint is difficult.
Elastic properties of many adhesives allow for flexibility in the joint and give it vibration damping characteristics.	Joint design is more critical than for many mechanical and thermal processes.
The ability to electrically insulate the adherends and prevent corrosion due to galvanic action between dissimilar metals.	Incompatibility with the adherends. The adhesive itself may corrode the materials it is joining.
The join may be sealed against moisture and gases.	Degradation of the joint when subject to high and low temperatures, chemical atmospheres, etc.
Heat-sensitive materials can be joined.	Creep under sustained loads.

12.6 Adhesive bonding

Modern adhesives are widely used for joining metal and composite components particularly in the aircraft industry. The advantages and limitations of adhesive bonding are listed in Table 12.6.

Figure 12.24(a) shows a typical bonded joint and explains the terminology used for the various features of the joint. The strength of the bond depends upon two factors:

- Adhesion.
- Cohesion.

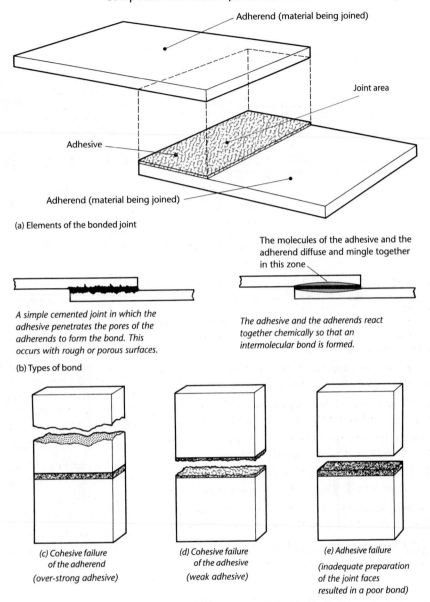

Adherend (material being joined)

Joint area

Adhesive

Adherend (material being joined)

(a) Elements of the bonded joint

The molecules of the adhesive and the adherend diffuse and mingle together in this zone

A simple cemented joint in which the adhesive penetrates the pores of the adherends to form the bond. This occurs with rough or porous surfaces.

The adhesive and the adherends react together chemically so that an intermolecular bond is formed.

(b) Types of bond

(c) Cohesive failure of the adherend (over-strong adhesive)

(d) Cohesive failure of the adhesive (weak adhesive)

(e) Adhesive failure (inadequate preparation of the joint faces resulted in a poor bond)

Figure 12.24 Adhesive bonding

Adhesion is the ability of the bonding material (*adhesive*) to stick (*adhere*) to the materials being joined (*adherends*). There are two ways in which this bond can occur and these are shown in Fig. 12.24(b).

Cohesion is the ability of the adhesive to resist the applied forces internally.

Adhesive bonds may fail in three ways and these are shown in Fig. 12.24(c–e). These failures can be prevented or minimized by careful design of the joint, correct selection of the adhesive, careful preparation of the joint surfaces and control of the working environment (cleanliness, humidity and temperature).

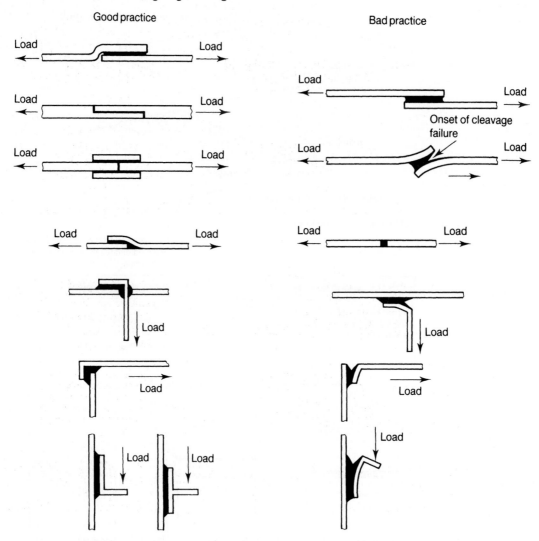

Figure 12.25 Suitable joints for bonding

12.6.1 The strength of bonded joints

No matter how effective the adhesive is and how carefully it is applied, the joint will be a failure if it is not correctly designed and executed. It is bad practice to apply an adhesive to a joint that was originally proportioned for bolting, riveting or welding. The joint must be proportioned to exploit the properties of the adhesive specified. Fig. 12.25 shows some suitable joints for adhesive bonding.

Most adhesives are relatively strong in tension and shear but weak in cleavage and peel. The shape of the joint is also important. Not only must it restrict the applied forces so that the bond is in tension or shear, it must also present the maximum possible area for the adhesive to adhere to. The ways in which adhesive bonded joints may be stressed are shown in Fig. 12.26.

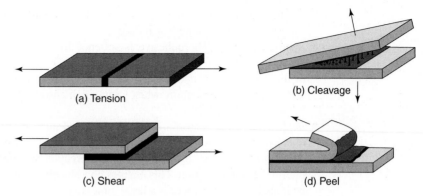

Figure 12.26 Stressing of bonded joints: (a) tension; (b) cleavage; (c) shear; (d) peel

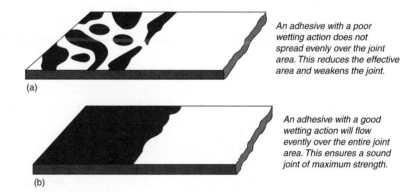

Figure 12.27 Wetting capacity of an adhesive; (a) an adhesive with a poor wetting action does not spread evenly over the joint area which reduces the effective area and weakens the joint; (b) an adhesive with a good wetting action will flow evenly over the entire joint area which ensures a sound joint of maximum strength

The joint surfaces being bonded together must also be carefully prepared. They must be physically cleaned to remove oil and dirt, and then chemically cleaned to remove any oxide film. The joint surfaces must be given a suitable texture to form a key by wire brushing, vapour or shot blasting or chemical etching.

The adhesive must 'wet' the joint surfaces thoroughly, as shown in Fig. 12.27, otherwise voids will occur and the effective joint area will be less than the designed area. This will weaken the joint considerably.

12.6.2 Thermo-plastic adhesives

- *Heat activated.* The adhesive is softened when heated and, when sufficiently fluid, is spread over the whole joint area. The materials to be joined are brought into contact, the adhesive adheres to them and, when cooled to room temperature the adhesive sets and a bond is achieved.

- *Solvent activated*. The adhesive is softened by a suitable solvent and a bond is achieved by the solvent evaporating. Because evaporation is essential to the setting of the adhesive, a sound bond is almost impossible to achieve at the centre of a large joint area, particularly when joining non-absorbent and non-porous materials as shown in Fig. 12.28.
- *Impact adhesives*. These are solvent activated adhesives which are spread separately on the two joint faces and then left to dry by evaporation. When dry, the treated joint faces are brought together whereupon they instantly bond together by intermolecular attraction. This enables non-absorbent and non-porous materials to be joined even when there is a large joint area.

Thermoplastic adhesives are based upon synthetic materials such as polystyrene, polyamides, vinyl and acrylic polymers and cellulose derivatives. They are also based upon naturally occurring materials such as resin, shellac, mineral waxes and rubber. Thermoplastic adhesives are not as strong as thermo-setting plastic adhesives but, being more flexible, they are more suitable for joining non-rigid materials.

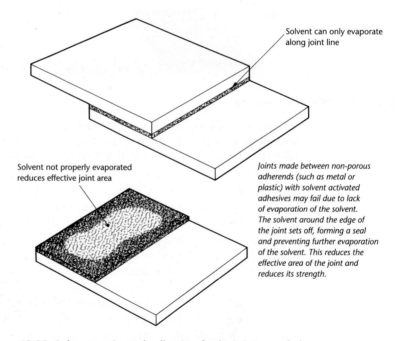

Solvent can only evaporate along joint line

Solvent not properly evaporated reduces effective joint area

Joints made between non-porous adherends (such as metal or plastic) with solvent activated adhesives may fail due to lack of evaporation of the solvent. The solvent around the edge of the joint sets off, forming a seal and preventing further evaporation of the solvent. This reduces the effective area of the joint and reduces its strength.

Figure 12.28 Solvent activated adhesive fault: joints made between non-porous adherends (such as metal or plastic) with solvent activated adhesives may fail due to lack of evaporation of the solvent. The solvent around the edges of the joint sets off, forming a seal and preventing further evaporation of the solvent. This reduces the effective area of the joint and reduces its strength

12.6.3 Thermo-setting adhesives

These are materials which depend upon heat to cause a non-reversible chemical reaction (curing) to take place to make them set. Once thermosetting adhesives have been cured they cannot be softened again by re-heating. This makes the strength of the joint less temperature sensitive than when thermoplastic adhesives are used.

The heat necessary to cure the adhesive can be applied externally by heating the assembly and adhesive in an autoclave. For example, this method is used to cure phenolic resins. Alternatively, the heat can be generated internally by adding a chemical hardener. For example, this method is used to cure epoxy resins. Since the setting process is a chemical reaction and not dependent upon evaporation, the area of the joint can be as large as is required to give the necessary strength.

> Care must be taken when working with adhesives as the solvents and the fumes given off by them are both toxic and flammable. The working area must be declared a 'no-smoking' zone

Thermo-setting adhesives are very strong and can even be used for making structural joints in high strength materials such as metals. The body shells of cars and stressed members of aircraft are increasingly dependent upon these adhesives for their joints in place of spot-welding and riveting. The stresses are more uniformly distributed and the joints are sealed against corrosion. Further, the relatively low temperature rise which occurs during curing does not affect the crystallographic structure of the metal.

Thermo-setting adhesives tend to be rigid when cured and, therefore, are not suitable for joining flexible (non-rigid) materials.

12.7 Adhesive bonding of metals

12.7.1 Anaerobic adhesives

These are high-strength adhesives widely used in the engineering industry. Anaerobic adhesives are single-compound materials that remain inactive when in contact with oxygen. The adhesives require two conditions to be present simultaneously in order to cure. That is, the absence of oxygen and in the presence of metal parts. These adhesives are widely used for threadlocking, threadsealing, gasketing and retaining (or cylindrical part bonding) at room temperature where ideal conditions exist for the adhesive to cure as shown in Figure 12.29.

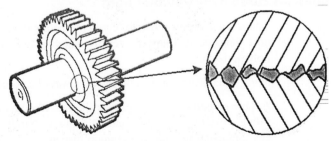

Figure 12.29 The adhesive flows into the spaces between the gear and the shaft where it has the ideal conditions to cure (absence of oxygen and metal part activity)

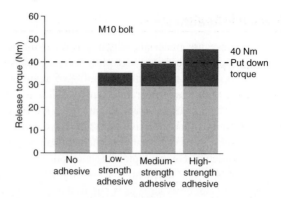

Figure 12.30 Graph showing release torques on an M10 bolt with various grades of adhesive where the Put down torque is 40 Nm

Copper, brass and plain carbon steel are 'active' as far as anaerobic adhesives are concerned and result in rapid curing at normal room temperatures. Stainless steel, aluminium and aluminium alloys, and plated products are 'passive' and require the use of an activator to ensure a full cure. As a liquid, the adhesive flows into the interstitial spaces between the male and female components of the joint where, devoid of atmospheric oxygen, it cures and becomes 'keyed' to the surface roughness. The curing process is also stimulated by contact between the adhesive and the metal surface that acts as a catalyst in the case of 'active'. Once cured to provide a tough thermoset plastic, the adhesive provides up to 60% of the torsional strength of the joint as shown in Figure 12.30. To ensure a successful joint the mating surfaces must be mechanically and chemically clean before applying the adhesive and any activator necessary – normally to the male component for convenience. Handling strength is achieved within 10–20 min of the joint being achieved and full strength is attained in 4–24 h depending on the gap between the mating components, the materials being joined and the temperature. The cured adhesives will resist temperatures up to 150°C and are resistant to most chemicals, although some higher temperature versions are available. Typical anaerobic adhesives from the Loctite product range and some applications are listed in Table 12.6.

12.7.2 Adhesives cured by ultraviolet light

These are adhesives that are cured by the application of ultraviolet (UV) light. The cure times depend on the intensity and wavelength of the UV light which initiates polymerization.

Note: Suitable eye-protection should always be aused when working with UV light.

Adhesives cured by UV light can be described as having high strength, high gap-filling capacity, very short curing times to handling strength, good to very good environmental resistance and good dispensing capacity with automatic systems as single-component adhesives.

The UV lamp is an essential part of the process and light sources can range from a simple bench-top open unit through to a fully automated conveyor system with

Table 12.7 Loctite product data

Anaerobic adhesives

Adhesives for retaining cylindrical parts

Loctite product	603	638	641	648	660
Diametrical clearance (mm)	up to 0.1	up to 0.25	up to 0.1	up to 0.15	up to 0.5
Strength	High	High	Moderate	High	Moderate
Handling strength at 23°C (min.)	10	10	30	10	30 (with activator 7649)
Temperature resistance (°C)	150	150	150	175	120

Adhesives for thread locking

Loctite product	222	243	2701
Diametrical clearance in thread (mm)	up to 0.15	up to 0.15	up to 0.15
Strength	Low	Medium	High
Handling strength at 23°C (min.)	15	10	10
Temperature resistance (°C)	150	150	150

Adhesives for gaskets/thread sealing

Loctite product	574	577	542
Flange sealing (gaskets)	Thread sealing (Coarse threads)	Thread sealing (Hydraulic fittings)	
Strength	Medium/high	Medium	Medium
Handling strength at 23°C (min.)	10	10	30
Temperature resistance (°C)	150	150	150
Pressure resistance	Steam pressure up to 5 bar	Up to burst pressure of pipe on hydraulic fittings	

several flood lights, incorporating special fixturing to hold the components in place during the curing cycle. UV light guides (or 'wand' systems) are often specified for small components as these units produce high-intensity light over a diameter of about 10 mm. These light guides ensure that the UV light is directed to the precise area for curing, thus minimizing glare and stray UV light. An example of a light guide is shown in Figure 12.31.

In some applications, it is not always possible to ensure that the entire adhesive is exposed to the UV light and products are available which will cure after the UV cycle. There are several options for the secondary cure as follows.

Heat

A secondary heat cure will ensure the full cure of many grades of UV adhesives. Cure temperatures are typically in the range 100–150°C.

Anaerobic

UV anaerobic grades are available for applications where a shadow cure is required. These anaerobic adhesives will cure through a depth of 0.2 mm due to contact with metal parts and the absence of oxygen.

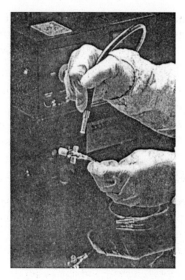

Figure 12.31 A flexible UV light guide unit for curing UV adhesives

Moisture

UV silicone products will immobilize after UV cure and then continue to cure over 24 h due to surrounding atmospheric moisture.

Surface moisture

UV curing cyanoacrylates will cure through several millimetres with UV light but also cure due to surface moisture in shadowed areas.

One of the main reasons why UV adhesives are used is the benefit of 'cure on demand' (i.e. the ability to cure the product exactly when required once, and not until, the components are fully aligned).

12.7.3 Adhesives cured by anionic reaction (cyanoacrylates)

Single-component cyanoacrylate adhesives polymerize on contact with slightly alkaline surfaces. In general, ambient humidity in the air and on the bonding surface is sufficient to initiate curing to handing strength within a few seconds. The best results are achieved when the relative humidity value is 40–60% at the workplace at room temperature. Lower humidity leads to slower curing, higher humidity accelerates the curing but in extreme cases may impair the final strength of the bond.

After adhesive application, the parts must be assembled quickly since polymerization begins in only a few seconds. The open time is dependent on the relative humidity, the humidity of the bonding surfaces and the ambient temperature. Due to their very fast curing times, cyanoacrylate adhesives are particularly suitable for bonding small parts. UV curing grades are also available to improve the cure throughout the depth of the adhesive and to cure any excess adhesive. Cyanoacrylate adhesives should be applied only to one surface and the best bond is achieved if only enough adhesive is applied to fill the joint gap. Activators may be used to speed the curing process and to cure any excess adhesive. A further benefit of cyanoacrylates is that they can be used with primers to bond lowenergy plastics such as polypropylene and polyethylene.

Features of cyanoacrylate adhesives are:

- very high shear and tensile strength;
- very fast curing speed (seconds rather than minutes);
- minimum adhesive consumption;
- almost all materials may be bonded;
- simple dispensing by good ageing resistance of single-component adhesives.

12.7.4 Adhesives cured with activator systems (modified acrylics)

These adhesives cure at room temperature when used with activators which are applied separately to the bonding surfaces. These components of the adhesive system are not pre-mixed, so it is not necessary to be concerned about 'pot life'. The characteristic properties of modified acrylics are:

- very high shear and tensile strengths;
- good impact and peel resistance;
- wide useful temperature ($-55°C$ to $+120°C$);
- almost all materials can be bonded;
- large gap-filling capacity;
- good environmental resistance.

12.7.5 Adhesives cured by ambient moisture

These adhesives/sealants polymerize (in most cases) through a condensation effect that involves a reaction with ambient moisture. There are three types of adhesive in this category, namely:

- silicones
- polyurethanes
- modified silanes.

Silicones

These materials vulcanize at room temperature by reacting with ambient moisture (RTV). The solid rubber silicone is characterized by the following properties:

- Excellent thermal resistance
- Flexible, tough, low modulus and high elongation
- Effective sealants for a variety of fluid types.

Polyurethanes

Polyurethanes are formed through a mechanism in which water reacts (in most cases) with a formulative additive containing isocyanate groups. These products are characterized by the following properties:

- Excellent toughness and flexibility
- Excellent gap filling (up to 5 mm)

- Paintable once cured
- Can be used with primers to improve adhesion.

Modified silanes

The modified silanes again cure due to reaction with atmospheric moisture and offer the following properties:

- Paintable once cured
- Good adhesion to a wide range of substrates
- Isocyanate free
- Excellent toughness and flexibility.

It has only been possible to give a brief review of the main types of these specialized adhesives and the reader is referred to the technical literature published by the **Henkel Loctite Adhesives Limited** for detailed information on adhesive types and applications.

12.7.6 Epoxy adhesives

These are based on resins derived from epichlorhydrin and bisphenol-A. They are characterized by low shrinkage on polymerization and by good adhesion and high mechanical strength. These are usually 'two-pack' adhesives that are cured by mixing the adhesive with a catalyst (hardener) immediately prior to use. Curing commences as soon as the catalyst is added. However, in some instances the hardener is premixed by the adhesive manufacturer and is activated by heating to (typically) 150°C for 30 min. Typical epoxy adhesives are found in the Loctite® Hysol® range. The substrate selector guide is shown opposite.

Typical Hysol® epoxy adhesives are as follows:

1. General purpose epoxy adhesives:
   - Multi-purpose adhesives for general bonding applications.
   - Ultra-clear adhesives for transparent bond lines.
   - Flexible adhesives for low stress bonding of plastics and dissimilar substrates.
2. Five-minute epoxy adhesives:
   - Rapid cure for a fast fixture time.
   - Ultra-clear adhesives for fast transparent bonding.
   - Emergency repair of metal parts like pipes and tanks.
3. Toughened epoxy resins:
   - Highest shear and peel strength.
   - Excellent impact resistance.
   - Adhesives for bonding high performance composites such as glass-reinforced plastics (GRP) and CFRP.
4. High-temperature epoxy resins:
   - Resistance to high operating temperatures.
   - UL94-V0 rating for fire retardant parts.
   - High thermal conductivity.
5. Metal-filled epoxy resins:
   - Rebuild and restore worn and damaged metal parts.

Loctite® Hysol®	General purpose Epoxies							Five-minute Epoxies							Toughened Epoxies						High temperature Epoxies					
	3421 A&B	3423 A&B	9481 A&B	9483 A&B	9484 A&B	9489 A&B	9533	3430 A&B	9455 A&B	3422 A&B	9450 A&B	3450 A&B	3455 A&B	3463 A&B	3425 A&B	9466 A&B	9463 A&B	9461 A&B	9464 A&B	9514	9491 A&B	9492 A&B	9502	9509	9493 A&B	9496 A&B
Steel	++	++	+	+	++	++	++	++	++	++	++	++	++	++	++	++	++	++	++	++	+	+	++	++	+	+
Stainless steel	++	++	+	+	+	+	++	++	++	+	++	++	++	++	++	++	++	+	+	++	+	+	++	++	+	+
Zinc dichromate steel	++	++	+	++	++	++	++	+	+	+	++	++	++	++	+	+	++	+	+	++	+	+	++	++	+	+
Hot dipped galvanised steel	++	++	+	++	++	++	+	+	+	+	++	++	++	++	+	+	++	+	+	++	+	+	++	++	+	+
Aluminium	++	++	+	++	++	++	+	++	++	+	++	++	++	++	++	++	++	++	+	++	+	+	++	++	+	+
Copper	++	++	+	+	+	+	+	++	++	+	++	++	++	++	++	++	++	+	+	++			++	++	+	+
Brass	++	++	+	+	+	+	+	++	++	+	++	++	++	++	++	++	++	+	+	++			++	++	+	+
Thermoplastics	+		+		+	+	+	+	+		+					+	++	++	++		+	+			++	+
Thermoset plastics	+		+	+	++	++	+	+	+		+					+	++	+	+						+	+
GRP/SMC composites	+	+	+	+	+	+	++	+	+		+	+	+			+	++	++	++	++		+	+	+	++	+
Rubbers	+	+	+		+	+		+	+	+					+	+	+	+	+							
Glass	++		+	++	+	+		++	++	+	+					++	+									
Ceramic	++	+		++	+	+	+	+	+	+	+	+	+		+	+	+	++	++	++	++	++	++	++	+	+
Masonry	++	+		+	+	+		+	+	+	+	+	+		++	+	+	+	+		+	+			+	
Wood	++	+	+					+	+	++	+	+	+		++	+	+	+	+						+	

++ Highly recommended
\+ Recommended

- Thermoplastics: ABS, Acrylic, PA, PC, PVC, SAN
- Thermoset plastics: DAP, Epoxy, Phenolic, Polyster
- Rubbers: Butyl, EPDM, Natural, Neoprene, Nitrile, SBR

Figure 12.32 Substrate selector guide

- Repair metal pipes and castings.
- Form moulds, tools, fixtures and models.
- Reduce sliding wear on moving parts.

Typical performance characteristics are shown in Table 12.7. The reader is referred to the comprehensive data published by **Henkel Loctite Adhesives Limited** who may be contacted at the address given in Appendix 3.

12.7.7 Redux process

A technique developed for bonding primary sheet metal aircraft components with a two component adhesive under closely controlled heat and pressure.

12.7.8 Bonded joints

It has already been stated that correct surface preparation is necessary for optimum bonding. Bond strength is determined to a great extent by the adhesion between the adhesive and the substrate. Therefore joint preparation should remove oil, grease, oxide films and dirt in order to ensure chemical and mechanical cleanliness. Mechanical surface treatment by grit-blasting or wire brushing ensures good adhesion.

In addition it is necessary to design the joint correctly. Adhesives are relatively strong in tension and shear but weak in cleavage and peel. Therefore the joint must be designed with this in mind and also provide an adequate surface area between the mating parts to ensure the required strength. That is, the joint must be designed with adhesive bonding in mind from the start and not just be an adaptation of an existing mechanical/welded joint. Fig. 1 below shows a number of correctly and incorrectly designed joints suitable for adhesive bonding.

The author is indebted to Mr Bob Goss, Senior Technology Specialist at Henkel Loctite Adhesives Limited for his assistance in compiling the above information on adhesive bonding.

General purpose epoxides

	General bonding	Clear bonding	Flexible bonding
Shear strength (GBMS)	17–23 N/mm²	19–23 N/mm²	22–25 N/mm²
Peel strength (GBMS)	2.7–2.5 N/mm	1.0–1.5 N/mm	1.3–2.0 N/mm
Operating temperature	Up to 100°C	Up to 80°C to 100°C	Up to 100°C

Five-minute epoxides

	Multipurpose	Metal repair
Shear strength (GBMS)	(Flowable) (Toughened) 16 N/mm²–20 N/mm²	(Gap filling) (Keadable stick) 6.8 N/mm²–23 N/mm² (Steel filled)
Peel strength (GBMS)	<1 N/mm–1.5 N/mm	<1.0 N/mm–1.5 N/mm
Operating temperature	Up to 80°C	Up to 80°C to 100°C

Toughened epoxides

	Extended working life	Medium working life	Heat curing
Shear strength (GBMS)	25–37 N/mm²	22–24 N/mm²	46 N/mm²
Peel strength (GBMS)	20–8.0 N/mm	10.0–10.5 N/mm	9.5 N/mm
Operating temperature	Up to 120°C	Up to 100°C	Up to 180°C

High temperature epoxides

	Room temperature curing	Heat curing	Fire retardant	Thermally conductive
Shear strength (GBMS)	15–20 N/mm²	29–35 N/mm²	20 N/mm²	17 N/mm²
Peel strength (GBMS)	4.2–5.1 N/mm	19–25 N/mm	2.1 N/mm	6.8 N/mm
Operating temperature	Up to 180°C	Up to 180°C	Up to 100°C	Up to 150°C

Metal-filled epoxides

	Steel filled			Aluminium filled		
	Putty	Pourable	Fast cure	Multipurpose	High temperature resistant	Metallic parts under friction / Wear resistant
Shear strength (GBMS)	20 N/mm²	25 N/mm²	20 N/mm²	20 N/mm²	20 N/mm²	20 N/mm²
Compressive strength	70 N/mm²	70 N/mm²	60 N/mm²	70 N/mm²	90 N/mm²	70 N/mm²
Operating temperature	Up to 120°C	Up to 120°C	Up to 120°C	Up to 120°C	Up to 190°C	Up to 120°C

Table 12.8 Typical performance characteristics for epoxy adhesives

12.1 Composite materials

a) Explain what is meant by a composite material and list the advantages and limitations of such a material.

b) Explain how lamination can be used to control the spread of cracks in brittle materials.

c) Describe how the laminated plastic called 'Tufnol' is manufactured, and list FOUR typical applications giving reasons for your choice.

d) Explain why plywood normally has an odd number of plies and explain how the properties of plywood compare with solid timer of the same thickness.

e) Select a suitable glass fibre and a suitable resin bond for moulding a small boat hull. Describe the stages of 'laying-up' such a moulding.

f) i) Explain how concrete can be reinforced with steel rods and wire mesh.

   ii) Describe what is meant by the pre-stressing of reinforcement and why it is required.

g) Given the data listed below, calculate:

   a) The matrix area fraction

   b) The reinforcement area fraction

   c) The tensile modulus for the composite

   d) The tensile strength for the composite

Average fibre diameter	0.005 mm
Number of fibres per strand	200
Number of strands	4800
Tensile modulus of polyester resin	3.8 Gpa
Tensile modulus of glass fibre	70 Gpa
Strength of polyester resin	50 Mpa
Strength of glass fibre	1450 Mpa
Diameter of component	25 mm

h) Explain what is meant by 'particle reinforcement' and how the particle reinforcement increases the strength and hardness of the composite.

i) Discuss the advantages and limitations of 'cermets' as cutting tool materials.

j) Explain how sintered metallic materials can be produced that cannot exist as true alloys.

k) i) Explain what is meant by the term a 'whisker'.

   ii) Explain how whiskers can be used to reinforce materials.

l) List the advantages of carbon fibres compared with glass fibres as a reinforcing material.

12.2 The manipulation and fabrication of polymers (plastics)

a) Describe the manipulation of rigid thermoplastic sheet by heat bending.

b) Describe the plastic sheet is heated locally in order to bend it.

c) With the aid of sketches describe how sheet plastic materials may be formed by:

   i) Vacuum forming

   ii) Blow forming

   iii) Pressing.

d) Describe the process of heat-welding thermoplastics, naming the most suitable plastics for this process and describing the heat source required.

e) Discuss the edge preparation required when heat welding plastics.

f) List the advantages and limitations of solvent welding compared with heat welding, paying particular attention to any safety precautions that should be taken.

g) Describe the process of friction welding thermoplastics and explain why it is a suitable process for joining medium and large diameter rigid plastic pipes.

h) Explain how plastics which are electrical insulators can be welded by:

   i) Electrical resistance heating

   ii) Electrical induction heating

   iii) Delectric heating. Name FOUR suitable materials for this process.

i) Briefly discuss the principles of ultra-sonic welding.

12.3 Adhesive bonding

a) With reference to adhesive bonding explain what is meant by the terms 'adhesion' and 'cohesion' and discuss the factors that affect the strength of a bonded joint.

b) Explain what is meant by the term 'impact adhesive' and state where such an adhesive type should be used.

c) List the advantages and limitations of heat activated adhesives.

d) List the advantages and limitations of impact adhesives.

e) List the advantages and limitations of solvent activated adhesives and describe the precautions that should be taken when using adhesives containing solvents.

f) Describe what is meant by a thermo-plastic adhesive and explain why such adhesives are widely used in the automotive and aeronautical industries.

Appendix A

Plasma arc welding and cutting

Job knowledge for welders – www.twi.co.uk

IngentaConnect On-line quality control in monitoring Plasma Arc Welding – www.ingentaconnect.com

IngentaConnect Stability of Keyhole in Plasma Arc Welding – www.ingentaconnect.com

Mig Welder, Tig Welder, Arc Welders and Plasma cutter – www.migtigarc.co.uk

MAS fact sheet – Plasma arc welding – www.mas.dti.gov.uk

Sensing weld pool surface using non-transferred plasma charge – stacks.iop.org

Science and practice of welding – Cambridge University Press – www.cambridge.org

Plasma arc welding: a mathematical model of the arc – www.iop.org

Space charge in plasma arc welding and cutting – www.iop.org

Plasma shape cutting (use of CNC) – www.torchmate.com

Electron Beam welding

Electron and Laser Welding – www.sws-trimac.com

The Welding Institute (TWI) – www.twi.co.uk

Electron beam welding in the oil, gas and chemical industry – www.twi.co.uk

Electron beam welding – www.mas.dti.gov.uk

Electron beam welding – www.fusion.org.uk

Laser Beam welding

Laser Beam Welding – MAS fact sheet – www.mas.dti.gov.uk

Computerization of laser beam welding – www.ingentaconnect.com

Laser beam welding of wrought aluminium alloys – www.ingentaconnect.com
Numerical simulation of weld pool geometry in laser beam welding – www.iop.org
Precision laser welding machines – www.precoinc.com

Automated welding

CNC welding and cutting – www.torchmate.com
Automated welding systems – www.weldlogic.co.uk
Automated welding machines and control systems – www.isotec.co.uk
Automation and Robotic welding for students– www.cranfield.ac.uk
CyberWeld: automated welding – www.cyberweld.co.uk

INDEX